New Earth Histories

Geo-Cosmologies and the Making of the Modern World

Edited by
ALISON BASHFORD,
EMILY M. KERN, AND
ADAM BOBBETTE

The University of Chicago Press
Chicago and London

The University of Chicago Press, Chicago 60637
The University of Chicago Press, Ltd., London

Published 2023
Printed in the United States of America

32 31 30 29 28 27 26 25 24 23   1 2 3 4 5

ISBN-13: 978-0-226-82858-9 (cloth)
ISBN-13: 978-0-226-82860-2 (paper)
ISBN-13: 978-0-226-82859-6 (e-book)
DOI: https://doi.org/10.7208/chicago/9780226828596.001.0001

Many contributors to this book gathered to discuss new earth histories in Sydney in December 2019, as fires engulfed the Australian continent. We would like to acknowledge the Gadigal people of the Eora nation, who are the traditional custodians of the land on which we met. The University of New South Wales contributed generously to that conference and to the ongoing New Earth Histories Research Program.

Library of Congress Control Number: 2023009341

♾ This paper meets the requirements of ANSI/NISO Z39.48-1992 (Permanence of Paper).

New Earth Histories

Contents

Part III New Elemental Histories

Part IV New Geo-Temporalities

Illustrations

Contributors

Alison Bashford is Scientia Professor of History at the University of New South Wales, Sydney, and founding codirector of the New Earth Histories Research Program. Previously, she was Vere Harmsworth Professor of Imperial and Naval History at the University of Cambridge. Her most recent book is *The Huxleys: An Intimate History of Evolution* (University of Chicago Press, 2022).

Adam Bobbette is a lecturer in political geology in the School of Geographical and Earth Sciences, University of Glasgow. His books include *The Pulse of the Earth: Political Geology in Java* (Duke University Press, 2023) and, edited with Amy Donovan, *Political Geology: Active Stratigraphies and the Making of Life* (Palgrave, 2018).

Melissa Charenko is an assistant professor at Michigan State University. Her work explores scientists' diverse understandings of climate. She received her PhD in history of science from the University of Wisconsin–Madison in 2018 and has been a fellow at the Max Planck Institute for the History of Science and the Institute for Historical Studies at the University of Texas at Austin.

Nigel Clark is a professor of human geography at the Lancaster Environment Centre, Lancaster University. He is coeditor, with Kathryn Yusoff, of a 2017 special issue of *Theory, Culture & Society* on "Geosocial Formations and the Anthropocene" and coauthor, with Bronislaw Szerszynski, of *Planetary Social Thought: The Anthropocene Challenge to the Social Sciences* (Polity Press, 2021).

Kathryn Dyt is currently a British Academy Postdoctoral Fellow in the History Department at the School of Oriental and African Studies (SOAS), University of London. Previously, she was a Past and Present Fellow at the Institute of Historical Research in London (2019–21). She has published research in the *Journal of Vietnamese Studies* and in the edited volume *Natural Hazards and Peoples in the Indian Ocean World*. Her forthcoming book with the University of Hawai'i Press is titled *The Nature of Kingship and the Nguyễn Weather-World in Nineteenth-Century Vietnam.*

Ruth Gamble is a senior lecturer of history and DECRA Research Fellow at La Trobe University, Melbourne. She is an environmental and cultural historian of Tibet and the Himalaya who has written two books about the relationship between sacred geography and Tibet's reincarnation traditions. Her forthcoming book, *Tears of the Gods: Life and Death by the Yarlung Tsangpo in Tibet*, is an environmental history of the upper Brahmaputra River.

Dan Hikuroa is an associate professor in Māori studies at Waipapa Taumata Rau / University of Auckland. He weaves together Indigenous and scientific ways of knowing, being, and doing, in collectively seeking solutions to our "wicked problems." He is the current UNESCO New Zealand Commissioner for Culture.

Jarrod Hore is a postdoctoral fellow at the University of New South Wales, Sydney and codirector of the New Earth Histories Research Program. He is a historian of environments, geologies, and photographies and the author of *Visions of Nature: How Landscape Photography Shaped Settler Colonialism* (University of California Press, 2022).

Emily M. Kern is an assistant professor in history of science at the University of Chicago. Previously, she was a postdoctoral research fellow in the New Earth Histories Research Program at the University of New South Wales, Sydney. A historian of the modern earth and paleosciences, she is currently at work on a monograph about the history of the search for the cradle of humankind.

Ruth A. Morgan is director of the Centre for Environmental History at The Australian National University. She is a coauthor of *Cities in a Sunburnt Country: Water and the Making of Urban Australia* (Cambridge University Press, 2022), and Bloomsbury will publish her forthcoming book, *Climate Change and International History.*

Zeynep Oguz is currently a senior postdoctoral researcher at the University of Lausanne's Laboratory of Cultural and Social Anthropology (LACS). Between 2019 and 2021, she was a postdoctoral fellow in environmental humanities at Northwestern University with a joint appointment at the Department of Anthropology. She received her PhD in Anthropology in 2019 at the Graduate Center, the City University of New York.

Sumathi Ramaswamy is James B. Duke Professor of History at Duke University. She has published extensively on language politics, gender studies, spatial studies and the history of cartography, visual studies and the modern history of art, and the history of philanthropy in modern India. Her publications include *Terrestrial Lessons: The Conquest of the Earth as Globe* (University of Chicago Press, 2017).

Alexis Rider is a postdoctoral fellow at the Institute of Historical Research in the School of Advanced Study at the University of London. She was awarded her doctorate in history and sociology of science at the University of Pennsylvania in May 2022. She is currently working on two projects: the role of ice in scientific conceptions of environmental change and an energy history of Ikea.

Natalie Robertson (Ngāti Porou, Clann Dhònnchaidh) is an artist and associate professor at AUT (Auckland University of Technology) who uses photography and moving image to advocate for ecologies and Māori cultural landscapes. She was awarded a doctorate in Māori studies at the University of Auckland in May 2022. Her PhD solo exhibition, *Tātara e maru ana—the Sacred Rain Cape of Waiapu*, responds to tribal aspirations for environmental reinvigoration.

Claire Conklin Sabel is a PhD candidate in history and sociology of science at the University of Pennsylvania. Her research also appears in *Gems in the Early Modern World: Materials, Knowledge and Global Trade, 1450–1800*, edited by Michael Bycroft and Sven Dupré (Palgrave Macmillan, 2019).

Anne Salmond is a Distinguished Professor of Social Anthropology and Māori Studies at the University of Auckland. She has written extensively about cross-cultural exchanges in the Pacific. Her most recent book is *Tears of Rangi: Experiments across Worlds* (Auckland University Press, 2017).

Perrin Selcer is an associate professor of history and the Program in the Environment at the University of Michigan. He is the author of *The Postwar Origins of the Global Environment: How the United Nations Built Spaceship Earth* (Columbia University Press, 2018). His current research focuses on the history of scientific stories about the origins of civilization at the end of the last ice age.

Raphael Uchôa is the Adrian Research Fellow at Darwin College, University of Cambridge, and affiliated to the Cambridge Faculty of History and Department of History and Philosophy of Science. He is an associate researcher at the ECO project in the Centre for Social Studies of the University of Coimbra and co-convenor of the working group "Science and Its Others: Histories of Ethno-Science" at the Centre for Global Knowledge Studies.

Foreword

For more than a decade now—ever since the idea of the Anthropocene was revived and renovated for our times by the Nobel-winning atmospheric chemist Paul J. Crutzen and other Earth System scientists, posing in turn new intellectual challenges to the human sciences—the interpretive debate on the Anthropocene has been stuck on the politics of naming the current geological epoch we may be living in. It is true, as this book itself testifies, the Anthropocene as an idea has become an integral part of the interpretive social sciences, equally popular and contested at the same time. Popular for it signals something about the scale of the overall impact on the planet of affluent humans and their technologies; contested because the term appears to conjure up and blame an undifferentiated humanity, the anthropos, for our current ills, while it is clear that the rich and affluent beneficiaries of capitalism bear far more historical responsibility for the destruction of the life-support system of the planet than humans who are impoverished and live on the margins of the global economy. Many commentators have therefore suggested that the root problem is capitalism and its colonial-imperial-racist origins, and, if anything, the current geological epoch should be called Capitalocene, Plantationocene, Econocene, or something similar, so that the constellation of causes behind the symptom of global warming can be rendered immediately visible. This has been a strong and effective critique, but its outlines are by now well worn.

What, then, makes the "earth histories" recounted in this book, called *New Earth Histories*, truly *new*? I think it is the fact that they effectively break out of the bounds of this older critique while also building on it. In addition, and more importantly, they move the debate forward by asking, with special reference here to the discipline of history, What can historians of modernity, capitalism, empires, colonies, science, race, and indigeneity do to historicize the elements of geological or geobiological thought that have come into the social sciences in the wake of this debate? In their recent book, *Planetary Social Thought: The Anthropocene Challenge to the Social Sciences*, Nigel Clark and Bronislaw Szerszynski made the powerful suggestion that the geological needs to be socialized.[1] The rich collection of imaginative and innovative chapters found in *New Earth Histories* provides an almost perfect demonstration of how that may indeed be done in so far as the discipline of modern history is concerned.

Yet, as readers of this book will discover to their pleasant surprise, the question about what makes the earth histories told here "new" does not admit of a single answer, for the accounts collected here adopt a diverse set of narrative, explanatory, and expository strategies. From the history of globes in circulation to the history of gemstones, from human pasts of volcanoes and water springs to the history of the last ice age, from the deep history of tectonic shifts to the modernity of the Anthropocene, from the history of empires to the Indigenous knowledge of the seas, rocks, and stars, and so on—the diversity of contexts within which the contributors to this volume situate their detailed and probing investigations is stunning in its variety. Every chapter is meticulously researched and lucidly presented. They have overlapping themes, but there is nothing reductive in their approach. One can see that the way European scientists, scholars, and administrators dealt with native or Indigenous knowledge varied from one context to another. As the editors say, with justification:

> Modern geo-cosmology emerged in part from the circulation of geologists within European empires collecting samples and fossils from terrestrial interiors and edges, depths and surfaces across the world: the "ultimate cosmopolitan science," as James Secord rightly describes. . . . While colonial geology was sometimes violently imposed on people and places, especially through the surveys, explorations, and extractions of "economic geology," it was also the site of transformative encounters with other knowledges. Recent work has analyzed how eighteenth- and nineteenth-century European geology and natural history were shaped by learned interaction with Chinese, Hindu, Indonesian, Polynesian, and Aboriginal cosmologies, among others. This was "a brokered world."[2]

The chapters of this book also make a methodological contribution. Taken together, they underscore the intellectual benefits of the historical method, of writing differentiated histories of modernity and the globe. There is no one universal narrative—either of the planet or of *Homo sapiens*—that trumps the entangled diversity of the worlds that modern humans have made. Indeed, one of the enduring lessons to take away from this book relates to what may be called the "colonial-cosmopolitan" nature of the science we often think of as exclusively Western (think of scientists' framing of the Anthropocene itself). This knowledge was "colonial" in the sense that the so-called scientific revolution in Euro-America cannot be separated from stories of empire-building, enslavement, European colonization and domination of different parts of the world, and the global dispossession of Indigenous peoples. But this knowledge was also "cosmopolitan" in that it resulted from difficult and fascinating exchanges between different knowledge systems of the world, albeit under conditions of Western domination. In other words, the cosmopolitanism in question was never completely benign. Sometimes, as in the story of the invention of gunpowder and modern explosions reconsidered here, what the West owes intellectually and materially to the non-West was never properly acknowledged, producing the effect of the "epistemic violence" that colonizers are often seen to have perpetrated on the colonized. But the colonizer's knowledge and that of the colonized do not always stand in a necessarily violent and binary relationship. Sometimes, as in the story of the Great Australian Artesian Basin, or of the volcanoes in Java, or of coal and the rise of geography and geology as disciplines in colonial Australia, the contributions of Indigenous knowledge are indeed acknowledged in the historical literature of empires, though it is true that they often get written over as the Western sciences undergo deep formalization in academic institutions.

As would be obvious from what I have already said, I firmly believe that this timely and stimulating book will do much to advance debates about the Anthropocene among historians and humanists generally. It will also offer attractive perspectives to scholars in cognate disciplines and to other Anthropocene scholars.

I felt honored when the editors invited me to write a short foreword for this marvelous volume. I very much hope that I have been able to convey to the reader some of the pleasure and excitement I experienced as I read this book. I congratulate the authors and the editors and wish them the success they so richly deserve.

Dipesh Chakrabarty
March 6, 2022
Chicago

INTRODUCTION

New Earth Histories

Alison Bashford, Emily M. Kern, and Adam Bobbette

New Earth Histories considers afresh the modern history of earth knowledges. The task is not small, as we seek to open a cosmopolitan conversation on multiple cosmologies, the many intersecting ways of knowing the earth's transformations, and the significance of geological time for the globe's multiple modernities.

We need look no further than the conception of Country held by Indigenous people in Australia for another way of thinking of earth, sea, land, and origins. Country means that kinship extends down to land and out to the cosmos. The night sky glimmers with ancestors, while landforms tell stories of how people and places came to be. The history of earth cannot be separated from the ancestral tie between land, people, plants, and animals.[1] It is unfamiliar to many. Yet, until quite recently, in conventional geological sciences too, rocks told the story of Genesis. The history of geology *is* the history of Judeo-Christian creation: strange stories about six days, lightness and darkness, dust and life, first peoples, and a great flood. Herein lies the conceptual richness of the history of geosciences and the possibility of new approaches. By broadening our field of view—thematically, temporally, geographically, epistemologically—what new ways of looking at the earth's history might appear? What new stratigraphies of knowledge and assemblages of meaning become visible in the history of the geosciences?

If we propose new earth histories, what do they succeed, intellectually and historiographically speaking? For decades, analysis of European ideas about the earth was core business in the history of science: from classical cosmology and medieval *mappa mundi* to disputes over stratigraphy and the age of the earth in the eighteenth and nineteenth centuries.[2] This is—classically—a history of ideas. But what ideas they were. The history of European geological thought over the seventeenth, eighteenth, and nineteenth centuries is made infinitely more intriguing and compelling because of complex doctrinal debates on the age of the earth and on its relation to the universal deluge and to biblical time. The English theologian William Whiston gives us his 1696 title to consider, *A New Theory of the Earth, from Its Original, to the Consummation of All Things: Wherein the Creation of the World in Six Days, the Universal Deluge, and the General Conflagration, as Laid Down in the Holy Scriptures, Are Shewn to Be Perfectly Agreeable to Reason and Philosophy*. The earth, time, water, fire, and human reason all in one title. *New Earth Histories* sits in that nexus, while confounding it with other ways of knowing altogether.

Geology *was* cosmology in the age of reason. Earth's archives told the story of divine truths. And yet its strata, fossils, and rocks were "active stratigraphies."[3] They told other stories as well, sometimes hidden ones—geologies, theologies, and cosmologies that needed to be uncovered, a changing age of the earth that was deeply part of the story of global modernity. This, too, is canonical historiography of the geosciences. Martin Rudwick has shown us how six days became thousands and then millions of years—how over a short set of decades at the end of the eighteenth century, time itself changed. The "limits of time" were stretched until they burst.[4]

Some of the earth archives that told such stories were geographically proximate to Whiston's British successors: in Wales and the Scottish islands, in the valleys and glaciers of the European Alps. Yet the earth's archives elsewhere in the world opened not just new geological ages but also larger and stranger temporalities, and radically different conceptions of the creation of the earth—ancient cosmogonies that were new to the geosciences. A colonial history of geology is becoming more familiar.[5] Modern geo-cosmology emerged in part from the circulation of geologists within European empires collecting samples and fossils from terrestrial interiors and edges, depths and surfaces across the world: the "ultimate cosmopolitan science," as James Secord rightly describes.[6]

While colonial geology was sometimes violently imposed on people and places, especially through the surveys, explorations, and extractions of "economic geology," it was also the site of transformative encounters with other knowledges. Recent work has analyzed how eighteenth- and nineteenth-century European geology and natural history were shaped by learned interaction with Chinese, Hindu, Indonesian, Polynesian, and Aboriginal cosmologies, among others. This was "a brokered world."[7] We are beginning to understand more precisely the content and implications of some of this brokered intelligence.[8] But can we learn more about earth-knowledge exchange over the long modern period, with wider instances and greater precision? That is what the chapters within *New Earth Histories* offer.

New Earthly Cosmologies

Western geoscience is conventionally afforded a special capacity to "universalize" and "globalize" a particular epistemology, but it is hardly unique in doing so. We begin by foregrounding a suite of other earth knowledges that are "cosmopolitan" in the base sense of other-than-European and in the more interesting sense of cosmological: ways of understanding earth-worlds, a globe, a universe. In doing so, we explore how the knowledge of earth is typically also the knowledge of heaven and the heavens. These were some of the other cosmologies that European geo-theology encountered, learned from, absorbed, rejected, misunderstood, or ignored. Here, also, lie histories of earthly expertise in times and places in which European geo-cosmologies were more or less irrelevant.

Sumathi Ramaswamy considers how the figure of the Goddess Earth, Prithvi, has been displaced by the impersonal and abstracted image of the spherical globe. As the world became increasingly geo-coded through the instruments and processes of modern cartography, Ramaswamy asks, Have Prithvi and her fellow gods truly been banished from the face of the earth? Her chapter shows how a process of displacement has instead transmuted into one of renewal. Public practices of religious nationalism in Hindu India ensure the gods are reinscribed and rejuvenated via the very instruments of scientific modernity that were meant to cast them into exile.

Humanities scholars have implemented a "planetary turn" in recent years,[9] driven partly by environmental and world historians who analyze large-scale geographical pasts, and partly within a long tradition of historical geography of globes themselves.[10] Ramaswamy has shown how worldwide consensus about the globe as a sphere was produced through the hard work of distributing physical models around the world.[11] This consensus, however, was patchy, transforming over time through bricolage, determined by materials at hand. Ramaswamy shows us how, following their creation and distribution, physical globes have been edited and revised over time through local conditions and imaginaries. Tracing such trajectories leads us to ask how radically different concepts of the earth's shape and universal placement have evolved. In Islamic, Christian, Ming, and Qing versions, celestial globes were spheres that surrounded another sphere; the earth and thus celestial globes were charts of the heavenly bodies observed from a position beyond fixed stars.[12] In classical Buddhism, earth is a sequence of concentric oceans and mountain chains, and the universe is in the form of a lotus.[13] Through what other spheres and shapes has the earth been comprehended?

Kathryn Dyt explains nineteenth-century Vietnamese comprehension of the cosmos, of the "celestial sphere" and "enveloping sky" that surround the earth like the hard shell of an egg enveloping the softer yolk-earth. Nine layers of phenomena become increasingly dense as they stretch from the earth itself outward to the sky-shell. This concept was derived from a Sinitic cosmology and therefore implicated Chinese emperors and divine imperial rule. For Nguyễn scholars and rulers in Vietnam, the sky told time, the future, and fortunes. It was the site of portents to be read, and Dyt shows how it was the Chinese world view that was subtly decentered in the process. In this part of Asia, before French colonization, it was the Chinese empire and its corresponding political cosmology that mattered.

Earthly cosmologies are often to be read in and through specific landforms in localized places. Ruth Gamble explains how the most famous mountain on earth, Chomolungma, became recast for some as "Everest." Chomolungma was sacred, her name short for the female deity Chomo Miyolungsangma. Non-Buddhists began appearing on the mountain from the 1920s, first in the form of the British Mount Everest Reconnaissance Expedition. They had been instructed by the Dalai Lama neither to hunt nor dig. But they did both, and the mountain brought her revenge. Gamble puts Chomolungma's subsequent history at the center of a twentieth-century geopolitics that involves Tibet, China, India, and Britain—and all the earth-and-mountain knowledges brought by these places and people. The brokering in this instance was far from bilateral, and extended well beyond the seemingly minor exchanges of people on the ground. Instead, it was—and continues to be—highly complex in geological, cosmological, epistemological, and geopolitical terms. Gamble explains just what the early nineteenth-century Great Trigonometrical Survey of India and the Chinese Academy of Geological Sciences surveys from the 1980s were literally overlooking, or else looking through.

The genealogy of "whole-earth thinking" is often told within the history of biosciences, linked to the intellectual history of ecology and of interconnected systems, as well as a new history of the species, the out-of-control part of that system. Yet "bio" and "geo" are the great twinned pair—really, conjoined twins. Thinking about the living and nonliving as part of one system, tracking them as indistinct, is the "geontology" that Elizabeth Povinelli elaborates.[14] It is also the tradition out of which "Gaia" was born. James Lovelock's original inquiry into life on Mars became a cautionary tale for life on Earth. It is hardly immaterial that belief and mythology gave the idea such an enduring life: Gaia, the goddess "Earth," who bore "Uranus" (Heaven/Sky), "Pontus" (Sea), and "Ourea" (Mountains). Gaia was primordial mother, who brought forth her children alone: this was an immaculate conception. Life and earth, *bios* and *geos*, turn out to have created the globe, the heavens, and the earth together.

Since Gaia's rebirth in 1972, the earthly biosphere has taken on great cultural and political purchase, now recognized as enveloped by new strata and materials, plastics and out-of-place ozones. In the Anthropocene, a key question is how to think *bio* and *geo* together—how to consider the earth as animate.

In "Think like a Fish," Anne Salmond, Dan Hikuroa, and Natalie Robertson present ways in which Earth System science and Māori knowledge of the ocean and the land that is Aotearoa New Zealand have conjoined. The great Pacific Ocean—Te Moana-nui-a-Kiwa—holds islands that are fish underneath a night sky that is the ocean, stars that are ancestors, and whales that are kin. They explain not just Māori comprehension but also emerging incorporations with other ways of knowing. In 2017, the Te Awa Tupua Whanganui River Act was passed by the New Zealand Parliament, recognizing a river as a legal being, not just with its own life but also with its own rights. Writing from the land fished up from the sea, Salmond, Hikuroa, and Robertson seek new oceanic histories rather than new earth histories.

New Geo-Theologies

When modern European concepts of the cosmos traveled around the world, they were often transformed and creatively appropriated by the cultures they encountered. Hindu and Buddhist concepts of Mount Meru as the axis mundi and center of the universe transferred to Africa and East and Southeast Asia, for example, and they often amalgamated with Christian and Islamic monotheism. Novel stories of sacred living mountains fused with Christian and Islamic ideas of powerful monotheistic deities. Geological knowledge drove accommodation and syncretism between world cosmologies. We might say that it was therefore never a "disenchanted" knowledge.

At the same time as Cambrian and Silurian disputes unfolded, and biblical time was being stretched beyond any possible Genesis; geology and religious cosmology were being comprehended in many other parts of the world in completely different terms.[15] On Taumako, in the Duff Island group in the Solomon Islands, for example, Moses's tablet is said to be an imposing rock face that spouts water from a hole Moses punctured with his staff. The island, though, long preexisted Moses's arrival; a boy once pulled it up from the ocean with a fishing line. White coral on the island's hilly peaks is evidence of its subaquatic origin.[16] Here, if not everywhere, the geological form of Taumako is the materialization of cosmopolitan histories, theological change made physical.

The quintessential Christian story about how the face of the earth was formed and animals distributed—the Great Flood—was always a cosmopolitan one. Zeynep Oguz examines the Great Flood, not only as told within evangelical Christianity but also in the Epic of Gilgamesh, the Torah, the Quran, and folk stories across Turkey and Kurdistan. To this day, the Flood remains a contested geological fact. Potential sites with the supposed remains of Noah's ark are subject to modern scientific stratigraphic and archaeological study. Claiming to live in the location of the ark unsurprisingly carries profound nationalist consequences. States can naturalize their contested status in the region by claiming their connection to the beginning of time. They inherit not only the biblical and Quranic tradition but also the catastrophes that shaped the world. It is a national political struggle over who gets to inherit ruins, and thus origins.

These collisions—scriptural exegesis meets lidar screening, Noachian fragment meets petrologic exploitation, and high-tech Prithvi—point to the persistent entanglement of the sacred and the earthly in *New Earth Histories*. We might ask how earth has been imbued with spiritual significance or consider the spiritual traditions that identify earthly sites as mediating between the material and divine planes. The creation and transference of sacred knowledge is the foundation of many of the oldest knowledge traditions. Frequently, deep knowledge of the sacred has been coincident with profound knowledge of the earth and its processes, a result of human efforts to mediate existence in uncertain metaphysical and geological terrain.

Non-Western cosmologies have shaped the conventional geosciences. The theory of plate tectonics, for instance, was not in any simple sense a Western scientific theory. In the 1960s, it became increasingly common for European and North American geologists to describe the theory as a "revolution" in the understanding of earth history. It has since become standard practice to highlight the contributions of Western geologists in the creation of the theory. Yet, as Adam Bobbette's chapter shows, the theory was also shaped by Javanese Islam on Indonesian volcanoes. In the late nineteenth century, colonial geologists in the Netherlands East Indies became increasingly worried about protecting the colony from volcanic eruptions. Their concern set in motion the earliest systematic studies of volcanism in the region, and it coincided with their increasing familiarity with Javanese Islam. Colonial scientists were literally taught how to look at the landscape and study its history from the perspective of syncretic Javanese Islam. Scientists then went on to translate these insights for American and European scientists, who subsequently developed the theory of plate tectonics. What would later be frequently narrated as a triumph of European and North American geosciences, and even a scientific revolution, in fact had its roots in Javanese Islam.

New geo-theologies point us to agents and agencies often overlooked by conventional histories of the earth sciences. When British geologist William Clarke surveyed New South Wales for resources in the 1830s and 1840s, he encountered Mount Wingen a coal seam permanently on fire. It was an important site for Kamilaroi people, whom Clarke consulted and learned that the fire had been burning for thousands of years. As Jarrod Hore tells us, Clarke's search for Christian revelation through modern geological science did not preclude his appreciation and frequent reliance on Indigenous earth knowledge. Clarke's oeuvre reminds us that Australian scientific geological knowledge is in no small part the product of Aboriginal memory transmission across generations.

The new stories contained here, then, remind us that the Western geological sciences are a continuation of theology by other means and that they have also been shaped by any number of non-Western traditions. And yet, new earth histories are not only about paying attention to the earth knowledges created by people and traditions who do not figure in the traditional historiography of the geosciences. These new histories also consider nonhuman agents, technologies, and elemental nature itself as part of making knowledge. If the old story of earth histories was about how a specific and materialist interpretation of the natural world came to shed other chronologies, cosmologies, and (perhaps above all else) theologies, and foreclose alternative possibilities, new earth histories demonstrate that multiple ways of knowing have long been able to coexist, even at the heart of the modern geosciences themselves.

New Elemental Histories

Vietnamese weather, Indonesian volcanoes, Aotearoa's fish-island that swims in the sea, and the "Burning Mountain of Wingen" each suggest a different history of and for air, earth, water, and fire.[17] *New Earth Histories* offers fresh accounts of the elements, of mutually constituting land, water, soil, fire, air, ores, minerals, rare earths, gems, and fossil fuels, found in the shallows and in the deep. How did the antique European and Islamic series—earth, water, fire, air—affect conceptions of the globe, of time, and of origins? How did the Chinese series, which includes wood and metal, shape those ideas differently? The Vedic Hindu series adds "akash"—space, void, or ether—to earth, air, water, and fire. It all throws Hippocratic "airs, water, and places" that constituted human health and ill-health into a new kind of comparison.[18] We might consider how elements were incorporated or disappeared, how they garnered the attention of outsiders and insiders, and how they were translated or misread.

Examining the histories of specific knowledges about earth elements makes it clear how knowledge has been commodified and tightly bound up with another globe-spanning cosmopolitan system, capitalism. As Claire Conklin Sabel describes in her study of gemstones and the mineral trade in early modern Southeast Asia, there is a deep and situated relationship between earth knowledge, mercantilism, and capitalism, a collision of geological and market forces that radically reshaped lived and material environments. The market for gemstones drove a desire among European and Asian commercial agents to better understand the circumstances of mineral formation and the broader composition of the earth that produced such riches. Sabel's narrative offers a new genealogy of geological knowledge in Europe that decenters the role of clashing Christian cosmologies and instead is reframed around connoisseurship, aesthetics, and the global commodities trade. Reincorporating elemental desires radically transforms the history of the early modern European earth sciences.

Ruth Morgan considers water, putting hydropolitics and biopolitics together to examine the history and geography of artesian groundwater reserves in the arid continent of Australia. Water, it turns out, is a compound whose geohistory requires consideration of the four antique elements together: air, water, earth, and fire. Morgan shows how "atmospheric territory" was made and claimed alongside the earthly territoriality of settler colonialism—and of subterranean territory and its hydro-resources. But did this precious water come from the sky? At least one geologist did not think so. J. W. Gregory considered artesian water to be the condensed vapors of molten rock deep within the earth, a "plutonic theory" derived from James Hutton. And yet, this was not the only way these watery environments were understood. When this distinguished chair of mineralogy and geology in Melbourne and then Glasgow sat down to write his geological text, *Dead Heart of Australia*, he began with an account of "how the Kadimakara came down from the skies."

> According to the traditions of some Australian aborigines, the deserts of Central Australia were once fertile, well-watered plains. Instead of the present brazen sky, the heavens were covered by a vault of clouds, so dense that it appeared solid; where to-day the only vegetation is a thin scrub, there were once giant gum-trees, which formed pillars to support the sky; the air, now laden with blinding, salt-coated dust, was washed by soft, cooling rains, and the present deserts around Lake Eyre were one continuous garden.[19]

In its myriad forms, water simultaneously speaks to *New Earth Histories*' concern with narratives of the elements and of deep time. As Alexis Rider discusses, the study of ice and the cryosphere preserved a record of the planet's deep past, revealing both the slow progression of past geological epochs and the rapid transformations sparked by anthropogenic climate change. Focusing on the nineteenth century, Rider shows how this temporal agency of ice was conceived and leveraged by researchers making claims about the age of the earth—and its eventual and inescapable future death as energy dissipated and the universe cooled. The future would bring an inevitable and final return of the ice. Scientists understood that the earth was becoming less "tropical." Stereotypes of the verdant, excessive, humid tropics as the cradle of biological life were contrasted with the cold dead landscapes of the frozen poles. The geography of European empires across the hemispheres furnished imaginaries of the future of the earth. The scientific preoccupation found cultural expression in the genre of the scientific romance of the mid-to-late nineteenth century. These works, usually associated with industrialization, technological change, and immediate social transformation, also reveal a persistent concern with environmental threats that operated on deep temporal scales. Ice, in these works, was not a fragile material or anthropogenic victim but a primordial force—the material incarnation of the approaching natural end of time.

Emily Kern takes the analysis of ice into the twentieth century, where she examines how glacial geochronology—a method of telling prehistoric time by the advance and retreat of Pleistocene glaciations—briefly connected geological time and human time in the 1920s and 1930s. New ways of seeing the earth as a climatological system enabled researchers outside the traditional territories of classical glaciology to index their own discoveries to Northern Hemisphere Pleistocene glacial events, in turn making visible a new geography and a new history of human prehistoric culture and migration. Glacial geochronology united the climatological and geological history of the Pleistocene with the chronology of the human Paleolithic—"ice time" bridging the gap between earth time and human time. Here, ice is not cast as destructive or apocalyptic; instead, it permits a new unification of human history across large continental divisions and the partial overwriting of racial hierarchies of teleological development.

Nigel Clark looks at fire, explosions, and gunpowder, using them as an occasion for engaging simultaneously with physical and epistemic violence. Clark conceives of the explosive force of gunpowder as a geological and planetary event, literalizing Joseph Needham's description of the invention of gunpowder as "earth-shaking." By thinking through the fact of gunpowder as a "novel anthropogenic fire"—prefiguring but in no way replacing the remnants of nuclear detonations as one of the proposed markers of the onset of the Anthropocene—and considering how gunpowder migrated from East to West, Clark argues for complicating the tenacious Western-centrism of the

Anthropocene narrative. Considering gunpowder's history as part of earth histories also makes apparent the linkages that run between the invention of gunpowder and the development of the internal combustion engine, another signal event in the history of our present runaway planetary heating. Attentiveness to the elemental series, in the oldest sense of the phrase, thus opens new vistas even in the most contemporary domains of the earth sciences.

New Geo-Temporalities

Time is our special object of inquiry in *New Earth Histories*. Chronos drives everything. The nomination of human-lithic relations in the Anthropocene has sparked fresh collaborations between geologists and historians. We are now in a moment when geological and historical time are part of one conversation (again).[20] But in truth, geologists have been historians from the beginning. We share time as both object and method of inquiry. Periodization is critical across our disciplines, substantiated through the earth's stratigraphic archives on the one hand and its paper archives on the other. *New Earth Histories* contributes to the disciplinary partnership between geology, paleontology, and history over the period that immediately precedes, and therefore defines, our own.

Although historians have become enamored with deep time, an everyday scale for geologists, the Anthropocene is not really deep history at all. As Alison Bashford has argued elsewhere, it is modern history.[21] On one measure, it is canonical modern history: the global history of the energy, economic, and ecological transformation called "the industrial revolution." We might also say that the Anthropocene was born alongside geology itself, with James Hutton's "Theory of the Earth" (1788). And yet, while the Anthropocene is modern history, its defining energy source—hydrocarbon-containing fossil fuels—puts us into quotidian contact with earth from hundreds of millions of years ago. Organisms from a prehuman earth energize the present. It is a dizzying vertigo of chronologies and geo-temporalities.

In the process of determining the "antiquity of man" and the age of the earth through field work on all continents, European geologists and natural historians encountered completely different concepts of time itself. Both India and Sri Lanka hosted significant geological inquiries in the mid-nineteenth century, for example.[22] But how did Hindu belief about a permanent cycle of time—that is, no beginning—impact geological and paleontological work driven by origins? And how did geological deep time engage with Australian Aboriginal temporalities in which "before" and "after" are often indistinct?[23] Cosmopolitan analysis of geo-temporalities invites not just consideration of long-term periodization or even the incorporation of deep time into the present. It also invites multiple comprehensions of short-term calendrical time.

Likewise, modern Western geochronology was dependent on spatializing the earth. There would be no modern globe without the long work of establishing longitude and latitude. The globe's grid lines map onto degrees that demarcate time zones and the celestial movement of the sun and moon. Cosmic mapping intertwined with mapping earth and debates about the order of calendars always implicated new possible shapes of the earth. Thai, Balinese, and Javanese calendars are derived from the Hindu lunar-solar Saka calendar, while the Islamic world relies on a lunar calendar. These divergent systems fuse in Southeast Asian Islam and create sacred days when ancestor spirits materialize in mountains and mines. The month of Suro on the Javanese calendar, for example, and the mid-seventh-moon month Yu Lan on the Cantonese calendar are periods when the boundaries between celestial realms, between living and dead, become porous and ancestors invade the present.

The creation of these calendars is inseparable from navigation and the need to chart movement in space. At the same time, these calendars invoke ideas about death and the invasion of the present by the past. In the Malay Islamic world, *kitab primbon* calendar manuals were used to determine auspicious days for entering mines, invoking metals and minerals, harvesting, or simply cutting one's finger nails.[24] Ours is a world of many calendars, ways of keeping time, and conceptualizing earthly space. Sometimes one calendar is fit, awkwardly, inside the other—a cosmos of multiple beginnings and endings. Yet it is often forgotten that the canonical event in modern geological history—"bursting the limits of time"—meant that living in multiple temporalities was entirely familiar to the founders of natural and earth sciences too.[25] When researchers on scientific expeditions encountered radically other ways of time-telling in the eighteenth century, they were themselves inhabiting a worldview that was sometimes a biblical six thousand years old, sometimes tens of thousands of years old. It was on the cusp of becoming millions of years old.[26]

Time was extended through global fieldwork and so was modern capitalism: geologists, surveyors, and their agents searched in and for coal seams, quarry pits, and oil wells. As the economic and political structures of modern societies became increasingly defined by digging into and distributing strata as resources, so too did cosmologies and geo-theologies change. The search for exploitable minerals and hydrocarbons has underwritten, literally and figuratively, significant portions of the canon of geological and, to an extent, paleontological knowledge. *New Earth Histories* collapses the difference between Triassic coal beds and the onset of modernity. We aim to make sense of these colliding regimes of geological time and also the experience of accelerating earth processes and human processes, measured and felt in melting glaciers, accumulating plastics, and warming climates.

In the midst of this boundary-crossing chronological telescoping, one powerful view of the universal past emerged from a limited set of cultural perspectives and national contexts. As Perrin Selcer discusses, the temporal framing of the Holocene—a global geological epoch—only made sense from the perspective of the North Atlantic, where the retreat of glaciers in that region 11,700 years before the present was clearly visible in the geological record and was apparently followed by new patterns of migration, hunting, and cultural production. But in other regions, this Holocene threshold was nowhere near as clear; in fact, in many other regions of the world, 11,700 years before the present is not marked by a significant shift in either climate or culture at all. Selcer shows that the geological debates around the origin of the Holocene is intimately linked to notions of the origins of modern humans and to debates around the idea of the so-called behaviorally modern human. Selcer addresses the problem of defining origin moments and transitional thresholds, from prehuman to full human, from deep past to geological present, and shows how these definitional solutions were far more modern and far more contested than we might expect.

Different concepts of deep and historical time carry an implicit, and sometimes explicit, politics. As Raphael Uchôa explains, the early nineteenth-century European naturalist Carl von Martius sought to counter dominant European modalities that described European peoples as historical and the Indigenous peoples of the Americas as "natural," without culture, and timeless. Instead, Martius argued that Indigenous peoples were possessed of a great and vast historicity, although subsequently had fallen into a state of "ruin," with the remains of past sophisticated knowledges about plants, animals, and the cosmos now maintained only piecemeal though shamanistic practices and local myth. By importing notions of catastrophe and deep time from geology into the ethnological sciences, as Uchôa shows, Martius extracted new meanings and new interpretations from Indigenous knowledge, forming a new vision of the potentialities of now-lost civilizations of antiquity.

History and geology meet in questions about the synchronicity of settlement and extinction events in late Pleistocene North America. Melissa Charenko discusses how theories about the relationship between human settlement and megafauna extinctions touched on colonial politics of Native Americans and settler scientists. One of the epistemic challenges was discerning synchronicity and causality on long time scales, a problem deeply implicated in politics of Native American history, sovereignty, and environmental activism. Charenko's chapter shows how Native American researchers critiqued modern geoscientists for privileging narrow ideas about evidence at the expense of Indigenous oral traditions and cosmologies. Native Americans, oral traditions contended, had always been present on the land; it was not them but climate change that extinguished the megafauna.

In recent years, the late Pleistocene and the Holocene have become familiar periods for modern historians' consideration, as Anthropocene urgencies seem to have thrown us all, rightly or wrongly, into planetary deep-time conversations. This has also prompted investigation of vastly older geohistories. Alison Bashford explores how the ancient southern megacontinent of Gondwanaland, which began to break up two hundred million years ago, has a modern political and cultural history too. It was named in the 1880s when the ancient *Glossopteris* fossil was found and co-located in Central India, southern Africa, and Australia, as well in South America and later Antarctica. And yet those deep-time fossils, combined with economic geology of the modern era, created the industries that extract (still) "Gondwana coal." The profoundly prehuman space and time of Gondwanaland becomes contemporary political geology in the Anthropocene. Gondwanaland, Bashford shows, is also a fabulous geography, to use Ramaswamy's term.[27] It is an imagined landscape that signals a faraway space-time, an "earth-without-humans," in which critical geographers and environmental humanities scholars are increasingly interested.[28] In its fantastic versions, Bashford suggests, Gondwanaland performs some of the cultural work that "wilderness" and wildness did in mid-twentieth-century high modernity, substituting not infrequently for Antarctica, now a peopled continent that can no longer be mobilized as pure or prehuman.

*

New Earth Histories suggests the depth and breadth of human engagement with the origins and meanings of the earth and cosmos—and, indeed, how existing scholarship has only just begun to comprehend and encompass this intellectual and epistemological diversity. We can begin to appreciate how fundamentally entangled contemporary earth knowledges—including the environmental sciences—are with much deeper and older cosmological and theological preoccupations. These cosmologies were sometimes highly localized, specific to a place—a topography, a climate, the regular movement of stars in the heavens—and a certain way of being in the world and making sense of humankind's situatedness within it. And yet, many of these cosmologies were also intrinsically cosmopolitan, emerging from encounters between different ways of knowing and out of relationships—collaborative, brokered or fraught, intimate or distant in time, space, and culture—among a varied set of knowledge-makers. By encountering other worlds, unfamiliar skies, and different ways of seeing, multiple geo-actors created the earth's history anew.

Martin Rudwick gave us his canonical history of geology as *Bursting the Limits of Time*. And yet, if we stretch geological thought to more places and people, if we expand temporalities, and if we think carefully with and through other cosmologies, might we burst time all over again?

PART I

New Earthly Cosmologies

1

Of Celestial Gods and Terrestrial Globes in Modern India

Sumathi Ramaswamy

> In a secular multicultural age, the image of the Earth is the nearest thing we have to an icon, a universal common property with shared meaning and, for many, spiritual resonance.[1]
>
> SHEILA JASANOFF, 2004

> Four billion years of evolution have given us a planet unsurpassed in beauty. We are part of it and through our eyes Gaia has for the first time seen how beautiful she is.[2]
>
> JAMES LOVELOCK, 2005

This chapter explores the trials and tribulations of an ancient divinity called Prithvi—Earth, personified as a sensuous, even sentient, female—as she confronts the demands and (dis)enchantments of mapped modernity in colonial and postcolonial India. As our planet came to be increasingly geo-coded through the various instruments, techniques, and processes of scientific cartography, the anthropomorphic form of Prithvi (also variously known as Bhu Devi, Bhoomi Devi, or Dharti Ma in the numerous languages of India) has had to compete and contend with a completely different—and newer—imagination of the earth as an inanimate but perfect sphere, impersonally and indifferently twirling daily about its own axis and annually around the sun, its surface variously marked with land formations and water bodies netted down by a mathematized albeit imaginary grid of latitudes and longitudes. The fundamental event of European thought, argues cultural theorist Peter Sloterdijk, is spheropoiesis. "The affair between occidental reason and the world-whole unfolded and exhausted itself in the sign of the geometrically perfect round form, which we still label with the Greek 'sphere,' and even more widely the Roman 'globe.'"[3] Insisting that "globalization began as the geometrization of the immeasurable," Sloterdijk reduces us mortals to "orb-creating and orb-inhabiting animals," in thrall to an enduring "orb piety."[4]

Thus, it is not altogether surprising in this reading that "the new image of the earth, the terrestrial globe, rose to become the central icon of the modern worldview."[5] Numerous others share this conclusion.[6] Indeed, if we were to rewrite Sloterdijk's thesis in the words of fellow German philosopher Martin Heidegger, the fundamental event of the modern age *is* the conquest of the world as globe.

Yet how complete and sweeping a conquest has it been, and where? I set out to respond to this question by considering several "scenes of world-imagining"[7] from one Elsewhere outside Europe, the originary home of the terrestrial globe, from where it was exported over a period of time when the planet for which it served as a thingly proxy was also taken over by the very peoples who had thought up the artifact with the help of "world hungry" disciplines like geography.[8] "The struggle over geography," Edward Said wisely noted many years ago, "is not only about soldiers and cannons, but also about ideas, about forms, about images and imaginings."[9] Inspired by this observation, I too anchor myself in this chapter to forms and images as I consider the shifting fortunes of an anthropomorphic imagination of the earth as it enters a modernity seemingly taken over by the spherical enchantment with our planet as "the geometrically perfect round form."[10]

Geo-Reverence in India before Europe

My entrée point is a beautiful painting completed in 2009 by a young Mithila artist, Shalinee Kumari (fig. 1.1).[11] Titled in English *Weeping Mother Earth Prays to the Sun God to Spare the Earth from Global Warming*, the painting casts Mother Earth as a beseeching bejeweled woman, her visible youthfulness belying the belief that she is a more-than-human being from remote antiquity whose existence in the region's imaginary has been traced by scholars to the earliest known Sanskrit texts dated to the late second millennium BCE. In these texts, named as Prithvi, Earth is "a stable, fertile, benign presence.... She is addressed as a mother, and it is clear that those who praise her see her as a warm, nurturing goddess who provides sustenance to all those who move upon her firm, broad expanse. The *Rg Veda* nearly always links her with the male god, Dyaus, but in the *Atharva Veda* and later Vedic literature she emerges as an independent being."[12] In the words of Minoru Hara, who follows her persistence and proliferation in subsequent generations of Sanskrit texts into the first millennium CE, "All the passages ... illustrate a life cycle of the earth as a woman. First, she is represented as a young, yet full-grown fair lady.... Then kings strive after her for marriage.... In the course of married

FIGURE 1.1. Shalinee Kumari, *Weeping Mother Earth Prays to the Sun God to Spare the Earth from Global Warming*, 2009, acrylic and ink on handmade paper, 76 × 56 cm. With permission of artist and image courtesy of the Ethnic Arts Foundation.

life, while she is enjoyed . . . by means of the various arts of love, she produces jewels and precious stones instead of offspring for her husband, as long as she is properly treated by her husband. . . . But once she becomes aware of the approaching calamity of her husband king, she trembles and is distressed and finally widowed. . . . Still, the earth is ever-lasting and remains always young. After her husband king dies, she starts again the same life cycle with another victorious hero, and thus is enjoyed one after another by thousands of kings during the course of a long history."[13]

From early on, Prithvi also became the subject of praise poetry. Among the most well-known of such hymns is the Prithvisukta (*Atharvaveda* 12.1), which reverences Earth across sixty-three verses as the mother of all beings. "It is she who supports us with her abundant endowments and riches; it she who nourishes us; it is she who provides us with a sustainable environment; and it is she who, when angered by the misdeeds of her children, punishes them with disasters."[14] Though gendered female, there is considerable ambiguity on whether Prithvi is solely or even predominantly humanoid and endowed with anthropomorphic form. Consider the praise of Prithvi in the second verse of the Prithvisukta: "The earth that has heights and slopes and great plains, that supports the plants of manifold virtue, free from the pressure that comes from the midst of men, she shall spread out for us, and fit herself for us."[15] Similarly, in a praise poem of thirty-three verses, titled *Bhustuti*, by the Tamil Srivaishnava theologian Vedanta Desikan (c. 1269–1370), Prithvi, referred to as Bhu, is declared to be "immeasurable" in her dimensions, her "expanse" impossible to fathom, even as a variety of female names are bestowed upon her: Acala ("steadfast"), Vasundhara ("bearer of all creatures,"), and so on.[16]

All the same, from the early centuries of first millennium CE, when artists of the subcontinent set out to visualize Prithvi, they inevitably turned to the female figure, especially the female human form (although occasionally, she is also pictured as a milk-bearing cow, given that "both of them milk out goods in accordance with the quality and capacity of their masters, if they are properly attended,"[17] a point to which I will return a little later). In such representations, unlike the multilimbed deities of the Hindu pantheon, Prithvi is generally depicted as a two-armed woman, holding a lily in her right hand, a sacred thread stretched across her torso. Unlike the fierce and single mother/goddess, she is rarely a stand-alone deity and is almost always paired with her divine husband, the super god Vishnu, to whom she is clearly subordinate. This is especially the case when, across India from around the middle of the first millennium CE, she becomes most visible in representations of her rescue from the clutches of a demonic other by Varaha, the boar incarnation (*avatara*) of Vishnu in his third reincarnation in our world to save it from oblivion. Although there is considerable variation in how she is visualized across various media and in how her cosmic rescue is imagined, often she is cast as a demure hapless woman, obviously in awe of the immense male animal god who is her savior.[18] Her capacity for independent agency is quite circumscribed, although not entirely nonexistent.

Nowhere in either the deep textual or elaborate visual imagination in India before Europe is Prithvi associated with the terrestrial globe. And yet in Shalinee Kumari's painting completed in 2009 (fig. 1.1), the female protagonist, who the artist names Dharti Ma in Hindi (literally Earth Mother), springs out of (or merges into) a globe, as she appeals to the (male) sun god to save our planet from the perils of global warming. Not the perfect sphere by any means of the dominant European imagination sketched out by Sloterdijk, the globe in the painting is certainly an expression of a modern, secular, and scientific planetary consciousness as it developed around the concept of our terraqueous planet imagined as constituted by five continents surrounded by bodies of water named oceans.[19] Indeed, the very presence of the globe in the painting confirms Prithvi's identity as Earth to the contemporary viewer (who is cartographically literate), giving her feminine persona a distinctive look, even as it allows her to be distinguishable from the generic Hindu goddess. It is worth underscoring—for what is to follow—that Indian though the artist is, the image of our planet is not centered on her country in her painting.

What is Europe's terrestrial globe doing thus in the company of an ancient "Hindu" goddess, even cast as a spherical prosthetic attached to her body? To answer this question, I need to take us back to a time when the terrestrial globe first arrived in India and then, in the course of a couple centuries of British rule over the subcontinent, became a part of the geographical common sense of those who went to school, studied geography, and learned to identify the earth not (just) as divine female but as a mathematized sphere.

Modern Earth in Modern India

As historians of cartography have correctly noted, knowledge of the earth's spherical form has not come to us naturally or easily, not even in the birthplace of this knowledge, Europe, where it became critical to the very definition of being educated, enlightened, *and* modern. The realization that our planet is spherical "is a residue of cultural activities, of watching ships come to us out of the sea for eons, of thinking about what that means, of observing shadows at different locations, of sailing distances, of contemplating all this and more at one time. It is hard won knowledge. It is *map* knowledge."[20] Materialized in novel objects such as the atlas, the map, and the terraqueous globe, this hard-won cartographic and geographical knowledge made its way to the Indian subcontinent slowly but surely from the closing decades of the sixteenth and especially in the seventeenth century, mostly through the mediation of Jesuit missionaries and Dutch, English, and French traders and ambassadors acting on behalf of modernizing European states. The passage was not easy or uninterrupted, nor was the landing necessarily soft for the earliest globes—triumphalist contemporary accounts notwithstanding—as they were often put to projects not intended for them by their early modern European donors. For instance, if we turn to another important scene of world imagining—the mighty Mughal empire, especially during the reigns of Jahangir (r. 1605–1627) and Shahjahan (r. 1627/28–1658)—the terrestrial globe came in handy for imperial artists seeking to portray their patrons as emperors of the world (*padshah*). Even while its form was taken on board, confirming as it did ancestral notions inherited from an Arabic-Islamic past of the earth as an immobile spherical entity resting within a cosmos of nesting spheres, the globe allowed imperial artists of the Mughal court to visually enact their patrons' claims to be "world-seizer" (Jahangir), "monarch of the world" (Shah Jahan), and "world-grasper" (Alamgir, a.k.a. Aurangzeb)—their very names signaling global aspirations for which this novel European object was quite efficacious. Thus, in numerous paintings, the emperor is painted holding the globe, or standing on it, his feet resting on land formations, variously delineated but rarely identifiable or named, Europe even frequently blurred or even disappeared. The emperor's body, elaborately bejeweled and sumptuously clothed and haloed, soars gloriously into the cosmos itself, frequently dwarfing the globe on which he stands or which he holds. In striking contrast to the European terrestrial globe, the Mughal artist frequently painted the object oriented to the east rather than to the north. Not least, the terrestrial globe's mapped surface allowed imperial artists to pictorially insist that it was not so much territory that mattered to the great emperor—as it apparently did on the European artifact—but the fact that lion and lamb lay together peacefully in the emperor's benevolent shade. Their "global" sovereign was thus someone who was more interested in justice and compassion rather than in naked territorial possession or the exercise of earthly power.[21]

The Mughals do not appear to have been interested in the terrestrial globe qua globe: it did not become the subject of scientific study, nor did it promote production of the object, as arguably happened in Ottoman Turkey and farther to the east, in Ming and Qing China, Tokugawa Japan, and Choson Korea. That changed when India progressively came under British rule from the later eighteenth century. Under the force of what I have called pedagogic modernity and with the help of its principal agents—the eager expatriate, the zealous missionary, the educated native, and the colonial state—the Empire of Geography, using the evangelism of modern cartography (and sometimes even of the Christian God), established the dominion of Modern Earth, a concept I borrow from Thongchai Winichakul.[22] Unambiguously, Modern Earth is spherical and occupies space without props of any sort. Modern Earth is also heliocentric, turning diurnally around its own axis and moving annually around the sun, causing eclipses and such. Not least, Modern Earth is objectified as the terrestrial globe and placed at the disposal of the learning child who is abstracted from its surface and expected to study it as a mandatory part of schooling—almost the very first lesson is to master its (correct) spherical shape, the only one admitted within the field of reckoning.[23]

In the Indian context, the votaries of Modern Earth had to contend with numerous other world-making knowledges, famously caricatured by the influential English historian and colonial administrator Thomas Babington Macaulay as "astronomy which would move laughter in girls at an English boarding school, history abounding with kings thirty feet high and reigns thirty thousand years long, and geography made of seas of butter and seas of treacle."[24] In addition to its anthropomorphic manifestation as Prithvi, and a theriomorphic imagination as Cow, the earth was dominantly conceived as a flat-bottom circular disk or plane (*gola*; *mandala*) in the venerable Sanskrit texts called the Puranas, whose spatial imaginaries were especially much maligned in colonial circles but which continued to exercise a hold on the populace. A rival body of texts called the Siddhantas, under the impact of Ptolemaic astronomy—the same system that was inherited by Arab geographers a few hundred years later and also became part of the imaginary of the Mughal imperial court in the seventeenth century—challenged the dominant discoidal view of the Puranas with its own conception of a stationary sphere, "only about 5000 yojanas in circumference, suspended in the middle of a sphere of fixed stars, around whose center the planets including the sun and moon were considered to move in tilted circular orbits."[25] Siddhanta texts often included complex instructions on how to construct the earthly sphere (*gola*), but such instructions notwithstanding, terrestrial globes were not really produced on their basis.

From fairly early on, and especially in the second millennium CE (likely in response to the growing dominance of Islam in the subcontinent), the Puranic and Siddhanta systems attempted various compromises, so that the obvious "formal" contradictions between the discoidal and the spherical—around which there was indeed much vitriolic debate, even outright ridicule by one of the other—never reached the brink. A material and visual materialization of this compromise is a beautiful artifact dated to 1571 CE in which the disk-shaped earth of the Puranas was wrapped around the sphere of the Siddhanta imagination.[26]

Educators in colonial India thus had their work cut out for them as they attempted to introduce Modern Earth into such a complicated and layered world. Their task was made even more difficult by the fact that, although different makes and sizes began to appear in different parts of India, the presence of terrestrial globes in school rooms and educational contexts—the principal site by the nineteenth century for their dissemination—was by no means assured. An administrator in 1870 described the condition that prevailed in many places: "The school house is an open shed. There are no maps, forms, chairs, tables, desks or globes or the usual apparatus of a school; or, if any they are of the rudest possible description. A round earthen pot serves for a globe."[27] Even in the absence of the globe, the foundational terrestrial lesson of the rotundity of the earth was taught with great ardor and enormous enthusiasm—with the help of any globular object at hand, such as a coconut, an orange, a wood apple, or an earthen pot—as the European preoccupation with our planet's sphericity became as well an Indian concern. As I have documented at length, over time, this enthusiasm found quite a bit of success in producing generations of Indians who too became enchanted with the proxy for our planet materialized in the terrestrial globe. So much so, in India's many languages, the subject and discipline that is called geography (or some variant) in most European languages, "earth writing," is translated as *bhugol*, "earth sphere."[28]

Importantly, as colonial rule gave way to rule by Indians, the nation-state also took up the cause of Modern Earth, bolstered as well by various "people's science" movements, which sought "to popularize scientific knowledge among the masses, to develop a scientific outlook among the masses; [and] to challenge the forces of supernaturalism, obscurantism, and superstition."[29] One measure of the success of these movements was a series of unusual surveys taken about fifty years after India secured freedom from British rule. The surveys were conducted by the New Delhi–based National Institute of Science, Technology, and Development Studies (NISTADS), which periodically collected and evaluated the Indian public's understanding of "science." They covered several areas of scientific knowledge, including astronomy, cosmology, and geography. Of particular importance to my argument is the fact that these surveys typically began with ascertaining the public's knowledge of the earth's shape. This was the very first question posed to respondents, and their answers to this question also led the reports on the state of scientific temper prevailing in the nation. Thus, the report of the 1995 survey declared that roughly three-fourths of Indians surveyed in that year "knew the correct shape of the earth," and this "*despite the absence of such notions in traditional structures*."[30] Equally important, 100 percent of those surveyed from southern India said that "the earth is round," and 91.7 percent of students across India responded similarly. By contrast, "49.3% of unskilled workers knew the correct shape of the earth," as did 52.5 percent of "housewives," or women who did not work outside the home. A "larger percentage of men [close to 80 percent] knew the correct shape of the earth as compared to women respondents [less than 60 percent]," and indeed, 30 percent of women surveyed responded with "don't know," as compared to 10 percent of men. While the general conclusion of the survey was that "a high degree of exposure to the formal education system" increased the chances of getting the right answer (hence the high percentages registered by southern Indians and by students), the role of mass media was also important, with 88.2 percent of newspaper readers, 85.6 percent of television viewers, and 81.2 percent of radio listeners returning the correct answer. In particular, the survey concluded that the state-run television network Doordarshan ("Distant Vision"), with its rotating globe logo, played an important role in this regard. The (secular) scientists of NISTADS who conducted these surveys were particularly gratified that over the years—from 1989, when they first started their project, to 2007—there was an incremental increase in the percentage of people polled who knew "the correct shape of the earth," with 81.8 percent responding with the right answer in 2007, thus confirming the scientists' conclusions about growing literacy rates and the role of mass media in consolidating the truth of terrestrial sphericity.

But even if one did not go to school in India, or study geography with its foundational lesson in terrestrial sphericity, from at least the late nineteenth century, the force of popular culture ensured the visibility of the terrestrial globe as it became manifest as image in a variety of media, ranging from matchbox and cigarette pack labels to advertisements for all manner of consumer goods. It appeared in nationalist posters and, from the 1930s, in cinema as well and, not least, television from the 1960s (as in the logo of the state-run Doordarshan). Nationalism's anticolonial image work—what I have elsewhere referred to as "visual patriotism"—drafted the terrestrial globe into the pictorial transformation of Indian leaders (like M. K. Gandhi, "the father of the nation," or Jawaharlal Nehru, who went on to become independent India's first prime minister) into worldly and world-renowned figures. As ubiquitously, the globe appears in patriotic visual culture, from at least 1907 but especially after the 1930s, in the company of Mother India, the emergent Indian nation imagined as goddess, queen, and mother. In the colorful chromolithographs and posters printed in the thousands and scattered across the length and breadth of the country, she appears standing on the gridded globe, or seated on it, showering her blessings on the map of India, which is inevitably centered on the globe.[31]

In the face of such concerted enthusiasm for Modern Earth, backed by the power of the modern state and its pedagogical project, the enchantment with secular science, and the reach of popular visual culture, has it been possible for the anthropomorphic imagination of Earth as a sensuous female to survive?

Down to (Modern) Earth

Another scene of world imagining, a striking print published around the late nineteenth century or the early years of the twentieth century (at the very height of the colonial period), helps us answer the question of whether the anthropomorphic imagination of Earth as a sensuous female has survived (fig. 1.2). A blue-gray Varaha—half man, half boar—emerges from the primeval waters, his four hands bearing the standard symbols associated with Vishnu. He is gloriously adorned in jewels, a crown atop his head. But arguably the most remarkable feature of this poster is the large terrestrial globe perched on his tusks, the landmasses and oceans marked in astonishing detail. As importantly, the globe is centered on the Indian subcontinent, its peninsular outline conforming to modern cartographic science's determinations rather than to Puranic conception of the same space as "Bharatavarsha." The poster seems to suggest that not only is Varaha rescuing Earth on this primeval mission but "India" as well. From the perspective of the long representational history of the cosmic act of Varaha rescuing Prithivi, a history that I have noted stretches back to the early first millennium CE, the poster marks a fundamental departure. Prithvi is not shown in her conventional manner as a more-than-human divinity; instead, her sensuous, sentient, and obviously female body makes way for a takeover by the inanimate spherical form of the terrestrial globe, the perfect orb of Modern Earth. This poster printed by the Ravi Varma Press—and quite likely designed by the famous artist himself, who started this firm—is not singular in the consequential displacement of Prithvi by the terrestrial globe: in fact, through the course of the twentieth century, this increasingly becomes the standard manner in which Varaha's cosmic rescue mission is visualized across numerous media, even in paintings and statues in contemporary temples where we would have expected the anthropomorphic earth to triumph.

As Denis Cosgrove writes in his magisterial study, *Apollo's Eye*, "To imagine the earth as a globe is essentially a visual act. . . . Such a gaze is implicitly imperial, encompassing a geometric surface to be explored and mapped,

FIGURE 1.2. *Varaha Avatar*, chromolithograph, late nineteenth century, published by Ravi Varma Press, Karla-Lonavla. Courtesy of Museum of Archaeology and Anthropology, University of Cambridge, catalogue number Z44986.15.

inscribed with content, knowledge, and authority."[32] To masterfully visualize the earth as an integrated totality, a spherical entity mesmerically suspended in the cosmos and gridded by a mathematical network of latitudes and longitudes, is also an act that requires an enormous leap of imagination that does not come naturally and must be learned. The artist of colonial India who painted the primeval Varaha with the modern terrestrial globe perched on his tusks did not arrive at this imagination overnight. From at least the mid-seventeenth century, the pictorial representations of Varaha—a popular theme with (Hindu) artists across the subcontinent, especially because of the widespread belief in Vishnu's periodic interventions in the affairs of the mundane world—bear the marks of the struggle on the road to the eventual triumph of the hard-won knowledge of the sphericity of the earth. Thus, in Pahari paintings from diverse eighteenth-century Rajput courts at Kangra, Basohli, Mankot, and Chamba, Varaha's twin tusks bear neither a sensuous Prithvi nor the gridded or mapped globe. Instead, they bear a landmass on which frequently are crowded forests and hills, mansions and temples, and various animals, especially the cow (which, as noted above, is also a stand-in for the earth). In a magnificently illustrated manuscript of the *Adhyatma Ramayana* commissioned around 1804, on the very cusp of the project of pedagogic modernity initiated around Modern Earth, is a watercolor in the Patna/Chapra style of Varaha bearing a small slice of the earth on his tusks. It is accompanied by an explanation for the painting's English patron: "After the creation of the world by Brahma, the giant Hirinyacsha rolled up the earth into a shapely mass, and carried it down to the seventh abyss. The God Rama [*sic*, Vishnu] assumed the form of a Boar and having slain the giant he brought up the earth on his tusks, and restored it to Brahma."[33]

All this is not to say that Prithvi as more-than-human female disappears from the visual landscape of modern India. Indeed, and fascinatingly, Ravi Varma—the artist who most likely produced the globe-bearing Varaha in figure 1.2—also painted, in 1880, a work in oil titled *Sita Bhumi Pravesh* (Sita's Entry into Earth). The painting recalls the last days of the heroine of one of (Hindu) India's most important ancestral narratives, the *Ramayana*. As summarized by anthropologist Christopher Pinney, "The painting depicts the final ordeal of Sita upon her return to Ayodhya. After enduring years of captivity by Ravan, her chastity is openly questioned by Kaikeyi, who had been instrumental in ensuring the banishment of Ram. Unable to endure this final insult, Sita appeals to Bhoomi Devi, the earth mother. A chasm opens at her feet and Bhoomi Devi takes her down into this as Ram looks on in astonishment."[34] What the otherwise astute Pinney misses in this description is that Sita, in some versions of her story, is considered the daughter of Bhoomi

Devi, hence the entire episode valorizes the return of the child to the mother. Pinney is also focused on the figure of the hapless Sita "who may have been read as a figure of the nation and its people, whose honor was also threatened by colonial interrogators."[35] I am, however, drawn to the other woman who is part of the mise-en-scène, Bhoomi Devi (a.k.a. Prithvi), beautiful and bejeweled, and a far cry from the inanimate terrestrial globe rescued by the mighty Varaha in the poster printed around the same time by the artist. All the same, if one did not know her complex history, or if Bhoomi Devi is not specifically named as such, there is very little to distinguish her as a being who personified our earth. Indeed, a colonial censor recognized this in the early years of the twentieth century when he noted, after observing all manner of innovations that had begun to accrue around the annual celebrations of the *Ramayana* in the town of Meerut, "a chauki [tableau] of Prithvi Mata [Mother Earth] is shown in the above procession along with those of other gods. . . . Prithvi Mata is exhibited as being the mother of Sita. . . . The educated classes probably call it or will call it Bharat Mata [Mother India], whatever she may be called by the pandits."[36]

Prithvi as a bejeweled goddess may also be spotted in the ubiquitous prints of Mother Cow, variously referred to as Gau Mata or Kamadhenu, in which she is frequently pictured as a two- or four-armed divinity, perched inside the belly of the bovine.[37] As noted earlier, the theriomorphic imagination of the earth as a cow also reaches into antiquity, and it is revived in the colonial and national modern in the context of a resurgent Hindu nationalism. Of course, it is only because she is named as such that we would even know that she is Prithvi/Bhoomi Devi, our Earth imagined as goddess.

Given such representational challenges, artists—long before Shalinee Kumari in 2009—turned, ironically, to the terrestrial globe to distinguish Prithvi from the generic Hindu goddess, such recourse also only possible, again ironically, with the consolidation of Modern Earth in the Indian imagination. I invoke two examples from two ends of the visual media spectrum. An artist called M. Ramaiah painted an image that was subsequently mass printed in 1953 with the title (in English) *Bhoodevi* (fig. 1.3). Bhoodevi is pictured simultaneously as a four-armed Hindu goddess *and* as a terrestrial globe contained by the graticule of latitudes and longitudes within which the peninsular outline of India is clearly demarcated. In the work, the artist is taking no chances, it seems.[38] Indeed, if we are to go by the title of the print, the viewer is called upon to see the gridded globe itself as Bhoodevi: there is a strategic ambiguity at play here that is worth underscoring. A few decades later, in the early 1980s, the creators of the comic book *Dasha Avatar* (The Ten Incarnations), in the widely read *Amar Chitra Katha* series intended for young readers, creatively

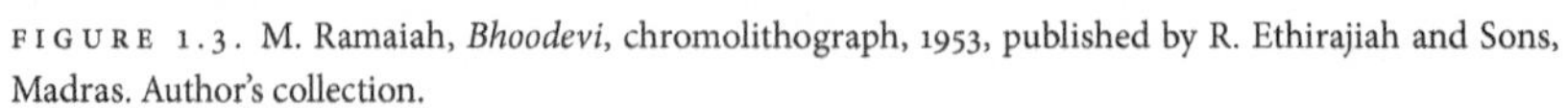

FIGURE 1.3. M. Ramaiah, *Bhoodevi*, chromolithograph, 1953, published by R. Ethirajiah and Sons, Madras. Author's collection.

FIGURE 1.4. Pratap Mulick, "Varaha Rescuing Bhoomi Devi," detail from *Dasha Avatar* (The Ten Incarnations) (Bombay: H. G. Mirchandani, 198?). With kind permission of Amar Chitra Katha Pvt. Ltd., Mumbai.

brought together the mathematized spherical and the feminized anthropomorphic, despite their essential incommensurability (fig. 1.4).[39] In their pictorial rendering of the cosmic rescue mission performed by Varaha, Mother Earth, also known as Bhoomi Devi, is rolled up inside a large globe, a damsel in distress.

With such antecedents, I take us back to my anchor image by Shalinee Kumari, an artist who was trained in an indigenous art tradition, where our planet is typically shown as a female divinity, and who was also educated in theories and forms of Modern Earth in graduate and postgraduate classes as she pursued advanced degrees in geography.[40] Not surprisingly, in this work and numerous others by the artist also featuring Dharti Ma (Earth Mother), the globe inevitably appears, confirming that the female being we are seeing is none other than Earth.[41] In other words, two incommensurable imaginations—each enchanting in its own way—coexist within the same frame, one supplementing the other.

In fact, it is not just Prithvi who comes to be explicitly associated with the terrestrial globe and the outline map of India; other gods of the Hindu pantheon do as well. They too are "carto-graphed," or fixed to a map form, as is the new goddess of the nation, Mother India herself, as noted above. In a 1927 poster by the artist Ghoting, the half-monkey, half-human Hanuman, with his trademark club-weapon, stands defiantly on a terrestrial globe (not suspended in a mathematized cosmos but immersed in the [primeval?] ocean). The outline map of India is clearly inscribed on the globe, even showing a couple of its principal rivers; Hanuman's legs straddle the mapped spread of India—its "geo-body"—even as his own body is carto-graphed by the globe and map. In another poster, published in Ahmedabad, and titled in English *Brahma Mahesh Vishnu* (with the phrase "One god in three forms" inscribed in Hindi), the artist T. B. Vathy places the Hindu trinity in the company of a terrestrial globe prominently centered on the Indian geo-body; Shiva's feet are firmly planted on the Indian map, making it his own. And in *Bansiwala*, "The Man with the Flute," that all-important Puranic deity Krishna sits firmly on a terrestrial globe on which, once again, the mapped shape of India (its geo-body) is distinctly outlined, with Krishna's feet resting on a lotus suggestively placed at its southern tip.[42]

In all such prints, going back to the earliest years of the twentieth century and continuing to the present, the divine bodies of Hinduism's many gods are unambiguously locked to the geo-body of the emergent nation, itself a product of colonial and scientific cartography.[43] Such pictures suggest that modern mapped knowledge is hijacked to ensure the survival of the gods rather than their exile, even as the gods themselves are used to popularize the image of the earth as an inanimate sphere, especially in the colonial period among a largely illiterate populace who did not go to school and learn the modern terrestrial lessons on offer through the discipline of geography. In other words, rather than modern science banishing the Puranas, Puranic knowledge becomes useful for disseminating pedagogic modernity's planetary consciousness centered on Modern Earth.

In turn, Hinduism's ancient gods and goddesses, including Prithvi, are transformed. Rather than freewheeling deities roaming an uncharted cosmos, they are carto-graphed, their bodies hitched to the terrestrial globe and pinned (down) to the outline map of "India." An entirely novel and innovative way of seeing these gods as carto-graphed divinities is thus inaugurated in the subcontinent. Most consequentially, through such a visual and cartographic act, Hindu gods are transformed into *Indian* gods, the geo-body of India nationalizing their divine bodies. With the help of modern cartographic instruments, Hinduism's ancient deities, rather than becoming irrelevant or redundant, are rejuvenated as members of the emergent nation's geo-body, lending it their aura, their powers, and, most importantly, their divinity. Correspondingly, the territory of India is sacralized, transformed from a geo-body on the impersonal face of the terrestrial globe into a hallowed land, a *punyabhumi* (sacred land) specially favored by the gods themselves. Denis Cosgrove has observed that the terrestrial globe with its geometric grid of latitudes and longitudes "universalizes space, privileging no specific point . . . extend[ing] a non-hierarchic net across the sphere."[44] Yet, in these god posters, India appears as the gods' chosen land: the images are centered on the outline map of the nation, which frequently appears as the only territory on the globe's surface, and they associate the divine bodies of these ancient deities with the modern nation's geo-body.

"The program of the Enlightenment," wrote Horkheimer and Adorno famously, "was the disenchantment of the world; the dissolution of myths and the substitution of knowledge for fancy."[45] Yet the contrary way scientific cartography is deployed in the proliferating god pictures of (Hindu) India demonstrates how the European Enlightenment project is undone at its (post)-colonial address by the revival of old myths and the return of fancy. Rather than heathenism being demolished as many a rationalist (European and Indian) had zealously hoped, the gods come back even more realistically, exuberantly, and potently, transformed from sectarian "Hindu" deities into nationalized "Indian" divinities through the mediation of the very scientific knowledge and objects that ought to have banished them from the lives and livelihoods of their devotees.

Geo-Theology in the Time of Modern Earth

The philosopher of science Bruno Latour has reminded us that we live in times (which he of course famously declared have never been modern) in which we see "a fabulous population of new images, fresh icons, rejuvenated mediators: greater flows of media, more powerful *ideas*, stronger *idols*."[46] Modern visual culture's carto-graphed Prithvi—which brings together the anthropomorphic and the spherical—is one "stronger idol," partaking of the powers and enchantments of two beguiling imaginations, mythic and modern. In this regard, she may not be alone. In the age of climate change, global warming, "green" ethics, and not least, (eco)feminism,[47] we are witness to the resurgence of other animated Earth entities, like Pachamama and Gaia, albeit in radically transformed terms, even in the writings of some heterodox scientists who have not hesitated to use the gendered language of anthropomorphism.[48] "Gaia is the great Trickster of our present history," says Latour.[49] Not surprisingly, Latour writes of Gaia that "we seem to have great difficulty housing her inside our global view, and even more difficulty housing ourselves inside her complex cybernetic feedbacks."[50] James Lovelock's Gaia theory has itself been rebuked by mainstream scientists for its vague mysticism and for flirting with metaphors of the divine (for which reason it was also embraced by some admirers). Latour, however, concluded his lecture, in which he too seemed to be "waiting for Gaia," as follows: "The idea, at once daring and modest, is that we might convince Gaia that since we now weigh so much upon Her shoulders—and Hers on ours—we might entertain some sort of a deal—or a ritual."[51]

Some years after Latour said this, the controversial Tamil politician H. R. Raja gave us a preview of what one such deal or ritual might look like in the Indian context. One hot summer, photographs of Raja "cooling" down the earth by pouring water over a mounted school globe—the master object of secular pedagogic modernity, as I have characterized it—circulated on social media, attracting both attention and attack.[52] Once again, Prithvi's female form appears to have conceded to the rotund (metal or plastic) sphere. But, all the same, it is worth noting in Raja's actions the transfer of rites and rituals from the sacral body to the secular object—placed on a lotus-shaped pedestal, no less, the habitual resting place for the goddess. Geo-piety has needed to resort to some unexpected salvational strategies in the time of Modern Earth in modern India. In turn, Modern Earth stands revealed for what it is—a fragile, all-too-human achievement that may be indispensable but is nevertheless inadequate in the face of other memories and other imperatives of being and belonging on our planet.

2

Living in an Eggshell

Cosmological Emplacement in Nguyễn Vietnam, 1802–1883

Kathryn Dyt

> The sky envelops the earth and the earth is like the yoke of an egg. There are nine layers of the sky . . . these layers are always in motion, and when you get to the ninth layer it is rigid, it is very firm. The shape of the earth is round, and it is located in the middle [of the sky] and does not move. . . . The sky moves around continuously and if it stopped moving for one second, the earth would fall down.
>
> ĐẶNG XUÂN BẢNG

In the text quote above, the nineteenth-century Vietnamese scholar-official Đặng Xuân Bảng compares the cosmos to an egg.[1] The earth, he writes, is like an egg yolk suspended inside a "sky-sphere" of nine whirling, stratospheric layers, with a hard shell-like surface at its exterior.[2]

Like many other nineteenth-century Vietnamese scholars, Đặng Xuân Bảng was schooled in Chinese history and philosophy.[3] His description of the cosmos paraphrases the writings of the Song dynasty philosopher Chu Hsi, whose ideas circulated at the Vietnamese court of the Nguyễn dynasty. Drawing on "celestial sphere" or "enveloping sky" theory (渾天) developed in the first century CE, Chu Hsi elaborated that the earth is sustained in the center of the sky-sphere because the vital energy (Chinese 氣 / Vietnamese khí), which animates the entire universe, moves in a rapid revolving motion.[4] According to this theory, there are nine distinct layers of sky, extending out from the earth to the stars, that rotate in increasing velocity. As Chu Hsi outlines, the speed of rotation produces a corresponding density in the atmosphere: "the inside layers are relatively soft; reaching towards the outside, they become gradually hard. I imagine when the ninth layer is reached, it simply forms a hard shell."[5]

The cosmic egg metaphor explained the spatial relationship between the heavens and the earth, and it also detailed the location of various objects and phenomena that could be observed between the earth and the ninth layer of the sky.[6] Celestial objects like the moon, sun, and planets were believed to exist on the outer, denser layers with the stars—composed of gaseous khí—located closest to the hard shell exterior. Meteorological phenomena, on the other hand, such as rain, wind, clouds, thunder, and lightning, were thought to occur in the softer, lower layers of the sky.[7] Celestial sphere theory also included cosmological ideas about the nature and origins of the universe. It explained the role of khí, the transcendent force, in creating the world and the idea that the objects on the distant sky-surface layer formed "patterns" (文/văn), which regulated the seasons and connected events in the heavens to those in the human realm.

Notions of a patterned cosmos in an eggshell blended the metaphysical and the earthly, but these notions were also anchored in a specific political and territorial reality. Within Sinitic texts, the shapes observed on the distant sky-shell were expounded in relation to Chinese (not Vietnamese) dynasties and territory. The cosmos and the earth were intimately bound up with the Chinese emperor, who was divinely appointed to rule, and patterns in the sky spoke to imbalances within his realm. More broadly, Chinese cosmologies and creation myths solidified the notion that "civilization" was embodied in the formation of the Chinese state and its systems of political organization. Such myths conceived of a world order that hinged on a dichotomy between the "civilized" (華/hoa) and the "barbarians" (夷/di). The "Treatise on Earth's Patterns" in *The History of the Han* (漢書/Han Shu) describes how the mythical-historical Yellow Emperor delineated the wild from the civilized realms in his journeys "Under-Heaven in all directions" at the beginning of time.[8] These mythologies supported China's claims to occupy the center of a geographical and conceptual "domain of manifest civility" (文獻之邦/văn hiến chi bang), a domain in which people accumulated "righteous energy" (正氣/chính khí) through engaging with Chinese ideas and texts.[9]

Sinic expositions about the sky provided the ideological underpinnings for the Chinese tributary system in which other countries could only become civilized through becoming vassals of China. The Vietnamese Nguyễn dynasty, which began with the reign of the Gia Long emperor (1802–1820), participated in what has been referred to as the "Chinese world order," or the "Sinosphere," meaning a region within the range of Chinese cultural influence.[10] Vietnam, like other countries in Asia in this period, was regarded by China as a "vassal state" (诸侯国/nước chư hầu). Nguyễn rulers were required to "request investiture" (請封/xin phong) from China to legitimate their reign and were required to offer regular tribute to the Qing court. Nguyễn literati like Đặng Xuân Bảng were immersed in a Chinese cultural world. Members of the nineteenth-century Nguyễn court were well versed in Chinese culture, and their literacy in Sinitic texts confirmed their elite "civility."

Through incorporating the known world into its systems of classification, Chinese cosmologies can be understood as "globalizing." In a recent article, Pennock and Power discuss how peoples living in Aztec Mexico and the late medieval Latin West subscribed to "globalizing cosmologies."[11] Such cosmologies, they explain, were globalizing due to their "capacity to envision their environment on a cosmic scale, to assign meaning to it, and to see their societies as significant agents in global time and space."[12] In pointing to the ways in which cosmological ideas translated to political action, Pennock and Power unmoor established narratives of "globe" and "globalizing" from a rigid embeddedness in European expansionism and Western imaginative capacity. Chinese cosmologies might also be understood to be globalizing in the way they challenged Western conceptions of the world order and the "globe," while also serving to prop up China's claims to cultural and political dominance. There are two aspects to this globalizing cosmology worth noting in this context. First, Chinese expositions on "All Under Heaven" and the tributary system saw European countries such as the Netherlands and Portugal relegated to the status of vassal states of China or even located outside of the Sinosphere and therefore beyond the orbit of civilization.[13] Second, while European notions of the terraqueous globe were integrated into Chinese cosmology from the seventeenth century onward, they were incorporated in a way that supported conceptions of the egg-like cosmos.[14] Notions of the terraqueous "globe" earth thus differed from predominately European views of it as a surface ripe for maritime endeavors and colonial occupation. Conceptions of the egg-yolk earth could not be detached from a larger embeddedness within an eggshell cosmos. Ingold's distinction between "globes and spheres" is useful here.[15] Whereas a solid globe can only be perceived from without, spheres are "perceived from within."[16] Within Chinese cosmology, the earth was a sphere nested within another sphere. The entire egg-like sphere of the universe was infused with the transcendent force of khí. It was this primal force—carried and enhanced by Sinic civilizing practices—rather than ships and technology, that demarcated the known world.

Nguyễn court divination practices, based on studying the skies, relied heavily on Sinitic texts to interpret the fate of the kingdom. Astronomers at the Nguyễn court scanned the sky-sphere for portents, or irregularities of nature, that pointed to disharmony. After making observations, they then consulted Chinese texts to reveal the meaning of celestial "patterns" and reported their findings to the Vietnamese emperor. But how did these texts, which connected cosmic occurrences to China, translate to a Vietnamese context? In other words, how did the Nguyễn court make sense of its place within the schema of the eggshell cosmos that mapped Chinese territory, history, and dynastic fortune onto the sky?

This chapter contributes to this book's move toward decentering European cosmologies by considering regional cosmological transfers within Asia, focusing on precolonial Vietnam from 1802 to 1883. This period spans across the reigns of the first four emperors of the Nguyễn dynasty—Gia Long (r. 1802–1820), Minh Mạng (r. 1820–1841), Thiệu Trị (r. 1841–47), and Tự Đức (r. 1847–1883)—before the French took control of Vietnam with the official formation of French Indochina in 1887. It was primarily Chinese, rather than European, modes of understanding the world and the universe that informed precolonial Nguyễn knowledge systems. When European ideas entered Vietnam in the nineteenth century, it was often via Chinese texts. Thinking about how elite members of the Nguyễn court conceived of their place in the world in relation to Chinese cosmology is important because it challenges simplistic narratives of the Sino-Vietnamese relationship as one of either passive acceptance or staunch resistance to Chinese power. As Kathlene Baldanza highlights in her insightful study of Vietnam and Ming China (1368–1644), there was often a much subtler power play at work in the relations between the two countries, which involved negotiation as well as a skillful "decentering" of the Chinese worldview.[17]

In this chapter, I argue that complex power dynamics were at play in the varied interpretation of portentous phenomena that fed into the Nguyễn project of cosmological emplacement. In reading the sky, Vietnamese diviners identified Vietnam within the revered Sinosphere and flexibly adapted Chinese cosmology in relation to their own political realities. In addition to the observation of the cosmos through the study of astronomical phenomena, Nguyễn court officials paid close attention to meteorological events. Importantly, Nguyễn meteorological knowledge was based on direct experience as well as the study of texts. So, this chapter also explores how the Vietnamese court understood its place in the world, and the world itself, through producing knowledge grounded in the phenomenology of the weather. Whereas interpretive frameworks provided in Sinitic texts were applied to distant astronomical phenomena, meteorological portents—occurring in the lower layers of the sky—were sensed and embodied and did not rely entirely on textual exposition. Indeed, localized Vietnamese understandings of the weather often meshed awkwardly with Chinese knowledge. Teasing out these two different modes of divination helps us better understand the complexities of the Nguyễn project of cosmological emplacement. Reflection on how climatic difference figured in Nguyễn attempts to interpret portents highlights the importance of the weather for solidifying the geographical space of Vietnam and for constructing China as a distant and "othered" locality.

The Court Observatory: Portents, Texts, and Training

Divination through studying the skies was highly esteemed at the Nguyễn court. Understanding heavenly patterns was important because the emperor's divine role was to ensure balance between the celestial and terrestrial realms. Portents or irregularities of nature, such as comets, eclipses, shooting stars, and unseasonal weather patterns, could be read as inauspicious. They might be interpreted, for instance, as a warning sign of an impending rebellion. Diviners kept watch for glitches or abnormal movements that pointed to disharmony and advised monarchs to carry out penitential measures to resecure heavenly favor.

The Observatory, known officially as "The Bureau for the Observation of the Sky" (欽天監/Kham Thiên Giám), was the court's central hub for identifying portents and interpreting their meanings. Working from a purpose-built tower positioned in the southwest corner of the outer citadel walls, members of the Observatory regularly collected astronomical and meteorological information through careful measurement, observation, and note-taking. If members of the Observatory noted anything out of the ordinary, they had to notify the throne immediately and cancel any celebratory events to show reverence to Heaven.[18] Any activity in the skies during Lunar New Year and transitional moments, such as solstices and equinoxes, was especially significant. During these times, members of the Observatory were instructed to "make hourly notes of phenomena such as clear skies, cloud, heat, wind, rain, and the movement of the moon, the sun and stars" and submit this information, together with an interpretation, to the throne.[19]

Some weather events and astronomical phenomena, including violent storms, strong winds, thunder, eclipses, shooting stars, and comets, were obvious causes for alarm. Others, such as a sudden change in temperature, the direction of wind, the spreading of fog, and the color of the sky at sunset, were more subtle but could play an equally important role in prediction. The interpretation of portents therefore depended on extensive knowledge about the stars, the movements of celestial bodies, and a sensitivity to the smallest changes in the skies and the atmosphere.

Members of the Nguyễn Observatory received specialist training in meteorology and astronomical concepts. By at least the time of Emperor Thiệu

Trị's reign, those seeking employment at the Observatory were required to undergo formal training in these areas. In 1842, Thiệu Trị announced that all members of the Observatory would be ranked based on rigorous examination on a range of subjects including feng shui, horology, and astronomy.[20] In 1856, Emperor Tự Đức introduced a three-year degree centered on astronomy and cosmological concepts for those seeking employment in the Observatory.[21] In their first year, students studied the constellations, the "Twenty-Eight Lunar Mansions" (二十八宿/Nhị Thập Bát Tú), and their associated stars. They moved on to study the "Three Enclosures" (三垣/Tam Viện) in their second year, and by the third year, students were expected to be able to accurately locate all the stars and the corresponding land under the influence of the stars.[22]

The training and texts given to students entering the Observatory underscores the value of Chinese knowledge to Nguyễn engagement with the skies. The book titles noted as mandatory reading for Observatory members, which are included in *The Official Compendium of Institutions and Usages of Imperial Vietnam*, are all Chinese.[23] Based on the titles for which information is available, we can see that the majority of these were written by authors during the Chinese Qing dynasty (1644–1912):[24]

1 *Correct Words and True Origins* (直竚真原), author and date unknown
2 *Precise Research on Climate* (月令粹編), Qin Jiamo, 1812
3 *Imperially Reviewed Collection of Studies on Ritual and Astronomy* (欽定儀象考成), multiple authors, seventeenth to eighteenth centuries
4 *Illustrated Description of Chime Clocks and Watches* (高厚蒙求), Xu Chaojun, 1799
5 *Elementary Collection on Divination Knowledge* (管窺集要) author unknown, Qing dynasty (1639–1912)
6 *Later Volumes on the Thorough Investigation of Calendrical Astronomy, Imperially Composed* (御制歷象考成), two volumes, author unknown, Qing dynasty (1639–1912)
7 *First Part of Music Composition Rules* (律呂上編), author unknown, Kangxi Era (1661–1722)
8 *Essence of Mathematics* (御制數理精蘊), multiple authors, eighteenth century
9 *Record on the Newly Built Astronomical Instruments for the Beijing Observatory* (新制靈台儀象志), author unknown, 1674
10 *Book of Questions and Answers between Zhu Xi and His Disciples* (朱子語類), thirteenth century
11 *Short Note on Physics* (物理小志), Fang Yizhi, 1611–1671
12 *The Mirror and Source of All Knowledge* (格致鏡原), Chen Yuanlong, 1717
13 *The Earth Explained in Book Form* (地球說書), author and date unknown

Beyond this core reading, as noted in passages in *The Veritable Records*, Observatory officials consulted a wide range of books that included discussion of astronomy and meteorology. Interpretations of portents were sought, for instance, in the Confucian classics and in Chinese dynastic histories ranging as far back as the Spring and Autumn period (722–481 BCE), through to the Qin (221–206 BCE), Han (206 BCE–23 CE), and Tang dynasties (618–907 CE). Books from China were therefore essential to the Observatory's ability to read and interpret celestial messages.

The Observatory's library was continuously updated through the arrival of new books. Some of these were collected locally from private archives. From the time of Ming Mạng, the Nguyễn court sought to amass books for its depleted state libraries and for the purpose of writing court histories.[25] In response to the call for books, in 1827 Hồ Quang of Thừa Thiên handed over a copy of the book *Astronomy and Chronology* (天元寶曆), and Hoàng Đa Trợ, deputy governor of Cao Bằng province, submitted a copy of *Celestial Signs* (乾象) to the Observatory.[26] Some books were personally carried back into Vietnam from China by members of the court.[27]

Some of the books included in the Observatory's list of core reading were produced during the reign of the Chinese emperor Kangxi (1661–1722). Mathematical, astronomical, and geographical knowledge during the Kangxi era greatly benefited from the influx of Jesuit missionaries into China.[28] Indeed, two of the books on the Observatory's official reading list—the *Later Volumes on the Thorough Investigation of Calendrical Astronomy*, composed by order of the emperor, and the *Record on the Newly Built Astronomical Instruments for the Beijing Observatory*—were in fact written by Jesuit missions during their stay at the Qing court.[29] Jesuits in the seventeenth century were also responsible for introducing the notion of the terraqueous globe to China. As Qiong Zhang's research shows, the Jesuit Matteo Ricci was credited with mapping ideas about the spherical earth onto the egg-yolk metaphor.[30] The nineteenth-century Vietnamese court, it seems, fully embraced theories of the "round earth." In autumn of 1826, the Emperor Minh Mạng commented that notions of the square earth were outdated and no longer had any currency.[31]

Contemporary scientific writings by European scholars continued to filter into Vietnam via China into the late nineteenth century. For instance, in 1879, Emperor Tự Đức ordered that a number of Chinese translations of Western books on the subjects of meteorology and geography be printed locally in Hải Dương province and "disseminated widely so that officials and scholars have access to them."[32] These included books by American Protestant missionaries in China, including *Treatise on Natural Philosophy* (博物新

編), penned by Benjamin Hobson in 1855, and *The Law of Storms* (航海金針), written by Daniel Jerome Macgowan in 1853.[33] Tự Đức considered such books to be "necessary for the study of the seasons."[34] Vietnamese engagement with Western literature on the environment through Chinese translations draws into question the conventional historiographical view that the adoption of Chinese knowledge in Nguyễn Vietnam was an index of conservatism, dogmatism, and rigidity. Rather, the influx of such books suggests that China also provided a gateway to European ideas for the Vietnamese court.

Celestial Courts and Vietnam's Place in the Cosmos

Gazing up at the remote, ever rotating sky-surface, Nguyễn officials used texts to decipher the objects they saw on the peripheries of Heaven, including the sun, the planets, the moon, and the stars. Following the instructions laid out in books, they drew correlations between the outer sky layer and human structures and activities on earth. Canonical texts in Chinese were read to reaffirm the imperial celestial model, which outlined how the political order of the kingdom of Vietnam corresponded to patterns found in the skies.

As events in the heavens were considered to be intimately entangled with the human realm, the Nguyễn court did not make a clear distinction between "astronomy" (i.e., the objective study of celestial bodies and their movements) and "astrology" (i.e., the impact of celestial bodies and their movements on human affairs and the natural world). The Vietnamese term for astronomy, *thiên văn học*, literally means "the study of heavenly patterns."[35] The notion that the celestial realms were "patterned" originated among the ancient people living along the Wei and Yellow River Valley in China and is also outlined in a section entitled "Teaching on the Patterns of Heaven" in the ancient text *Masters of Huainan* (淮南子/*Hoài Nam Tử*), which describes the origins of the universe.

According to the Chinese texts studied by Vietnamese astronomers, star groups, or asterisms, formed a correlative cosmological system that directly linked to people, places, and objects on earth. A large area of the sky was imagined as a celestial court that contained star groups representing the emperor, the royal family and imperial bureaucracy, and real-life personalities of the court. The concept of the stars as a celestial court with terrestrial correspondences was developed during the Han dynasty (202 BCE–220 CE) and was elaborated in the ensuing centuries with reference to the Chinese imperial court.[36]

Within Sinitic texts, the patterns observed on the distant sky-shell were deciphered in relation to Chinese dynastic fortune, but Vietnamese astronomers imported the celestial court and used it as a framework for divining events in their own kingdom. Nguyễn astronomers reread Chinese astral roadmaps as Vietnamese ones, with a celestial Vietnamese emperor at its North Star helm.[37] Similarly, the asterisms closest to the North Star were taken to be representative of the people at the Nguyễn court who were closest to the emperor. Like the imperial palace in Peking, the sky around the North Star was understood to be divided into "Three Enclosures," which corresponded to the Vietnamese imperial palace in Hue. Each of the Three Enclosures moved progressively further away from the Celestial Emperor North Star: first was the "Purple Forbidden Enclosure" (紫微垣/Tử Vi), then the "Supreme Palace Enclosure" (太微垣/Thái Vi), and finally the "Celestial Market Enclosure" (天市垣/Thiên Thi).

The Purple Forbidden Enclosure included star groups representing court members in the inner court sanctum—for example, the crown prince, concubines, high-ranking civil officials, and imperial historians. It also included areas of the palace and objects, such as a kitchen, a guest house, a bed, and a canopy for providing shade. A map (fig. 2.1; see table for an explanation in English) of these star groups is shown in Phạm Phục Trai's *A Recitation of the Essentials for Enlightening Children* (1853).

Asterisms belonging to the celestial court provided important reference points for assessing portentous activity. On the fifth lunar month of 1861, the appearance of white mist and a comet in the area of a number of these core star groups alarmed Nguyễn astronomers. The *Veritable Records* explains:

> The white mist originated from the area of the Purple Forbidden Enclosure and spread to the area of stars which represent the states of Han and Wei. The leftover vapours reached the border of the first star of the Celestial Ford asterism. . . . A long comet appeared in the northeast on the hour of the rooster during that month. It looked like raw cotton and had a shape like a cloud being spat out. Its light spread for more than two *trường* (approx. six metres) and pointed towards the west of the Purple Forbidden Enclosure. It ascended one degree each night.[38]

This description of white vapor and a comet uses stars as a framework for assessing the positions of these sightings and emphasizes their danger. The fact that the phenomena appeared within a key area of the celestial palace—the Purple Forbidden Enclosure—where the emperor resided was particularly worrisome. In an urgent petition to the throne, court diviners consulted their texts and connected such appearances as ill omens, which included epidemic outbreaks, bouts of drought, and skirmishes on the Cambodian border. They advised Tự Đức to "take heed and think of Heaven in all that you do."[39]

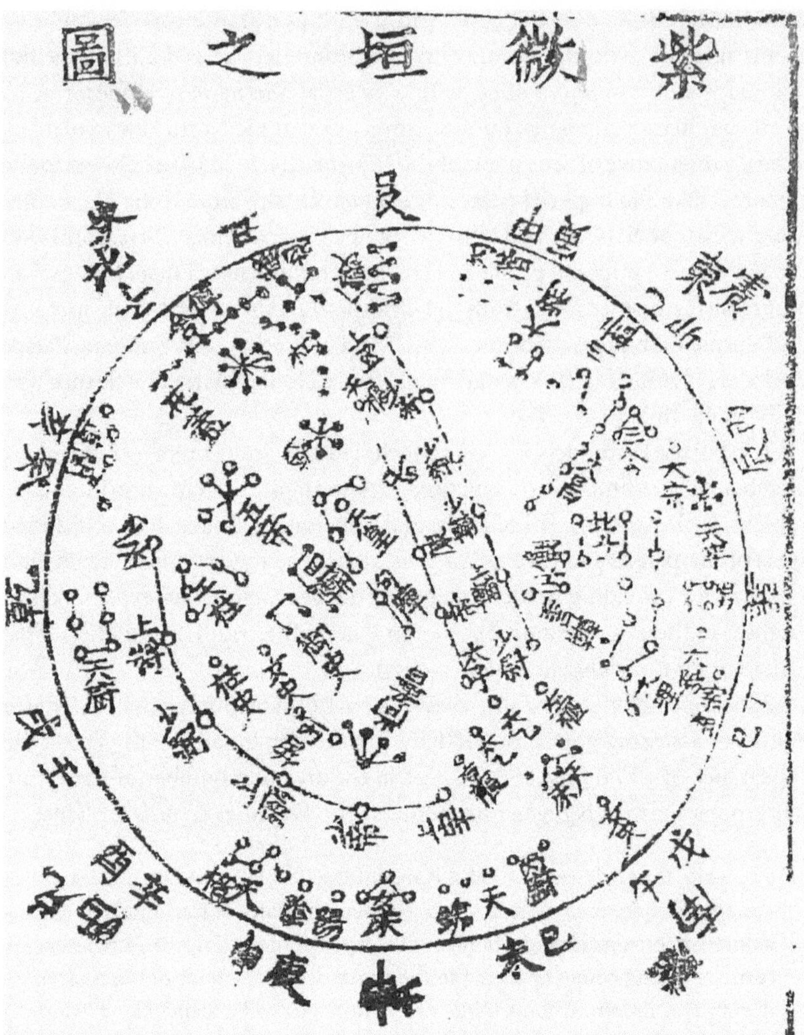

FIGURE 2.1. Map of the Purple Forbidden Enclosure in *A Recitation of the Essentials for Enlightening Children* (Khải Đồng Thuyết Ước) by Phạm Phục Trai, 1853. Courtesy of the National Library of Vietnam (R. 562 NLVNPF-0617).

In the passage above, Vietnamese astronomers link stars with the ancient states of Han and Wei, which existed during the Warring States period in China (475–221 BCE). The aligning of the stars to geographical localities—a technique known as "field division" (分野/phân dã)—had long been practiced by Chinese astronomers.[40] This system of astral-terrestrial correspondences projected different parts of China's geography onto the skyscape, thus en-

TABLE 2.1. Explanation in English of figure 2.1

Left map	(Inner) Center	Right map
天廚 (Celestial kitchen)	太子 (Crown prince)	床天/天床 (Celestial bed)
舍傳/傳舍 (Guest house)	帝 (Emperor)	天牢 (Celestial prison)
杠 (Canopy support)	柱史 (Royal archivist)	三師 (Three instructors)
棓天/天棓 (Celestial flail)	尚書 (Royal secretary)	

abling the observer to pinpoint the exact location of imminent danger. The multitude of star groups under the umbrella of the Twenty-Eight Lunar Mansions played a particularly important role in field division. During the early millennium, an elaborate system of correspondences was established between the Twenty-Eight Lunar Mansions and the various states, and it became more sophisticated during ensuing dynasties.

The system of field division was steeped in a Chinese geographical reality. For the Vietnamese astronomers who interpreted the cosmos through Sinitic texts, the geographical locales linked to the stars were Chinese towns and provinces, many of which were centuries old and no longer in existence.[41] For instance, the river of light that dominates the night sky, the Milky Way, was imagined as an astral replica of China's geographical Yellow River. Vietnamese readings of the stars as sites of an ancient Chinese past often defied a linear temporality. It was not uncommon for Vietnamese astronomers to cite dynastic histories and texts spanning over twenty centuries in one reading, and thus to see in the stars both the ancient states of the Warring States period (475–221 BCE) and Daxing, the capital of the Sui dynasty (589–618 CE). Understanding the stars as a repository for overlapping pages of Chinese history was underpinned by the Observatory's reliance on the texts they used to interpret the remote, outer layer of sky. Removed from the realm of human experience, court astronomers depended on texts stretching back as far back as the Spring and Autumn period (722–481 BCE) to find answers to the cosmos.[42]

Nguyễn astronomers were able to relate courts in the sky to political figures within their own kingdom. For them, celestial movements spoke to a Vietnamese, rather than a Chinese, Son of Heaven (the title denoting a divine mandate to rule). It was harder, however, to uproot Chinese cosmology from its embeddedness in Chinese territory without dismantling the system entirely. In the Sinitic texts Nguyễn astronomers read, the cosmos was predominantly a Chinese landscape: it reflected a Heaven that supported Chinese historical feats and territorial claims. Vietnamese diviners were forced to work within this framework to find their own place in the stars. The only texts that mentioned Vietnam were produced during the period of Chinese rule over Vietnam from 111 BCE to 980 CE. Within these texts, Vietnam is represented in the stars as a province of China and hence is placed within a relatively meager portion of the cosmos.

Furthermore, not all Chinese texts attributed the geographical space of Vietnam to the same star group or portion of the sky, which caused headaches for Nguyễn astronomers. During the nineteenth century, a number of scholars grappled with the problem of determining which celestial space governed Vietnam by sifting through ancient texts. Trịnh Hoài Đức, who served as manager of the Observatory between 1812 and 1817, tackled this issue in the opening section of the *Gia Dinh Citadel Records* (*Gia Định Thành Thông Chí*) penned in 1820. Trịnh Hoài Đức notes that the chapter "Embrace of the Mandate of Heaven" in *The Spring and Autumn Annals* places Vietnam under the influence of the Ox Mansion (牛宿/sao Khiên Ngưu).[43] The "Treatise on Geography" in the *History of Han*, on the other hand, places Vietnam under both the Ox Mansion and Girl Mansion (女宿/sao Vụ Nữ). Trịnh Hoài Đức explains that *The Classic of the Stars*, compiled during the Warring States period, is more specific, placing Vietnam under the first and second asterism of the Ox Mansion.[44] After sifting through other texts, including the *Masters of Huainan*, the *History of the Tang*, and *Records of the Grand Historian*, Trịnh Hoài Đức finally concludes that Vietnam is governed by the third star of the Southern Dipper (南斗六星/sao Nam Đẩu), which is positioned underneath the Well Mansion (井宿/sao Tỉnh) and on the periphery of the Ox Mansion.[45]

The *Unified Gazetteer of Đại Nam* (*Đại Nam Nhất Thống Chí*), which was published in 1882, explores Vietnam's place in the cosmos in even greater detail: it discusses the astrological divisions corresponding to the terrain of each province in the kingdom.[46] Building on the earlier work of Trịnh Hoài Đức and drawing on information contained in the *Book of Tang*, the text concludes that Vietnam is under the governance of different star groups depending on the region. It concludes that the area extending south from the "Five Passes" (Ngũ Lĩnh), which refers to some of the mountainous regions along the present-day China-Vietnam border, is under the rule of both the Wing Mansion (翼宿/sao Dực) and the Chariot Mansion (軫宿/sao Chẩn).[47] Whereas the Vietnamese provinces of Sơn Tây, Hưng Hóa, Tuyên Quang, Lạng Sơn, and Cao Bằng, which are located at a higher longitude, lie in the realm of the Eastern Well Mansion (井東/sao Đông Tỉnh) and the Ghost Mansion (鬼/sao Dư Qui).[48] And much further south, the three provinces of Hà Tiên, Vĩnh Long, and An Giang are noted as being under the governance of the Ghost Mansion and the star Thuần Thủ (鶉首).[49]

Such texts indicate that the question of which stars corresponded to the territory of Vietnam was never conclusively resolved. Depending on the text consulted, Vietnam's celestial space was located to various stars either in the northern or southern quarter of the sky. Court records give no indication that Nguyễn astronomers tried to re-correlate the entire cosmos to Vietnamese territory or that they raised questions about why one country should have a monopoly on Heaven, although this did happen in some other countries within the Sinosphere, such as Japan.[50] The Nguyễn elite who worked at the Observatory, it seems, never doubted that the cosmos was inextricably bound up with the "patterned" geographical space of China. Instead, through textual readings of the skies, the Nguyễn court supported its claims to belong to a Chinese world and determined whether its actions accorded with that world's foundational principles. For Vietnamese astronomers, the elements of the outer sky layer referred to in Chinese books were the most important. Sky-surface divination was thus highly referential: diviners confined their attempts to orient Vietnam's place in the cosmos to the stars referred to in canonical texts.

Divination practices pertaining to the remote layer of the sky, which adhered to rules and celestial roadmaps enumerated in texts, were a form of cosmological emplacement. However, this was not the only mode by which Vietnamese sought to know their place in the world. The skies in Vietnam were not only read, they were also experienced. While the court's interpretation of the weather did make use of text-based knowledge, meteorological phenomena in the lower layers of the sky were more complex, dynamic, and unpredictable than astral movements, and court knowledge about the weather emerged experientially through inhabitation. Let us now turn to considering how formulaic divination of weather patterns, like thunder, had to be adjusted in light of Vietnam's specific climate and through drawing on locale specific knowledge.

Meteorological Portents and Lived Experience

In the first lunar month of 1835, the Minh Mạng emperor queried a particularly loud burst of thunder. He summoned the manager of the Observatory, Nguyễn Khoa Minh, for an explanation. Nguyễn Khoa Minh consulted his books and reported:

> According to the *Essentials of Astronomy*: "when there is thunder in the West, half-ripened crops will be destroyed by caterpillars." The book also says that "thunder in the first month means that people who live in the area where there is thunder will not have any rice." In my humble opinion, the book is refer-

> ring to the Qing dynasty in China. The Qing dynasty is located in the extreme north, where the yin atmosphere is dense and therefore thunder develops and dies down in the eighth month. But our dynasty is established in the south, where the yang atmosphere is dominant. So thunder dies down in the tenth month and builds up in the twelfth month. The recent bout of thunder appeared in the west, this is where Siam is located. This country has recently severed its friendly relations with us, showing us hatred and ingratitude. Heaven will never forgive them and they will have to pay for their crimes.[51]

Here, Nguyễn Khoa Minh first struggles to select a single meaning for the recent thunder from the multiple interpretive possibilities provided in his text. He then questions whether the interpretation of thunder in the *Essentials of Astronomy* can be relied upon owing to the differences in climate and ying-yang balance between China and Vietnam. Finally, he connects the thunder to Heaven's wrath toward Siam. Nguyễn Khoa Minh's report was written just after a period of armed conflict between Siamese and Vietnamese forces, between 1831 and 1834, which was sparked by disputes over Cambodian territory.

Linking meteorological portents to recent events seems to have been a strategy that enabled the Observatory to negotiate the different interpretive possibilities presented in texts. The texts used for deciphering meteorological portents presented nuanced readings of their meanings and often ascribed various prognoses for phenomena with the same, or very similar, characteristics. This was especially the case with thunder. Thunder was a particularly ambivalent and multivalent sign. If heard in places of political significance, like the imperial palace, it was thought to be particularly ominous.[52] But thunder was not always a bad omen. Volume was key. *Observing Astronomy* (看天象), for instance, explains that if "gentle" thunder is heard on New Year's Day, "the rice and the crops will be abundant."[53] In the twelfth lunar month of 1863, the Observatory director, Trần Tiễn Thành, reported hearing thunder that was "gentle and refined" and concluded that this was a "good omen."[54] Such favorable prophecies stand in stark contrast to the meaning attached to loud thunder within the *Astronomical Vestiges* (一蹟天文家傳), which explains that "if there is loud thunder [during this time], a lot of people will die, the harvest will be reduced to half and many of the animals will perish."[55]

The difficulty of interpreting meteorological phenomena was often met with dismissal of texts based on climatic differences between Vietnam and China. Nguyễn diviners, and even emperors themselves, regularly pointed to this climatic difference and insisted that Vietnam's weather needed to be studied on its own terms, through careful observation, rather than with sole reliance on Chinese texts. When members of the Observatory presented their interpretation of thunder in a report to the throne in the first lunar month of 1833, for example, Minh Mạng lambasted his officials for mindlessly following the divination manual rather than taking into account the different weather patterns between the two countries: "If we only read ancient books without examining our own climate, then it is not the fault of the books—it is your own fault."[56] He made this point again in 1837 in relation to thunder in the tenth lunar month:

> On the day of the Cat there was thunder and the emperor summoned his officials and said: "Thunder in the tenth lunar month, why is that? Usually there is no thunder at this time of the year because the khí from the earth and the sky cannot meet. This is widely known and we have a proverb: 'In the tenth month, thunder subsides and in the twelfth month thunder arises.' . . . Our climate is not like the climate in China [and] reading good or bad omens in natural phenomena is like trying to read something very profound. It is impossible to understand these phenomena on their own terms, much less by looking to countries who are separated by thousands of leagues and do not share the same climate."[57]

In a similar way, when Minh Mạng received a report about a recent bout of thunder from a member of the Observatory in the first lunar month of 1835, he drew on his knowledge of local climatic rhythms, this time assessing that the thunder was normal and not portentous: "Having thunder in the South in the first lunar month is a normal weather occurrence . . . we do not have to consult divination texts to understand it."[58]

As I have discussed elsewhere, passages such as these served to demonstrate the emperor's superior understanding about the skies and reinforced his imperial authority.[59] In addition to asserting Ming Mạng's position as supreme ruler, however, such passages also stress the importance of observation and climatic variation over a reliance on Chinese texts. In stark contrast to astronomical phenomena, meteorological divination consolidated a sense of living in a geographical space that was "other" to that of China, because the climatic conditions of the two countries were clearly distinct.

Weather proverbs, such as the one cited by Minh Mạng in his musings on the possible explanations for thunder in 1837, often appear within official discourse as an alternative to knowledge contained within divination texts. Vietnamese proverbs containing forecasting information are derived from locally observed weather patterns, and by using a proverb to assess a bout of thunder, Minh Mạng distinguishes between the space of Vietnam and the "outside" space of China. As discussed by Sarah Strauss in her discussion of proverbs as a form of "weather wising," proverbs are not usually "meaningful to those outside" because of the way that they refract a shared experience of the weather.[60] In other words, the inherently place-based nature of weather proverbs solidifies a sense of living "here" (inside) rather than "there" (outside).

To aid divination practices and foretell severe weather events, the Nguyễn court also drew on proverbs about animal and insect behavior and agricultural rhythms. In the eighth lunar month of 1839, Minh Mạng correctly predicted a flood through knowledge derived from a local proverb about bird behavior. In an audience at court, he explained to his official Phạm Thế Trung:

> There is a proverb that says, "we only have to look at the birds that fly back to the mountain to foretell a flood." If we think about it, when there is a flood the waves in the ocean become really choppy and the steam evaporating from the water has a really pungent smell. All the birds that feed along the seashore will fly back to the mountain, so I think this proverb makes a lot of sense. I predict that in autumn this year the flood will come around the twenty-fourth or twenty-fifth day.[61]

When the predicted flood materialized, Minh Mạng expressed relief that it had occurred late in the harvesting season so the crops were not damaged.

Court officials were encouraged to seek out local, place-based knowledge about the weather to assess whether an event was portentous. For example, when the crops in Phú Vinh, in the Thừa Thiên area of Huế, suffered pestilence in the first lunar month of 1832, Minh Mạng sent out the local governor, Trương Phúc Đĩnh, to gather information from local farmers. The farmers informed him that the "water pests" were not harmful to crops but were in fact a sign of a good harvest and were normal after a cold spring. The farmers recited a proverb as evidence of this fact: "the rice pests are harmful, but if we see the water pests then the harvest will be good." Sure enough, after some rain the pests disappeared and a good harvest followed.[62]

Local climatic information was crucial for assessing if a weather event was portentous. Climate sections in geographical texts compiled throughout the nineteenth century include an abundance of weather proverbs with specific relevance to local areas. For example, the section on Quảng Bình province within the *Unified Gazetteer of Đại Nam* includes proverbs about reeds and local weather patterns. In the following proverb, the reed flowers appearing late in the season were understood as a warning of flood: "If by the tenth month no flowers have appeared, it will rain heavily for the rest of the year."[63] Other proverbs from Quảng Bình similarly linked weather to particular outcomes: "If it rains on the fifth day of the fifth month, parasites infest; if it rains on the sixth day of the sixth month, vermin abound."[64]

The importance of weather phenomenology, which was prioritized over Chinese textual knowledge, is evident in court officials' attempts to seek out local knowledge. Proverbs about the weather for different locales permeate official court records and geography texts. These proverbs give a sense of how understandings of meteorological phenomena in the lowest layer of the sky were wedded to local experience and regional climatic differences. Divination practices in nineteenth-century Vietnam thus drew on the lived experience of inhabiting the environment and local perceptions of natural phenomena.

Interpretations of the weather also enabled the court to go beyond notions of the "world" outlined within text-based, sky-surface divination. As with Nguyễn Khoa Minh's interpretation of thunder as an indicator of Heaven's wrath toward Siam, discussed at the beginning of this section, meteorological phenomena provided the court with a means to read regional politics in relation to a Vietnamese center. Unlike the stars, which affirmed the centrality of Chinese civilization, the weather enabled the Vietnamese court to envisage their own territory on a larger geopolitical scale—a scale that encompassed their regional neighbors—and to imagine celestial support for Vietnamese political endeavors. Weather played a significant role in the conceptualization of interregional politics beyond divination practices. Places deemed less civilized, like Vietnam's own "vassal states," were depicted as climatically poorer. Within court discourse, countries like Cambodia and Laos, as well as remote, forested regions like the central highlands in Vietnam, were often portrayed as hotbeds for miasma, insalubrious climate, and disease.[65]

Conclusion

The Nguyễn court emplaced itself in the cosmos through observation of the skies and divination practices. On one level, Nguyễn court divination practices affirmed Chinese political order. The remote outer layer of the sky acted like a charter that was "read" by Nguyễn officials to locate themselves vis-à-vis the great empire to the north and to affirm the court's rule over a civilized kingdom within a patterned cosmos. When deciphering astronomical phenomena, the Observatory adhered closely to esteemed Sinitic texts, which maintained distinctly Chinese understandings of the world and supported China's "globalizing" territorial assertions. Reference to classical astrological texts gave the dynasty its moral foundations, even while complicating efforts to locate the kingdom within the known geographical realm.

While emplacement in relation to China was important, Vietnamese diviners also reimagined Chinese courts in the sky with relevance to their own kingdom. They flexibly adapted Chinese cosmologies, interpreting astral movements as heavenly messages intended for Vietnamese emperors and the Nguyễn court. Crucially, the skies in Vietnam were not only "read" through textual knowledge, they were also experienced. Whereas interpretive frameworks provided in Chinese texts could be applied to distant astronomical phenomena, weather events in the lowest layer of the eggshell cosmos were more reliant on Vietnamese experience of the environment. In divining meteorological phenomena, the court drew on a reservoir of accumulated sensed and observed local knowledge. The Nguyễn court's methods of divining the skies, therefore, moved fluidly between the interpretation of texts and knowledge gained from experience of the weather. This was a dynamic hermeneutic approach: empirical evidence was weighed against canonical texts and experiential knowledge of meteorological phenomena. Climatic differences between China and Vietnam led diviners at the Nguyễn court to question, and even dismiss, explanatory frameworks offered in texts. Through such interpretive moves, the lived experience of environmental events afforded the Vietnamese a distinctive mode of cosmological emplacement. For the Nguyễn court, a sense of place was gained, in part, through the phenomenology of the weather. Vietnam's unique weather environment enabled the court to conceive of geographical difference from China and to subtly decenter Chinese globalizing cosmologies by positioning Vietnam at the heart of its own geopolitical sphere.

3

The Mountain's Many Faces

How Geologists Mistook Chomolungma for Everest

Ruth Gamble

In 1921, Dzatrül Rinpoche noticed British "Sahibs" traveling past his retreat hut in the Rongpu ("Upper Canyon") Valley toward the mountain he knew as Chomolungma, and they, the Sahibs, knew as Everest. Both Rongpu Valley and Chomolungma were sacred to Tibetans and Sherpas who lived at the mountain's base.[1] The Sahibs were the first non-Buddhists to enter the valley, and Rinpoche was concerned about their presence. In the short term, he was worried that their disrespect would disturb the goddess after whom the mountain was named, Chomo Miyolungsangma, "the wholesome, immovable ox." Disturbed goddesses were known to create earthquakes, avalanches, and other calamities. He also recognized the Sahibs' "back-the-front views" (*log lta ba*) as a long-term threat to the site's sacredness.[2] In his *Guide to Dza Rongpu*, Rinpoche explained that sentient beings maintained sacredness through their perception.[3] There were many ways to look at the mountain as a sacred space, and he encouraged this plurality of visions. But those with completely back-the-front views could not see this sacredness. A person with a "back-the-front" view could not only not see the sacred but also held onto another—materialist or creationist—view of the world with prejudice. Rinpoche was worried that the arrival of one Sahib portended the arrival of more.[4] And the more people who arrived with a back-the-front view of Chomolungma, the less sacred Rinpoche's valley would become.

The Sahibs were members of the British Mount Everest Reconnaissance expedition. The Dalai Lama's government in Lhasa had granted them permission to enter the region under political duress but demanded they agree to strict prohibitions against hunting and digging. Alexander Heron, an employee of the Geological Survey of India and the first geologist to visit Everest, kept his urge to hunt in check, but he could neither abide nor understand the Tibetans' prohibition against digging. Locals reported hammering sounds, noted the expedition's extraction of rocks, and made an official complaint to the Lhasa government. When there was an outbreak of scarlet fever in Lhasa, Heron's actions were blamed, and the Tibetans banned geologists from future Everest expeditions.

The British accepted this rule and then broke it. Between 1922 and 1938, they sent seven climbing expeditions to the North Face. These expeditions included scientists working as climbers, and four expeditions included geologists, the most represented scientific field. The expedition leaders saw a clear link between the work of geologists and that of climbers. Climbers needed geologists to determine the best climbing route, and geologists needed to climb to reach the mountain's rocks. In the pursuit of the summit, expedition leaders granted geologists leave to surreptitiously dig into the earth, extract rock samples, and desecrate sacred sites. Both the climbers and the field of geology dismissed the Tibetans' beliefs about extraction as superstition. They insisted that the study of the mountain's age, structure, economic worth, and climbing feasibility was not only imperative but a fundamental good. As imperial soldier, president of the Royal Geographical Society, and chairman of the Mount Everest Committee, Francis Younghusband stated, they believed "the spirit of mankind, as a whole, has also been affected for the good by these undaunted efforts to reach the earth's highest summit."[5]

Younghusband and his climber-scientists had, in effect, created—and were determined to "conquer"—an imperial mountain that was disconnected from the locals' sacred mountain. The imperial mountain, Everest, was conceived in 1856 when the Great Trigonometrical Survey of India calculated that Peak XV, as they then called it, was the highest point above sea level. Unable to gain entry to either Tibet or Nepal, the surveyors determined from afar that it had "no true native name"[6] and named it for their recently retired chief surveyor, George Everest. Once it was deemed an important object of knowledge, Everest became an inevitable site of imperial conquest. Its geological and geographical secrets had to be known, and it had to be climbed. Geologists began conducting expeditions in the broader Himalaya and, during the latter part of the nineteenth century, used rocks transported from the region to Europe to develop their science. Their early twentieth-century expeditions to the mountain were an explorative extension of the field's abstract construction of the mountain.

More significantly for Chomolungma's people, the British imperial idea of the mountain outlasted the empire that created it. The British expeditions left behind the soft (disciplinary knowledge and political access) and hard (roads, ropes, and scientific equipment) infrastructure that supported subsequent and contemporary geological studies of the mountain. This infrastructure enabled the two states that claimed sovereignty over the mountain in the 1950s, the Kingdom of Nepal and the People's Republic of China (PRC), to intensify their administration of the region. Like the British, they both renamed the mountain. The Chinese transcribed its local name as Zhumulangma,[7] and the Nepali state called it Sāgarmathā. For decades, the mountain's border was also a Cold War divide. Western scientists continued to conduct research and mountain climb on the Nepali side, dependent on underpaid Sherpa labor, Sherpa deaths, and Sherpa beliefs. When the PRC began conducting large-scale geological surveys of the North Face in 1975, its scientists worked within Chinese Communist Party rules that disallowed the acknowledgment of nonmaterial sacredness. The post-1975 geological surveys collated the bulk of the mountain's geological data. This data enabled geologists to describe the mountain's deep past while simultaneously speaking with the authority of cutting-edge, contemporary science. Through numbers and models, they made the mountain's longer-than-human story comprehensible to humans. In the process, they detailed a new origin story for the mountain. Dependent on the British construction of a global mountain, the geologists' new story predated and further marginalized local visions of the mountain.

As Dzatrül Rinpoche foresaw, the climbing-scientists' version of the mountain, their Everest, had become globally predominant, overshadowing but not erasing his community's Chomolungma. This community has maintained Dzatrül Rinpoche's view of Chomolungma. Since their initial encounters with British expeditions, the Tibetan and Sherpa communities have adapted their visions and rituals to accommodate new political (primarily on the Tibetan side) and economic (primarily on the Nepali side) regimes. Within these frameworks, many Sherpas and Tibetans have added the geological view of the mountain to their communities' already multiple views of it. Geologists' engagement with these plural local visions is possible, but it requires more knowledge about Chomolungma, and greater reflection on the field of geology's involvement in its erasure.

This chapter is about these different visions and their symbiotic material realities. It follows the encounters and frictions between them from the nineteenth century to the present, concentrating on the pre–World War II period, when the rules of engagement had not been hardened by nation-state geopolitics.

The Mountain Goddesses from the Lost Ocean

Rinpoche's vision of Chomolungma was created through his community's millennia-long relationship with the mountain. Most of this relationship was conducted across the plains of the Bumchu (upper Arun) River basin to the mountain's north. Humans have occupied these plains for between 3,900 and 6,700 years, primarily in irrigated river settlements around the villages of Dingri, Shelkar, and Kharta.[8] The plains' habitants were not climbers and generally tried to avoid mountains unless they needed to cross passes for trade or to use high pastures for herding or hunting. There is, however, much archaeological and textual evidence that mountains played a significant role in their world. Theirs was a *much-more-than-human* world in that it was occupied by not only humans and animals but also supernatural beings who lived in tiered abodes. Deities occupied the heights—both the sky and the mountains—along with nondomesticated animals and birds. Humans, domesticated animals, and land-bound spirits lived in the valleys. The subterranean and subaqueous zones were the homes of other classes of spirits. In this worldview, all mountains, and particularly snow mountains, were sacred realms. Activities that would morally or materially pollute them, like mining and digging, would bring the deities' wrath, causing illness and geological disasters.[9]

The oldest extant stories of the Central Himalaya's deities are those told from the perspective of the Buddhist kings and yogis who "tamed" them between the sixth and twelfth centuries. This taming required the deities'—often forced—conversion and their promises to renounce blood sacrifices and protect Buddhists. In these tales, two groups of goddesses are said to live in the Central Himalaya: the Twelve Stabilizing Deities (female deities that guard the passes to Tibet) and the Five Sisters of Long Life.[10] As goddesses, they are all given the honorific title *Chomo*, meaning "goddess," "princess," or both. As snow mountains, they were also called *Gangkar*, "white snow."[11] The deities' abodes were not fixed. People tended only to view the mountains—and, therefore, the goddesses—from afar in groups, and the assignment of individual goddesses to particular mountains changed over time.

Along with the taming of the gods, the introduction of Buddhism to Tibet brought with it other ways of understanding the mountains. Because Buddhism held that the material world was only provisional—which is to say understood through the senses but not existent outside them—those who promoted it were allowed multiple explanations of their world's material formation.[12] The only cosmological parameter was that nothing was permanent. Within these bounds, the Tibetans blended traditional mountain stories with Indic Buddhist cosmologies. The most influential of these was the cosmology outlined in the *Abhidharma*, which evoked vast temporal and cosmographic scales with few premodern counterparts. Along with this acceptance of large scales for existence, the *Abhidharma*'s most well-known earth-formation story was one that would find resonance in later geological theories: that continents rose out of the world ocean.[13]

Over generations, the Tibetans added sites associated with yogis' life stories to their sacred conception of the mountains. Many such sites in the Central Himalaya were associated with two of the Tibetan Plateau's most famous yogis, the imperial-era semimythical Padmasambhava and the eleventh-century poet Milarepa. One of Padmasambhava's most sacred sites is the Pelmotang Plain, eighty-six kilometers from Chomolungma, where he tamed the Twelve Stabilizing Deities.[14] He was also said to have visited Rongpu Valley and converted its flesh-eating demonesses to Buddhism.[15] Milarepa was a reformed murderer who spent much of his life in repentance, performing yoga in the Central Himalaya's uninhabited valleys.[16] Some of his most sacred sites are just west of Chomolungma in the Rongshar, Labchi, and Nyalam Valleys. During his stay in the valleys, Milarepa is also said to have practiced sexual yoga with the mountains' Five Sisters of Long Life, whose relationship with Milarepa increased their fame. The Five Sisters of Long Life came to be associated with the mountain Tibetans call Chomo Tseringma and the Nepalis call Gauri Sankar.[17] One of these sisters, the southern, yellow deity who rode a tigress and enriched those who respected her, was associated with the mountain at the end of Rongpu Valley. Her full name was Chomo Miyolungsangma; on the Tibetan side of the mountain, this name was shortened to Chomolungma (fig. 3.1).[18]

From the fifteenth century onward, the sacred site network began to include a series of "hidden lands" that offered shelter to Buddhists seeking refuge from war and strife.[19] There was much from which they needed shelter. The Little Ice Age (c. 1300–1850) lasted for an extended period in the mountains, destabilizing the climate.[20] Central Asian groups raided and sometimes ruled the Plateau from the time of the Mongol Empire (1206–1368) until the twentieth century. The fifth Dalai Lama, Ngawang Lozang Gyatso, and his Ganden Podrang government eventually consolidated their rule over most of the Plateau's central river valleys in the seventeenth century, but they did so with the help of first Mongol rulers and then the Manchu dynasty (1644–1912), both of whom were known to displace and occasionally massacre the Ganden Podrang's competitors.[21]

The Sherpas (whose name, *Shar pa*, means "easterners") entered the "hidden lands" on Chomolungma's southern side as refugees between the fifteenth and nineteenth centuries. Their histories claim they fled their homelands in eastern Tibet to escape Mongol troops, arrived in south central Tibet to find

FIGURE 3.1. The Five Sisters of Long Life. Chomo (Miyo) Lung (Sang) ma is in the top right corner. Goddess of Mount Everest, Tseringma, Tibet, 19th century, pigments on cloth. Rubin Museum of Art, gift of Shelley & Donald Rubin Foundation, F1996.10.2 (HAR 433).

more fighting, and continued south of Chomolungma into hidden valleys where they made a new home.[22] There they practiced agriculture, herded yak, and became involved in Tibet-Kathmandu trade.[23] They also developed a complex sacred geography that included but was not focused on Chomolungma, who lay hidden behind closer mountains.[24]

Chomolungma was not, however, utterly devoid of attention. North of the mountain, the number of meditators living in the Rongpu Valley increased until, by the late nineteenth century, Dzatrül Rinpoche began long-term retreat there. In 1901, he founded the Dza Rongpu Monastery and supported a sizable all-season monastic community. Shortly after this, he wrote a *Guide to Dza Rongpu*, in which he praised Chomolungma and likened her abode, the mountain, to a "crystal stupa."[25]

Despite developing a protectorate relationship with the Nepali rulers in 1772, the Sherpas were mostly left to govern themselves.[26] They were fortunate. Between the mid-seventeenth and nineteenth centuries, the emerging Nepali state, the Tibetans, and (in the 1790s) the Tibetans' Manchu backers fought several wars over trade and territory disputes in valleys just to their west.[27] Although these wars did not affect the Sherpa or meditators directly, they did lead to the region's high passes being recognized as Nepali-Tibetan border posts. This outcome would later impact the region profoundly.

Magic Mathematics

Chomolungma's story might have unfolded much like other mountains in the region but for the Great Trigonometrical Survey of India.[28] The Survey started its work in South India in 1802 and George Everest expanded it when he became surveyor general of India in 1818. Its members mapped India using large theodolites, precision optical instruments that measured angles between points on horizontal and vertical planes. When Everest retired in 1841, the Survey had mapped India from its southernmost point to the foothills of the Himalaya. His successor, Andrew Waugh, head of the Survey between 1843 and 1861, surveyed the Himalaya.[29]

As the British had only limited control of the mountains and were banned from Nepal and Tibet, they relied on "Pundits," Indigenous surveyors who traveled into Tibet and Nepal disguised as traders and pilgrims. Even they had trouble seeing Chomolungma. Kumaoni Hari Ram mapped the Bumchu River in 1871 but did not see the mountain.[30] The Sikkimese surveyor Rinzin Namgyal traveled through northern Nepal and southern Tibet between 1884 and 1885, but only saw Chomolungma from a distance.[31] Confined to India, British surveyors used multiple theodolite sightings of the Central Himalaya from hill stations and plains to calculate the mountains' heights. They first observed "Peak B" in 1847 and assumed it had a lower elevation than either Kangchenjunga in Sikkim and Nanga Parbat in the Western Himalaya. But in 1852, the Survey's chief computer, the Bengali mathematician Radhanath Sikdar, determined that mountain, renamed Peak XV, was the world's highest mountain at exactly 29,000 feet (8,839.2 meters). So that Sikdar's calculation would not be mistaken for a rounded estimate, Waugh added two feet to the number, making it 29,002 feet (8,839.8 meters).[32]

It was Survey policy to give mountains their local name, but the surveyors were too far away to determine Peak XV's name. Darjeeling locals called it "Deodungha" (Dev Ḍhuṅgā), meaning "god's stone," a generic term for sacred mountains, much like *Gangkar* in Tibetan. Surveyors also mistook Chomo Tseringma's mountain, Gauri Sankar, for Chomolungma.[33] After making inquiries, Waugh decided that it had no local name[34] and determined it should be named for his predecessor as surveyor general, Everest. Colonel Tanner, reporting for the Survey in 1883, noted that "Everest has all the appearance of a very moderate hill, not in the least imposing and hardly picturesque . . . (it is made) interesting only by trigonometrical operations."[35]

The topographical survey of the Himalaya led to renewed interest in its geology.[36] The British government had been conducting small-scale excavations in the mountains since the early 1800s[37] and shipping rock fossils from the mountains back to the Indian Museum in London. Its surveyance rate increased dramatically with the establishment of the Geological Survey of India in Calcutta (Kolkata) in 1851. This survey's initial remit was to find coal to fire an industrial revolution in India, but it also conducted scientific explorations. These explorations significantly increased the flow of Himalayan fossilized rock to London.[38]

In 1862, the Austrian geologist Eduard Suess observed these fossils and "noted the stratigraphic correlations of Triassic fauna between the Himalaya and Europe."[39] Suess was a longtime student of the European Alps and combined this knowledge with his study of Himalayan fossils to develop ideas about mountain formations. Until Suess, most geologists had argued that mountains were caused by volcanoes or steady shrinkage of the earth's surface. Suess, by contrast, maintained that mountain building—or orogenesis—created volcanoes. His 1893 article "Are Great Ocean Depths Permanent?" claimed that the asymmetric movement of rock layers caused by the earth's cooling led to slow, lateral, localized contractions, such as folding and overthrust, which also moved ocean basins.[40] One such movable ocean basin, he argued, was the Tethys Sea (later the Tethys Ocean), which once stretched from the eastern end of the Himalaya to Spain before orogenesis displaced it. He named this new ocean for the mythology at its western end:

> The folded and crumpled deposits of this ocean stand forth to heaven in Thibet [*sic*], the Himalaya, and the Alps. This ocean we designate by the name "Tethys," after the sister and consort of Oceanus. The latest successor of the Tethyan Sea is the present Mediterranean.[41]

He extrapolated on this idea in his most extensive work, *Das Antlitz der Erde* (1883–1901), translated into English as *The Face of the Earth* and published between 1904 and 1908.[42] In this work, he laid out his theory of buckling lands, disappearing oceans, and spasmodic orogenesis.[43] His disappearing Tethys Sea—reminiscent of the *Abhidharma*'s disappearing ocean—had a profound impact on Himalayan geology. In 1935, geologist John Auden, brother of the poet W. H. Auden, named the plains that separate Tingri and Shelkar from Chomolungma, "the Tethys Himalaya."[44]

Suess accomplished this reformation of the mountains' geological model without visiting them. Martin Rudwick called Suess's work "the outstanding example of (the) globalization of geology."[45] It also represented the most abstracted understanding of the mountain. Suess did not engage with local knowledge from the Himalaya to form his vision of it, and he was spared the hardships and cultural confrontations of being there. The next group of Himalayan geologists, by contrast, were determined to contend with both, and they transformed how Himalayan geology was conducted, not always to its betterment.

Rock Climbing

In the first half of the twentieth century, European, primarily British, geologists began attaching themselves to imperial expeditions into unexplored parts of the Nepali and Tibetan Himalaya. The stated purpose of these expeditions was to maintain British India's defenses and trade routes, but gradually their focus shifted to climbing or "conquering" Everest, a feat that would symbolize British supremacy. As with much British empire-building and scientific exploration across South Asia, Indigenous labor carried out much of their physical labor. In the mountains, as Sherry Ortner described in detail, the Sherpa performed most imperial labor.[46]

The first surveyor to breach Nepal's and Tibet's ban on Europeans was Henry Wood, who was permitted entry to the Kathmandu Valley in 1903. From Kathmandu, he could see Gauri Sankar (Peak XX, Chomo Tseringma) clearly, but not Everest (XV), and disentangled the two mountains of the Five Sisters of Long Life in the Survey's imagination.[47] Later that year, the euphemistically titled "Tibet Frontier Commission" invaded Tibet. Headed by Francis Younghusband, this expedition killed many Tibetans and occupied Lhasa for nearly a year. Its geopolitical purpose was to open Tibet to trade and stop purported Russian influence. While in Lhasa, the British forced the Ganden Podrang government to host British delegations in Lhasa and Shigatse, and the British continued to occupy the Chumbi Valley until 1908.[48]

Charles Ryder, a geographer, and Henry Hayden, a geologist, were members of the Younghusband expedition. Hayden noted that the north side of the Himalaya contained sedimentary rocks that had "been converted by heat and pressure into gneiss" and came in "at least two varieties, a foliated rock composed essentially of quartz, felspar, and biotite (black mica), and a younger non-foliated form containing, in addition to quartz and felspar, white mica (muscovite), black tourmaline, beryl, and various accessory minerals." But he was forced to admit that "the most important mass of all, the Everest group, is still a blank on our geological maps."[49] Both surveyors also insisted that there was no local name for the mountain. In his 1904 report on the invasion, Ryder claimed, "Nowhere could we hear of any local name for Everest."[50] Writing three years later, Hayden agreed, stating, "After 50 years of controversy no true native name has been produced for Mount Everest: each of those suggested has in turn been shown to be inapplicable, and the evidence that no such name exists is overwhelming."[51]

World War I interrupted the region's exploration, but soon after it ended, Younghusband and other members of the Mount Everest Committee began planning an "assault" on Mount Everest.[52] The 1921 Everest Reconnaissance Expedition that Dzatrül Rinpoche noticed from his retreat hut was the vanguard of this offensive. It was made possible by the intervention of Charles Bell. He had been the British political officer in Sikkim when the thirteenth Dalai Lama, Thubten Gyatso, had been in exile there. The Dalai Lama invited Bell to Lhasa, and on an extended visit to the Tibetan capital in 1920, Bell convinced the Ganden Podrang government to allow a British Everest expedition. Notably, the Dalai Lama had not heard of the mountain and did not know it lay within Tibetan territory. Nevertheless, many Tibetans saw the proposed expedition as a threat to both their sacred sites and independence.[53] Bell assured them that the British would respect their sacred sites and refrain from invading them again. There is even some evidence he suggested the British would help Tibet against the Chinese. The passport the expedition was granted alludes to this new relationship, using the Tibetan word for "ally" to describe the British.[54] It also contains a Tibetan name for the mountain that

Bell claims to have heard in Lhasa, Lho Cha-ma-lung, which he translated as "the southern district where birds are kept." It is probably a Lhasa Tibetan calque of Chomolungma. The origins of this name for the mountain are unclear, but Bell's use of it even encouraged some Tibetans to promote this as the mountain's name.[55]

The Everest expedition set off shortly after it gained the Tibetan passport. It was led by Charles Howard-Bury, a writer and naturalist. Alexander Heron was part of his team. In his book on the expedition, Howard-Bury noted that the people of "Rongbuk Valley" were unhappy about their arrival as it would "disturb and distract their meditators." He also noted that locals had "heard vaguely of the fighting in 1904, and they imagined that our visit might be on the same lines. They imagined, too, that all Europeans were cruel and seized what they wanted without payment."[56]

While other expedition members assessed possible climbing routes, Heron surveyed "the Tibetan portion of the Arun (River) drainage area, with, in the West, the headwaters of the Bhotia Kosi and its tributaries."[57] His report, published as an appendix to Howard-Bury's tome on the expedition, noted the following:

> Geologically this area is divided into two broad divisions: (a) Tibetan and sedimentary, (b) Himalayan and crystalline, a distinction which is clearly displayed in the topography resulting from the underlying geological structure, for to the North we have the somewhat tame and lumpy mountains of Tibet contrasting with the higher, steeper, and more rugged Himalayas on the South.

North of the mountains near Tingri, in the Tethys Himalaya, Heron saw evidence for Suess's lost sea, and in the south near the mountain, a band of crystalline rock. He did not take rock samples from Everest itself but wrote a short description of it.[58] In his longer report, he noted that other mountains in the area were made of biotite gneiss, a hard metamorphic rock that did not erode quickly, and that there were no rocks of commercial interest (fig. 3.2).[59]

As noted in the introduction, Heron also caused an international scandal after his digging caused locals to send a telegram to the Tibetan prime minister, Lönchen Shédra Penjor Dorjé, in Lhasa. The lönchen passed on his displeasure to Charles Bell in a telegram that accused the expedition—and primarily Heron—of digging in sacred ground and carrying away "precious stones, turquoises from the sacred valley of Rongshar and rubies from Shegar Dzong." The lönchen condemned the expedition as "an excuse for digging earth and stones from the most sacred hills of Tibet, inhabited by fierce demons, the very guardians of the soil" and warned that, as a result, "fatal epidemics may break out amongst men and cattle."[60] The concurrent

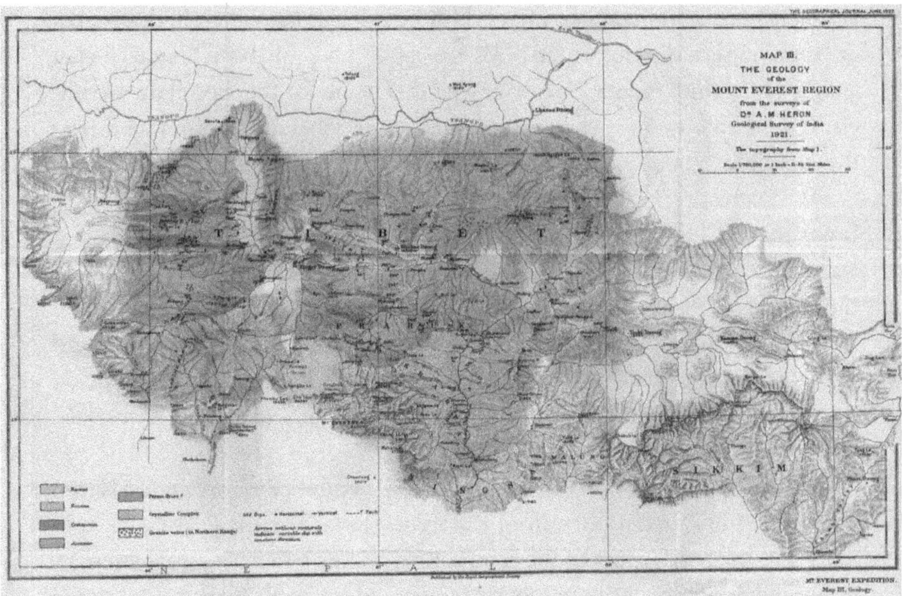

FIGURE 3.2. Alexander M. Heron, "Geological Results of the Mount Everest Expedition, 1921," *The Geographical Journal* 59, 6 (June 1922): 418–31; map at p. 480. Royal Geographical Society.

scarlet fever outbreak in Lhasa increased their fears. In a letter to the Royal Geographical Society, Heron dismissed the Tibetans' objections as superstitions and pleaded "'not guilty' to the charge of being a Disturber of Demons." He further wrote, "I did no mining and the gentle hammer tapping which I indulged in was, I am sure, insufficient to alarm the most-timid of the fraternity."[61] The Tibetans and Bell were unimpressed by Heron's defense and issued a ban on geologists.

There were no geologists on the next expedition, headed by Charles Bruce, but this group was described in greater detail in Dzatrül Rinpoche's biography. In this text, Rinpoche interrupts his descriptions of his meditational visions to note his displeasure at the Sahibs' arrival.[62] Developing on the importance of pilgrims' view, which he had described in his *Guide to Dza Rongpu*, Rinpoche stressed the importance of entering sacred sites with the proper perspective and intention. Those who were incapable of respectfully enacting sacred sites' visualizations, he explained, would degrade its sacredness.[63] The climbers, he went on to suggest, would also be the first of many outsiders with "back-the-front" views, and he did not particularly want to meet them. "If you meet with one person with back-the-front views," he wrote, "you will not stop the rest of them coming."[64] Nevertheless, because the leaders of Dingri and Shelkar insisted that he meet with the Sahibs, he resigned himself to their presence and agreed to meet them.

Rinpoche's record of the meeting describes his baffled tolerance for the visitors' bizarre behavior. "Where are you going?" he reports asking Bruce. "This is the highest mountain in the world," Bruce replied. "If we arrive at its peak, the English Government will give us a large payment and high honours." Rinpoche recalls trying to dissuade him by explaining that the valley was "bitingly cold and extremely difficult for those who are not Buddhist (yoga) practitioners. What is more, the local earth deities are powerful and uncompromising." Bruce ignored his warning but returned several weeks later, asking him to perform prayers for seven Sherpas who died in an avalanche. Rinpoche was even further disquieted when a group of local teenagers nearly died trying to retrieve the expeditions' remnant supplies. Reflecting on their behavior, he noted his "intense compassion for people who undergo such suffering for no good reason."[65] Bruce took Rinpoche's photo (fig. 3.3). Rinpoche

FIGURE 3.3. Dzatrül Rinpoche and attendants, Rongpu Monastery. Charles Granville Bruce, *The Assault on Mount Everest, 1922* (New York: Longmans, Green, 1923), 78. Courtesy of Chronicle / Alamy Stock Photo.

looks at the Sahib's camera evenly, and the monks around him are equally unimpressed. It could look to the casual observer as though this photograph was staged. But those used to spending time in monasteries recognize the arrangement of the monks around Rinpoche as a common social formation.

After meeting Bruce, Rinpoche retreated into meditation and refused to bless later expeditions. The British took advantage of his retreat and began once again sneaking in geologists. Noel Odell was a geologist-climber on the 1924 expedition. He added to Hayden's report by documenting Everest's upper strata and the "great belt of crystalline rocks [that had] developed along the main axis of the Himalaya."[66] He also suggested correctly that the trans-Himalayan rivers may predate the mountains.[67] Tellingly, his geological reports note that he took rock samples from Rongpu and nearby Rongshar, a sacred Milarepa site, entry to which was forbidden. This incursion may have been overlooked because of George Mallory and Andrew Irvine's death during the expedition, and a concurrent scandal about monks performing rituals as entertainment in London.[68]

After the climbers' deaths, the Ganden Podrang suspended permits to the mountain for nearly eight years until a team led by Hugh Ruttledge attempted the climb in 1933 and 1936. Geologist Lawrence Wager accompanied Ruttledge on both expeditions and provided a geological report for his leader's book.[69] His report begins by noting the "objection of the Tibetan Government" to his profession and suggesting, seemingly unaware of the cultural proscriptions against digging, that "perhaps the geologist is feared because he is often the forerunner of commercial exploitation."[70] He noted what he called the "Limestone Series" at the top of the mountain and the "Yellow Slabs" within it.[71] He agreed with Odell that the region's rivers predated the mountains and produced both a series of photographs and a detailed geological map that depicted their geology.[72]

This period of geological exploration of the Central Himalaya represented both breakthroughs in geological surveying and appalling precedents for geologists' practice: seven Sherpas had died, and sacred sites had been desecrated. But toward the end of this period, two Swiss geologists conducted an innovative survey and presented an improved (but by no means perfect) paradigm for working with local communities. Augusto Gansser-Biaggi and Arnold Heim were independent experts in Alpine geology who produced the Himalaya's first relief map.[73] Their geological findings, drawing on the work of Odell, Heron, and Wager, noted that "the highest mountain of our globe is weathered out of the normally stratified back part of the greatest known thrustfold."[74] However, unlike the British geologists, they acknowledged the contribution their Sherpa guides, Paldin and Kirken, made to their research.[75] Perhaps because of their close working relationship with the Paldin and Kirken, Gansser-Biaggi and Heim also became the first Europeans to use the mountain's Indigenous name. Throughout their writing, they called it "Choma Lungma (M. Everest 8982 meters)."[76]

Borders in the Sky: Tectonics and Geopolitics

Most histories of post–World War II Asia describe this period as "postcolonial." European empires were disbanded, and nationalist movements across the region demanded self-determination for those previously colonized. Decolonization is often said to have been accompanied by the "indigenization" of Asian science.[77] Rather than gains, the people of Chomolungma experienced the further erosion of their independence. The Kingdom of Nepal increased its control of the mountain's southern side and slowly opened it to climbers and scientists.[78] In 1953, Tenzin Norgay and New Zealand climber Edmund Hillary summited the mountain in a British expedition. As the world focused on Hillary and Norgay's achievement, the mountain's geopolitics were changing profoundly. The People's Liberation Army fought their way onto the Tibetan Plateau during the 1950s, and Tibetan refugees, including the young fourteenth Dalai Lama, Tenzin Gyatso, streamed over the mountain passes into Nepal and neighboring Sikkim, Bhutan, and India. The mountain became a Cold War border and a space of Communist-Capitalist contention.[79]

Cold War competition combined with the growing acceptance of, first, continental drift and then tectonic plate theory made the Himalaya a site of intense geological interest. The geologists stitched together a new story of Everest. It began when the pieces of Gondwanaland that became India broke off, tracked north, displaced the Tethys Sea, and subducted under the Eurasian Plate. The force of the subduction buckled and melted the Eurasia Plate's lithosphere, allowing magma to spill onto the Plateau's surface. In geological time, this happened relatively recently, sixty million years ago, and it remains an unfinished event, marked by frequent seismic disturbances.[80] These are young mountains, restless with earthquakes and landslides, and still rising. Evidence of the Tethys Sea was found in the sedimentary rocks of the southern Tibetan Plateau, particularly its southern section between the Central Himalaya and the Indus-Yarlung Tsangpo Suture Zone.[81] The Central Himalaya rose highest because it was composed of a mix of ancient seabeds and the magma that seeped upward as the lithosphere buckled, creating metamorphic rocks. This was the crystalized rock the early geologists noted. It became known as the "High Himalayan Crystalline Sequence."[82]

The region's new states established national geological surveys: the Republic of India inherited the British Indian Geological Survey, the Chinese Academy of Geological Sciences was founded in Beijing in 1959, and in Nepal, the Nepal Geological Survey was founded in 1967.[83] But for most of the 1950s, 1960s, and the first half of the 1970s, these organizations were poorly funded and concentrated their efforts on commercially exploitable surveys for mining or engineering projects.[84] Until 1979, all geological studies of Everest were conducted on the mountain's southern slopes, and most of its geologists were Westerners. None were Sherpas. The Chinese Academy of Geological Sciences made preliminary maps of Rongpu Valley in 1966–1968 and 1975.

As China's economy and influence grew after the 1980s, the region's geopolitics and geological studies changed again. In 1979, as China opened, Taiwan's Academia Sinica organized an international meeting of geologists that produced *A Scientific Guidebook to South Xizang (Tibet)*.[85] In recent years, the Chinese Academy of Geological Sciences has begun surveying and surveilling the mountain's north side and contributing to Nepal's Department of Mines and Geology. The majority of Nepali exploration had focused on mining and engineering projects rather than scientific discovery.[86] But China has begun sponsoring large scientific studies of the region's geology. Their approach to knowledge-gathering on the Plateau in many ways resembles British knowledge accumulation; they need knowledge to govern a region about which they have little prior knowledge.

The Mountains' Many Views

Few Tibetans and Sherpas have been involved in the PRC's geological surveys. The first Tibetan geologist, Dorje, began working in the 1990s but recently left science to become a high-ranking government official in the Tibetan Autonomous Region.[87] There is no record of other Tibetans following in his footsteps. As Tibetan students come from low socioeconomic backgrounds and must conduct all their studies in the Chinese language, it has been difficult for them to win competitive places in science education.[88]

But the Sherpas' ability to welcome multiple views of the mountain and carve out a resistant socioeconomic space for their community despite national and international interference in their affairs has been impressive. Their culture preserves generational memories that meld terrestrial transformations—earthquakes, glacier movement, avalanches, landslides, seasons—with their human histories and cosmologies. They have also accepted parts of the mountain's geological history, and they benefit from its global fame. The Sherpas' integration of ritual into their climbing economy, through the regular performance of forgiveness rites, demonstrates their pragmatism.[89]

The Sherpas' pragmatism and acceptance of multiplicity could represent a model for geological studies. Contemporary geologists are not compelled to accept the implicit claims of universalism and the imperialistic truth claims that underpinned the work of their nineteenth-century forebears. The future does not have to repeat the past, and there are some signs of hope that it will not. Over the past few decades, there has been a significant increase in Sherpas' living standards, a change in fortune that many attribute to the wealth-giving Chomolungma.[90] This economic growth has enabled some young Sherpas to study internationally, and among them are two young women completing PhDs in geology in the United States.[91] As Indigenous people from the developing world and scientists, these young women walk in many worlds, and they already know how to see both Everest and Chomolungma.

4

Think like a Fish

New Oceanic Histories

Anne Salmond, Dan Hikuroa, and Natalie Robertson

In a foreword to a recent book on climate change in the Pacific, His Highness Tui Atua Tupua Tamasese Ta'isi Efi, former head of state of Samoa, urged his readers to adopt a perspective based on *va tapuia*—sacred relationships between the cosmos, ancestors, humans, and animals.[1] He suggested that we might think about climate change from the vantage point of other life forms—perhaps a dog, the ocean, the stars, trees, a bird, or a fish—and explore Pacific worlds patterned by existential interconnections between people and other beings. Likewise, a generation ago, the Tongan scholar Epeli Hau'ofa conjured up a vision centered upon the Pacific, the largest geographical feature on the planet: "Oceania is vast, Oceania is expanding, Oceania is hospitable and generous, Oceania is humanity rising from the depths of brine and regions of fire deeper still, Oceania is us. We are the sea, we are the ocean."[2] Hau'ofa noted that most Western views of this great sea are based on a terrestrial vision:

> Continental men, namely Europeans, on entering the Pacific after crossing huge expanses of ocean, introduced the idea of "islands in a far sea." From this perspective, the islands are tiny, isolated dots in a vast ocean. Later on, continental men—Europeans and Americans—drew imaginary lines across the sea, marking the colonial boundaries that confined ocean peoples to tiny spaces for the first time.[3]

This imperial view of the Pacific Ocean, with its gridded control of space-time, also divided the earth's seas, one from another. In fact, our planet is ruled by one great, interflowing ocean, with its circulating winds and currents, migratory fish, and surging tidal rhythms (fig. 4.1). As the United Nations' *Second World Ocean Assessment* notes, this great sea "covers more than 70% of the surface of the planet, and forms 95% of the biosphere."[4] Rather than Earth, our planet might more accurately be called Sea—and instead of "New

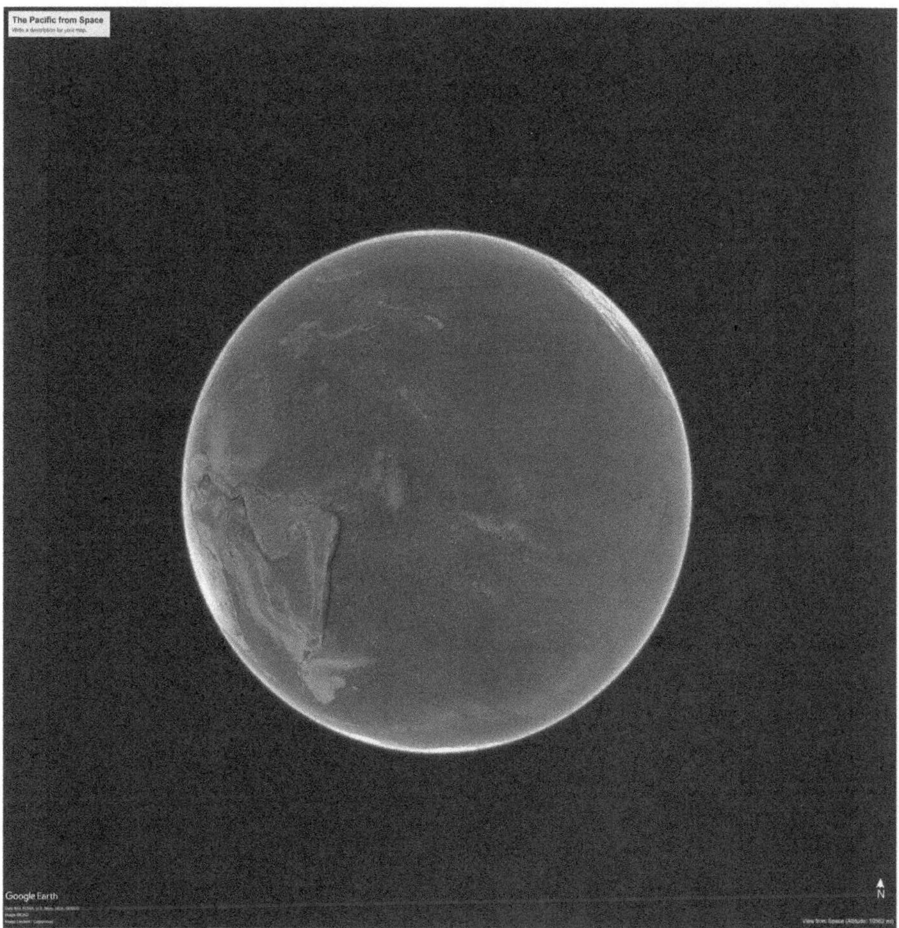

FIGURE 4.1. Pacific Ocean viewed from space—altitude: 10,562 miles (16,998 km). With permission from Google Earth. Data: SIO, NOAA, US Navy, NGA, GEBCO; Image: IBCAO; Image: Landsat/Copernicus.

Earth Histories," we might speak of "New Ocean Histories," a shift away from terrestrial framings and anthropocentric visions.

In world history, the ancestors of Pacific Islanders were the first to invent blue water sailing. They crossed the wide ocean following winds, currents, migrating whales, land-roosting birds, the sun, and star paths—successions of stars that rise above the horizon at night, marking the bearing of particular destination islands. Sailing was at the heart of their cosmological visions. According to Tahitian chants, at the beginning of the world, star ancestors sailed across the sky in their canoes, fishing up new stars and constellations. Star pillars were placed above voyaging *marae*, ceremonial centers on particular islands. As island navigators sailed through the night, watching the stars in the sky mirrored in the ocean below, they were retracing the sky voyages of their star ancestors.[5]

The first European explorers who followed the star navigators into the Pacific saw the sea differently. They had been sent in a race to claim new lands for their monarchs. Using instrumental observation, they measured and gridded the world, tracing coastlines and producing logs and charts as proof of their "discoveries." While they were also people of the sea, watching the sun and the moon and weathering storms, their mission was one of mastery—not just of the Pacific itself but of islands and their peoples. Since colonial times, a struggle between "thinking like a master," in which the sea and its inhabitants are understood as resources for human uses, and "thinking like a fish," one oceanic life form among many, has been ongoing in the Pacific, and around the planet.

Western and Tahitian Ways of Understanding the Sea

ANNE SALMOND

As it happened, the first close exchanges between island navigators and European cartographers began in Tahiti in April 1769, when the HMB (His Majesty's Bark) *Endeavour*, a Royal Navy vessel commanded by James Cook and carrying a Royal Society party of artists and scientists, arrived at the island.[6] Soon after, a leading star navigator and high priest from Ra'iatea, Tupaia, joined up with the Royal Society party.[7] Fascinated by their astronomical and artistic work, Tupaia often sketched with the ship's artists, and he guided James Cook and Joseph Banks during a boat trip around the coastline of Tahiti.

When the *Endeavour* party set sail three months later, Tupaia decided to go with them, hoping that Cook would help him to free his home island from Borabora invaders. During the passage through the Society Islands, Tupaia acted as the *Endeavour*'s pilot, calling up favorable winds and guiding the ship safely through passages in the coral reefs. Although Cook refused to attack the Borabora warriors, Tupaia took the *Endeavour* to Taputapuatea on Ra'iatea, the great voyaging *marae* and headquarters of the 'Arioi cult, where he had trained, leading the party through the rituals of arrival. The 'Arioi were artists, performers, warriors, and devotees of 'Oro, the god of fertility and war. They traveled long distances on their ceremonial voyages from island to island,[8] and Tupaia was a leading 'Arioi—a high priest, artist, and *fa'atere*, or star navigator.

The seas the 'Arioi voyagers sailed were very different from those charted by eighteenth-century Europeans. According to early Tahitian accounts, the ancestors of the ʻArioi saw the Pacific Ocean as a flat plane, joined around the edges of the horizon by the great arching bowl of the sky and crossed by sea paths among clusters of the known islands. It was also a *marae*, a sacred place where people went to cleanse themselves in times of spiritual trouble. The islands in this sea were fixed on a rock, Te Tumu or the "rock of foundation," and below this rock, and beyond the layered arches of the sky, there was Te Pō, a cosmic darkness inhabited by the gods and ancestors.[9] As star ancestors crossed the ocean in the sky at night on their star canoes, they were mirrored on the sea below. A great shark swam in Te Vaiora (Living Waters, or the Milky Way), while his *ata* (shadow), the blue shark, swam below in the ocean. The island of Tahiti was a great fish,[10] hauled up by the first voyaging ancestors. As they sailed across the sea, they sometimes followed migrating whales, the *ata* of Taʻaroa, the creator ancestor.[11]

In Polynesia, voyaging itineraries were commonly recited as island lists, typically including the bearing of the destination island from a particular point on the home island, the succession of stars rising on that bearing throughout the journey (the *rua*, or star path), the duration of the journey in *pō* (nights), the zenith star (*ʻaveiʻa*) that marked the position of the destination island, and information about each island.[12] During the party's passage through the Society Islands, Tupaia worked closely with Cook and Molyneux, the ship's master, giving them the names of about 130 islands and sharing some of his navigational knowledge. As a result of these conversations, Cook arrived at a remarkably contemporary conclusion about the exploration and settlement of the Pacific:

> In these Pahee's [pahi] . . . these people sail in those seas from Island to Island for several hundred Leagues, the Sun serving them for a compass by day and the Moon and stars by night. When this comes to be prov'd we Shall be no longer at a loss to know how the Islands lying in those Seas came to be people'd, for if the inhabitants of Uleitea have been at Islands laying 2 or 300 Leagues to the westward of them it cannot be doubted but that the inhabitants of those western Islands may have been at others as far to westward of them and so we may trace them from Island to Island quite to the East Indias.[13]

When the *Endeavour* set sail from Raʻiatea, Tupaia urged Cook to head to the west, where he said there were plenty of islands that he had visited in a journey that took "10 to 12 days in going thither and 30 or more in coming back."[14] Despite Tupaia's entreaties, however, Cook set his course south, intent on resuming his search for Terra Australis.

During the long days that followed, Tupaia often sat in the great cabin

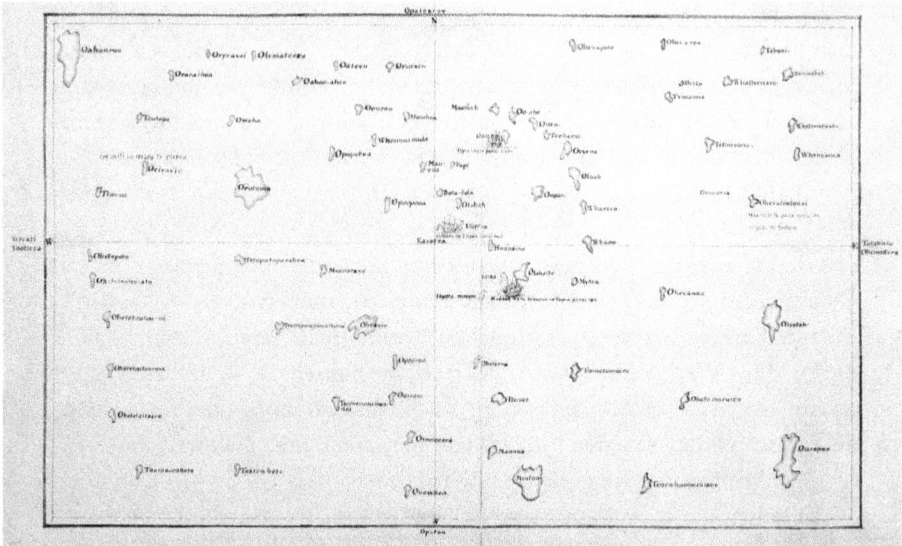

FIGURE 4.2. Tupaia's map, 1770. British Library Board BL ADD MS 21593c. Courtesy of the British Library.

with Cook and the ship's officers, helping to draft a remarkable map of the Pacific Ocean (fig. 4.2). It seems that Cook, Molyneux, or perhaps Isaac Smith, Cook's nephew who assisted him in drafting the ship's charts, first laid out the Society Islands, using their own charts (to which Tupaia had already contributed) as a basis. Tupaia then named and placed a series of other islands on the chart in arrays based on voyaging itineraries and island bearings that he had memorized during his training and travels as a *fa'atere* (star navigator).[15] On his list, Cook marked the names of twenty islands Tupaia said he had visited, including at least ten in the Society Islands,[16] Niue, two islands in Eastern Samoa, Rurutu (Orurutu), and two in the Cook Islands.[17]

Like Tupaia, James Cook was a deep-sea sailor, skilled at crossing the world's ocean and reading its signs. He was also a distinguished hydrographer, and the chart on which he and Tupaia worked together was a cosmological compromise, based largely on Western cartographic conventions. The navigational instruments used for such journeys were computational devices in which generations of voyaging experience were embedded.[18] In the *Endeavour*'s charts, the ocean—normally gray or blue-green, the home of birds, fish, and whales, surging with tides and currents, ruffling or roaring in the wind—was transformed into a static, white, two-dimensional expanse, gridded by lines of latitude and longitude and mathematically partitioned and measured.

Cook's mission was to discover and document new lands, with a particular focus on coastlines. Near harbors or lagoons, the depth of the coastal seabed was measured with lead soundings, carefully recorded on his charts. In a process of instrumental observation, the blurred, shifting liminal zone between land and sea was reduced to a simple line. In the ship's log, the official record of the voyage, space and time were also gridded. On board the *Endeavour*, for instance, ship's time was measured by an hour glass and corrected at noon when the sun was at its zenith, and the hours for each day were recorded in the first column in the ship's log. Also at noon, the angle between the horizon and the sun was measured with a sextant, establishing the ship's latitude; its longitude was established by comparing ship's time with Greenwich time, using a nautical almanac and mathematical calculations. The speed at which the ship was traveling was measured with the "chip log," a triangular piece of wood dragged behind the ship on a knotted rope, and recorded as "knots" per hour in the next two columns of the log; course or direction, determined by reference to dead reckoning and the compass, was recorded in the fourth; and the direction of the wind in the fifth column of the log.[19]

This process of cartographic and textual gridding and control was linked with colonial expansion.[20] Cook had instructions from the Admiralty to claim any new lands he might "discover" for the British Crown, a prelude to their dispossession.[21] At this time in Europe, the sovereignty of the Crown (or *imperium*) was held to extend not only over the land but also to about three nautical miles from the coastline, or within cannon shot, although property rights (*dominium*) could be granted within that limit. While cartographic conventions precisely split land from sea, political interests extended control over land out into the ocean.

When the *Endeavour* arrived in Aotearoa in October 1769, these divergent visions of the sea shaped what unfolded. In Tūranga-nui-a-Kiwa, the first harbor they visited, according to ancestral stories, the people thought the ship might be Waikawa, a sacred island off the end of the Māhia peninsula, floating into their harbor. When Cook followed his Admiralty orders and took possession of "New Zealand," marching the marines ashore to set up a British flag, the first encounters on land and sea ended in the shooting of Te Maro, Ngāti Oneone rangatira (chiefly leader), and eight other local Māori. During their six-month circumnavigation, Tupaia handled most of the exchanges with Māori, often warning Cook and his men to be on their guard. The presence on board of a high-priest navigator from the ancestral homeland, Ra'iatea, and from Taputapuatea, a famed voyaging *marae*, transformed this journey. Many Māori thought that the *Endeavour* was Tupaia's ship. When the *Endeavour* arrived at Waikawa, off the end of the Māhia peninsula, a sacred island and the site of a school of ancestral learning, *tohunga* (priestly experts) chanted and warriors in canoes threw spears at the hull of the vessel.

In the Bay of Plenty, a large canoe carrying sixty warriors came out from Whangaparaoa, in Te Whānau-ā-Apanui waters, and circled the ship, with a priest reciting incantations as the crew performed a *haka* (ceremonial war dance). When the crew of the *Endeavour* fired a cannon loaded with round shot overhead, the warriors and the priest fled back to the land. Off the tip of the Coromandel Peninsula, another ancestral voyaging site known to Māori as Te Tara-o-te-Ika-a-Māui (the jagged barb of Māui's fish), warriors in two large carved canoes threw stones at the side of the vessel. When Tupaia warned them to stop this aggressive behavior, insisting on the Tahitian idea of the sea as a great *marae*,

> they answerd [*sic*] him in their usual cant "come ashore only and we will kill you all." Well, said Tupaia, but while we are at sea you have no manner of Business with us, the Sea is our property as much as yours.[22]

The idea of "property" introduced here must have come from Joseph Banks, Tupaia's translator, since in Polynesia, the sea was an ancestral realm. While islanders would guard their coastlines from strangers, these were ritual challenges, asking them to identify themselves and their *atua*, and to make known their intentions.

On the whole, Cook respected these provocations, retorting with warning shots rather than shooting the warriors. In his instructions to Cook, the president of the Royal Society had insisted that people in these new lands had the right to defend their own territories, including their coastal waters. Later, this same understanding underpinned the promise in the English text of the Treaty of Waitangi, signed between the leaders of various *hapū* (kin groups) and Queen Victoria in 1840, that Māori would enjoy "full, exclusive, undisturbed possession of their Fisheries and other properties . . . so long as it is their wish and desire to retain the same in their possession."[23]

As accomplished deep-sea navigators, James Cook and Tupaia had much in common, yet their understandings of the ocean were very different. On one hand, Cook was part of an imperial contest to chart and claim "new" lands across the Pacific for his monarch. On the other hand, Tupaia—like the islanders they met in Aotearoa—engaged with the sea and its inhabitants through ancestral kin networks. Stars and constellations, islands and archipelagos, people and their canoes, and fish were existentially interconnected in an oceanic world in which sky above and land below (at the *pae*, or horizon), and land and sea (at the *tai*, or shoreline), were interfolded and fluid. These different knowledges of the ocean, coastlines, sea, and sky continue to unfurl and clash today.

Māori Understandings of the Ocean

NATALIE ROBERTSON / NGĀTI POROU, CLANN DHÒNNCHAIDH

In the millennia before European arrival, the seafaring people of Te Moana-nui-a-Kiwa lived in intimacy with the ocean. Hinateiwaiwā—the moon—"weaves the cycles of time and tides."[24] Cosmogonic narratives connected the immense spaces of stars above and ocean below, while genealogies across the sea recalled the first voyaging ancestors. In Aotearoa New Zealand, as in Tahiti, islands were seen as fish, swimming through the ocean.

The North Island, for instance, is known as Te Ika-a-Māui (Māui's fish), hauled up from the sea by the hero ancestor Māui. In 1871, the *tohunga* (scholarly expert) Mohi Ruatapu of Tokomaru Bay chanted the story of Māui pulling up his great fish:

<table>
<tr><td>Taku ika, taku ika</td><td>My fish, my fish.</td></tr>
<tr><td>I akuakuna, i akuakuna–</td><td>That cannot be moved, cannot be moved</td></tr>
<tr><td>I akoakona, [i] akoakona,</td><td>It's moved, moved,</td></tr>
<tr><td>Ka hapahapai he toka-nuku,</td><td>It's lifted up, an earth rock,</td></tr>
<tr><td>Ka hapahiipai he toka-rangi,</td><td>It's lifted up, a sky rock!</td></tr>
<tr><td>Tau tika, tau tonu,</td><td>Float straight up, keep going,</td></tr>
<tr><td>Tau tonu koe kite hiipai-rangi e!</td><td>Keep going to the sky-lifting![25]</td></tr>
</table>

When Te Ika-a-Māui emerged from the sea, Mount Hikurangi in Tairāwhiti (the East Coast of the North Island) was the first place to surface. As Māui hauled up the great fish, his *waka* (canoe) became stranded on top of Hikurangi, where its prow can be found in petrified form by Roto Takawhiti, the small lake nestled below two peaks representing *mana tāne* and *mana wāhine*, the ancestral powers of men and women. Most Ngāti Porou *whakapapa* (genealogies) descend from Māui. We live on a fish-island that still swims in the sea.

In Ruatapu's account, Māui's canoe is named Te Tuahiwi-nui-o-te-rangi. This is also a name for the Milky Way, referred to as the canoe of Māui or his fish, Te Ika-a-Māui. The night sky *is* the ocean. Stars can be seen in brilliant clarity from the summit of our ancestral mountain Hikurangi, sailing through the sky. Māori astronomers Pauline Harris and her colleagues identify many other names for the Milky Way, including Te Mangaroa, Mangoroa, Te Ika-a-Māui, Te Ika Nui, and Te Ika Roa, names for sharks and other fish. The star bodies are also fish, swimming through the night sky.[26]

In Ruatapu's version of the Māui story, his *aho* (fishing line, line of descent) is called Te Aweawe-o-te-Rangi. "Aweawe" has several meanings, but all circle around ideas of influence or reach. Names ending in *rangi* usually indicate the sky or heavenly realm. *Aho* are woven strands that connect earth and sky, linking earthly and celestial ancestors and knowledge. In 1944, the Ngāti Porou scholar politician Apirana Ngata described how ancestral place names are understood:

> This class of names, which is termed reminiscent, will be found everywhere along the New Zealand coastline with the addition of the peaks and ranges, which were first to raise themselves to the vision of the voyagers.[27]

The phrase "to raise themselves to the vision of the voyagers" offers insights into an ancestral way of understanding the relationship between land, sea, and the navigators on board their double-hulled *waka hourua*. Mountain peaks and ranges raised themselves for the explorers to see, indicating geographical agency. This is in keeping with the Oceanic navigational vision in which the *waka* stands still while land is pulled toward it on an invisible *aho*, or fishing line. Ngata also explained that in the *takiaho* method for reciting *whakapapa*, fishing terms are used to connect people as "fish" strung along an *aho*—in a line on the cord. Ancestors are also understood as fish, hauled up out of the ocean.

A map of Aotearoa turned upside down shows the fish-like shape of Te Ika-a-Māui—known since colonization as "the North Island" (fig. 4.3). Oriented north and south along its backbone, and east to west, fin to fin, it resembles a giant *whai repo*—one of the enormous long-tailed eagle rays found off East Cape. In our oral histories, the many hapū of Tairāwhiti recognize that both land and the people come from the ocean. Western knowledge systems also acknowledge this. In geological terms, it is not very long since this island emerged from the sea. Te Riu-a-Māui / Zealandia, the continent of which Te Ika-a-Māui is a part, is still largely submerged underwater, and in deep time, moves around in the ocean.[28]

Just as islands could *be* sea creatures, so could people. Oceanic people could think *as* fish, because over the millennia, blue ocean navigators had forged close relationships with their fellow ocean-going travelers. Observations of migratory patterns of birds, fish, and marine mammals led star navigators to form close bonds with other species. In Māori, for instance, whales are known as Te Whānau Puhā, "the family that expel air."[29] As creatures that crossed the ocean and breathed air, the same as humans, these iconic cetaceans were revered. In that family, Paikea—the southern humpback whale

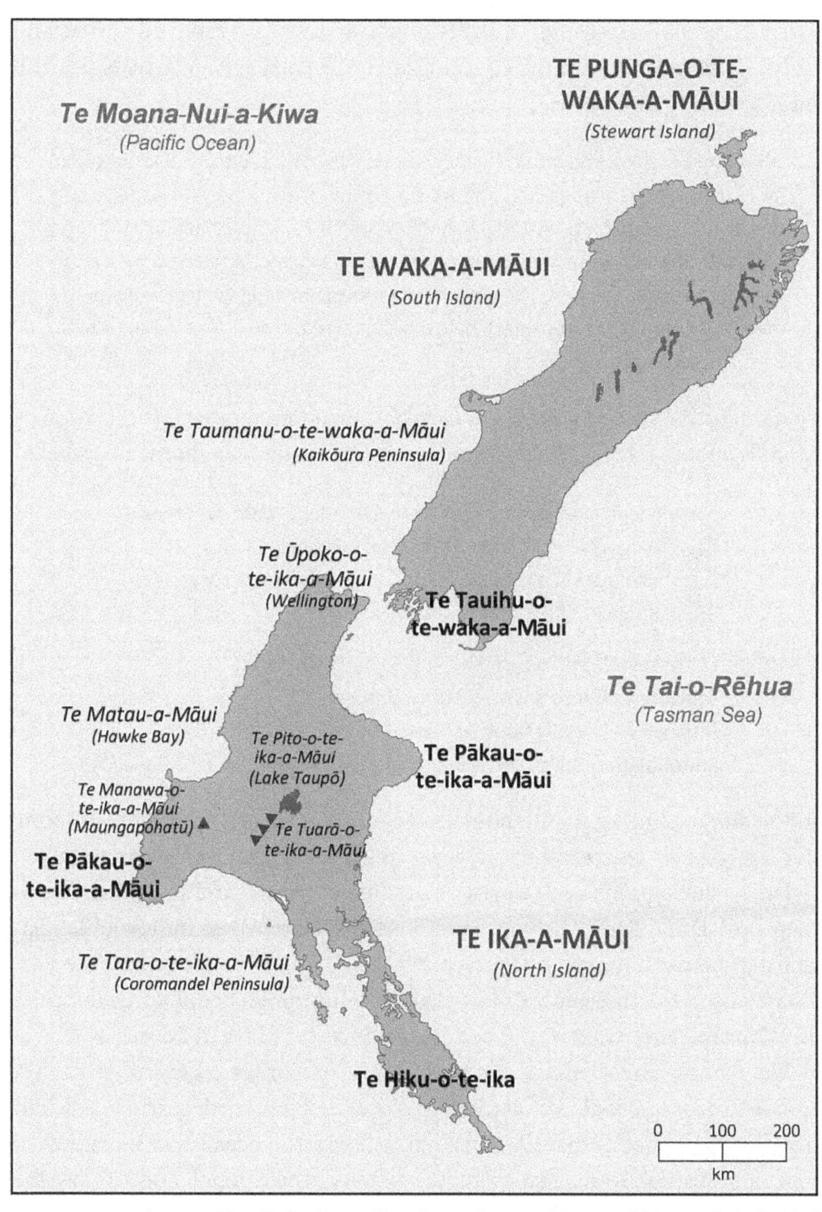

FIGURE 4.3. Aotearoa New Zealand map aligned according to a Māori worldview with the head of Te Ika-a-Māui upward. Concept and design by Dan Hikuroa; draft by Charles Hendtlass.

(*Megaptera novaeangliae*), with its white and deeply grooved throat and belly—is associated with the voyaging ancestor Paikea. As Wayne Ngata and Billie Lythberg write,

> It is commonly accepted that [Paikea] alone survived a marine disaster called Te Huripūreiata through his mobilization of his marine ancestors, his family of whales, who helped him reach Aotearoa. Paikea is described as riding on the back of a whale, or transforming into a whale, and is referred to accordingly as he tahito, he tipua, he taniwha, he tohorā, he tangata, he tekoteko—an ancient being, an extraordinary being, a denizen of the deep, a whale, a man, a sentinel for his people.[30]

Te Karakia Whakakau a Paikea, a *karakia* (incantation or chant) recorded by Mohi Ruatapu, records the ancestor's transformation from human to whale:

> Ka whakaika au i a au, ē Hiki-tai ē, Hiki-tai ē, Hiki-tai ore wae!
> Ka whakapakake au i a au, ē, Hiki-tai ore wae!
> Ka whakataniwha au i a au, ē Hiki-tai ē, Hiki-tai ore wae!
> Tū-maunga e, tīkina mai au, kawea ki uta![31]
>
> I turn myself into a fish, Hiki-tai, Hiki-tai ore wae!
> I turn myself into a whale, Hiki-tai ore wae!
> I turn myself into a taniwha, Hiki-tai, Hiki-tai ore wae!
> Standing mountain, come and take me to shore!

In this shape-shifting chant, the ancestor *becomes* Paikea, not simply thinking like a migratory cetacean but invoking "his *whakapapa* connection to *tohorā* (whales) through his mother (the matrilineal breath) and the mutuality of their breath and flesh (Tis your breath, and mine also), an indissoluble relationship between human and whale."[32] Upon arriving in Tairāwhiti, the East Coast of the North Island, Paikea married Huturangi, daughter of the chief Te Whironui and Araiara, and became an ancestor of Ngāti Porou.

Far from seeing whales as ancestors, however, Europeans saw them as a resource for commercial exploitation. In the early nineteenth century, as whaling ships swarmed into Te Moana-nui-a-Kiwa, the ocean was transformed into an industrial fishing factory that decimated whale populations. Today, the Paikea Southern Hemisphere humpback whale population has not recovered from near extinction levels, while the relationship between Māori and whales has been rendered predominantly symbolic.

In Tairāwhiti, an intimate relationship between Tangaroa and people still survives in the case of the *kahawai*, a migratory fish. In 1901, Tiimi Waata Rimini, an elder of Ngāti Awa and Te Arawa tribes, relayed a narrative about kahawai as the children of Tangaroa, ancestor of the ocean. After his son

FIGURE 4.4. Natalie Robertson, *Kahawai, Waiapu Ngutu Awa (River Mouth), Te Tai Rāwhiti (East Cape)*, 2014, photograph.

drowned in the Mōtū River, a man named Pou set sail and searched for his son in Hawaiki, the homeland. When Pou arrived at the home of Tangaroa, a "fountain of fish" seething with kahawai, he asked whether the sea ancestor had taken his son. Tangaroa denied it. In disbelief, Pou invited the sea ancestor to attend his son's *tangi* (funeral).[33] When Pou returned to his own region, the ocean of Toi-te-Huatahi, it was summer, and he told his people to make a great net. When the net was ready, the people looked out to sea and saw Tangaroa and his attendants—a huge shoal of kahawai—approaching, swimming inside Whakaari (White Island). As the shoal approached the mouth of the Mōtū River, Pou ordered his people to cast the great net. Thousands of Tangaroa's children were caught and fed to the crowds that attended the *tangi* to mourn Pou's son.[34]

The Mōtū River flows from the western side of the Raukūmara Peninsula while Waiapu River flows from the eastern side (fig. 4.4). As Rimini explained, when the kahawai arrive at the mouth of East Coast rivers every year, the story of Pou is reenacted. In order to lift the *tapu* (sacred restriction) of Tangaroa, a chiefly youngster was taken to catch three kahawai, which were offered to Pou and the high chief of the region. By acknowledging the mana of Tangaroa, Māui, and Pou, these "first fish" rituals guarded the fertility of the ocean. Rimini described catches of twenty or thirty thousand kahawai at a time in the ocean of Toi-te-Huatahi (Bay of Plenty). Rimini said that when the kahawai shoals arrive at the mouth of the Mōtū River, "the fishing lines thrown out on one side of the river and the other are as close as the telephone lines in Wellington. The fish there are as thick as if packed in an oven. If a paua shell should be thrown out on the shoal, it would remain on the surface."[35]

In 1923, Tūtere Wi Repa wrote an article for the *Gisborne Times* that also discussed the importance of kahawai fishing in Te Tairāwhiti. He mentioned "sacrifices to the residing deity, Pou" that are "dictated by variations in local physical conditions."[36] According to Wiremu Kaa, food gathering activities at Waiapu on the coast are still sacred.

> They were never performed or enacted lightly. These were the teachings we were exposed to by our *pakeke*, to be careful to respectful. If you do it right, you'll feel right and you'll be alright. "Ma te pai, te pai e utu, he pai to mai ka pai koe, te kino to mai ka kino ko koe."
>
> If you do it wrong, it will bite your bum. They would say it all in Māori: "Ngaua tautau, teeth will get onto your bum. Niho ki to whero—niho te kai o tautau."[37]

Kaa's message is that risky environments, such as tidal river mouths, are not sites for play or leisure—"*Nō te kahawai tēnā papa takaro*, that's the kahawai's playground, not yours! *Hoki mai ki uta.* [Come back to the shore.] *Tākaro* [play] on the shoreline, not in the water—*nō te kahawai tena kāenga*. That place belongs to the kahawai. Be a human, don't pretend to be a kahawai." He insists that we must always be mindful of the only home that the fish have—be aware that for the *kahawai*, "the *moana* [sea] is their home, and don't go in and spoil it."[38]

In his statement to the Waitangi Tribunal, established to address breaches of the Treaty of Waitangi, signed between Queen Victoria and Māori chiefs in February 1840, Ngāti Porou elder Tate Pēwhairangi, from Tokomaru on the East Coast, also cited ancestral insights into the ocean:

> Our tipuna had so much knowledge about their environment. Whether it be fishing by the signs and at certain times of year or when to plant, grow, harvest certain crops, they knew. They knew the fish were fat when the manuka was in bloom. They knew the significance of the kowhai in bloom.
>
> They used the Milky Way Te Ika o Te Rangi, and the phases of the moon for fishing and planting. They were a very knowledgeable people and worked in harmony with their environment. There was a use of the karaka berry, nikau, and other natural plants by the whānau living in those areas.[39]

Tautini (Tini) Glover, who was raised further along the coast at Ūawa, discussed the relationship between the stars and the prevailing winds: "[My grandfather] taught me about the pātiki (flounder) in the Milky Way. He said the direction the tail of the pātiki was facing was where the wind would come from. At night he would look up at the stars and say [t]here will be a southwesterly tomorrow. The next day would bring a wind just as he said." Tini also observed that, although he had learned about it from his grandfather, there was little application for the connection between stars and weather patterns in modern times.[40] Tini's lament may have been premature, however. Among Māori astronomers and environmental scientists, there is a resurgent interest in understanding the impact of astronomical cycles on land, sea, plants, and animals. Earth Systems scientist Dan Hikuroa describes both *pūrākau* (ancestral stories) and *maramataka* (astronomical calendars) as rigorous systems of codified environmental knowledge. He argues that these can be "both accurate and precise, as they incorporate critically verified knowledge, continually tested and updated through time."[41] Inspired by the revival of ocean voyaging and a care for the sea, many young Māori are once again studying and using knowledge of the ocean, *mātauranga moana*.

New Oceanic Histories

DAN HIKUROA / NGĀTI MANIAPOTO, TAINUI, NGĀTI WHANAUNGA

In "thinking like a fish," Te Moana-nui-a-Kiwa, the Pacific, is one expansive ocean, the great connector, in which the islands themselves are fish, moving through the water, mirrored above by the stars, which swim or sail across the sky ocean. Because the sea was so often crossed by our forebears, including Kupe, the great blue water navigator, some Māori regard it as the main *marae* of their ancestors, just as Tupaia asserted generations earlier. In 1955, Ngā Puhi elders lodged an application with the Māori Land Court for title to Te Moana-nui-a-Kiwa, the Pacific Ocean. The claim was based on "rights from Tangaroa, as descendant of Rangi and Papatuanuku; the act of Maui-tikitiki-a-taranga in fishing the island from the sea; on Kupe through his voyage to the island across this ocean, and his naming of points on land alongside it; and through human blood which Kupe smeared on his face when fishing the island from the sea."[42]

This claim was triggered by a recognition that the ocean was being damaged by extractive practices (including bottom trawling, overfishing, and mining for minerals), smothered with sediment from the land, and poisoned by the dumping of human and industrial waste. More recently, *waka hourua* (double-hulled open-ocean voyaging canoes built based on traditional designs) have been voyaging across Te Moana-nui-a-Kiwa, their crews using ancestral navigation techniques to demonstrate their deep knowledge of the ocean and their ongoing habitation of it. These activities are reestablishing *whakapapa* (genealogy) connections back into the Pacific, as well as rekindling relationships with the ocean as a living being, kin to human beings.

Over the last decades, however, Te Moana-nui-a-Kiwa has been the focus of a rush for control over resources and territories involving external powers, intertwined public and corporate interests, and Pacific Island states. Across the Pacific, the high seas, or mare liberum[43]—that part of the ocean that falls outside the exclusive economic zones, an expanse free to all nations but belonging to none—has been shrinking as nation-states seek to expand their terrestrial sovereignty out from beyond their coastlines. This is a kind of oceanic enclosure, extending the process of cartographic control that Cook and other European explorers initiated. In settler nation-states such as New Zealand, this justified the appropriation of Indigenous sea spaces, at first out to cannon shot—roughly three nautical miles beyond the coast—and then, in 1982 under the United Nations Convention on the Law of the Sea (UNCLOS), out to twelve nautical miles for territorial waters, and to two hundred nautical miles for an exclusive economic zone, thus placing these waters outside the "common heritage of mankind."

UNCLOS has two paradoxical purposes. On one hand, it defines what is "national," what is "ours," and hence the opposite, what is "not ours," by expanding territorial sovereignty out into the ocean. In this way, the sea is divided into exclusive economic zones that are managed as though they were land, charted and partitioned, as opposed to mare liberum, or "international waters," a view close to Tupaia's contention that the ocean belongs to everyone. On the other hand, it recognizes that the oceans are a single, complex integrated ecological system—a step toward "thinking like a fish," except that the vision is still anthropocentric. The objective of UNCLOS was to ensure that the ocean "continued to sustain present and future generations and that its uses contributed to peace, security, and the equitable development of peoples."[44] To realize this, a new legal principle was required, "one which claimed all ocean space as a commons (belonging to all humankind), and placed it under an international commons management regime, for the benefit of all."[45]

The problem is that, under UNCLOS, the marine environment is partitioned, with seas within national jurisdictions separated from each other and excluded from that overarching vision. In addition, although it focuses strongly on equity and sustainability, UNCLOS is limited by its human exceptionalism. The "all" in the "benefit for all" refers only to humans—fish, for example, are defined as "resources." As well, fish and other marine species do not recognize the lines that nation-states draw in the ocean; for instance, they swim beyond protected areas, or from exclusive economic zones into international waters, often resulting in the failure of management regimes based on this kind of thinking. "Thinking like a fish" would quickly tell us that this way of framing the sea is unlikely to lead to healthy oceanic outcomes.

Under New Zealand settler state law, where "inshore fisheries were a public right; they were a commons, a free for all in both the literal and figurative sense,"[46] fish stocks became severely depleted. In response to this "tragedy of the commons," in 1986 the government introduced the "world's first comprehensive privatised fishery management system,"[47] which set quotas for particular species, with the right to catch these quotas being bought and sold on the market. Māori, who had enjoyed the bounty of the oceans through generations, suddenly found themselves excluded from undertaking activities they had practiced since arriving in Aotearoa New Zealand. This new regime saw value realized only once the fish caught were turned into money, severing the kinship-based relationship with fish and fisheries, and the consequent responsibilities that came with invoking "user-privileges."

From the outset, Māori kin groups hotly disputed the Crown's right to introduce this regime. Under the Treaty of Waitangi, the *rangatira* (chiefly leaders) and *hapū* (kin groups) had been guaranteed "the full exclusive and undisturbed possession of their Lands and Estates Forests Fisheries." In 1985, northern kin groups appealed to the Waitangi Tribunal, arguing that they had never ceded control over their ancestral fisheries, and the Tribunal (and the courts) agreed.[48] In response, the Crown gave the claimants 20 percent of the quota for all new species set under the Quota Management System and the cash to purchase 50 percent of one of the country's largest fishing companies. The commodification of fish quota and its management by *rūnanga* (kin group corporates) has proved controversial, however, with kin-based relationships with fish and the ocean fundamentally disrupted. Many Māori fishers have been forced out of the industry and quotas sold off to non-Māori interests.[49] Although some allowance was made for "customary harvest," and subsequent legal battles wrestled some "use rights" back for Māori, these have proved difficult to manage, with practices at times aligning more closely with making money than feeding people, treating fish as kin, or taking care of Te Moana-nui-a-Kiwa.

A similar, controversial clash between Māori kin groups and the Crown centers upon the foreshore and seabed. In ancestral times, use rights tangled across the coastline, from gardens and fernroot diggings to eeling pools and birding trees, shellfish in sandy beaches, channels in reefs, and offshore fishing grounds. Families activated *whakapapa* (kin networks) to gain access to particular *mahinga kai* ("food working places") at particular times of the year, with accompanying responsibilities to look after them—although this was never understood as "ownership." As Natalie Robertson has described, it was a matter of kinship, of existential interlock with these places. Detailed calendars, or *maramataka*, containing seasonal information on fishing and instructions on when, where, and how to target specific species were developed. If particular species or areas became depleted, a *rāhui* or protective *tapu* was placed over them. Other measures to care for coastal ecosystems included the periodic flushing of *hāpua* (coastal lagoons), reseeding shellfish beds, and harvesting practices that ensured that breeding stocks were protected.[50]

These complex, overlapping networks of kin links with land and sea were fragmented after European settlement, when the land was divided from the ocean, surveyed, cut up into "blocks" (bounded areas with bounded lists of "owners"), and commodified. At the same time, successive governments assumed that, as in Britain, the "foreshore and seabed" was "owned" by the Crown, and they took control over these coastal zones. While kin groups protested, they were ignored until the Waitangi Tribunal was established in 1975 and, from 1985, began to report on historic breaches of the Treaty of Waitangi. Over that same period, Māori kin groups returned to the courts to uphold their rights under the Treaty.

In the case of the foreshore and seabed, the saga began in the Marlborough Sounds, at the northern end of the South Island. Although the local tribes had repeatedly applied to the local District Council for licenses to farm mussels in their ancestral *rohe* (territory), none were granted. Finally, in frustration, in 1997 they applied to the Māori Land Court to recognize their customary rights over the foreshore and seabed in the Sounds, where their claim was upheld. The case was referred to the High Court, where the judge reversed the ruling, and then to the Court of Appeal, where the judges ruled unanimously that upon the signing of the Treaty, the Crown had acquired only a radical right or *imperium* over the sea with the acquisition of sovereignty. Citing the doctrine of Aboriginal title, they ruled that unless the rights of *dominium* had been legally extinguished, they remained with Māori kin groups, and that this was also the case with the foreshore and seabed. Furthermore, they argued, the distinction in English common law between land above the high-water mark and land below it did not apply.[51]

The judges referred the case back for the Māori Land Court to determine whether the Marlborough *iwi* (tribes) had customary ownership of the foreshore and seabed in their ancestral territories.[52] Fearful that the *iwi*'s legal challenge would be upheld, the government hastily wrote and passed the

Foreshore and Seabed Act 2004 to ensure public (Crown) ownership of the foreshore and seabed. This was repealed and replaced with the Marine and Coastal Area (Takutai Moana) Act 2011, providing Māori kin groups the opportunity to exercise their customary interests. The conditions are so stringent, however—requiring uninterrupted relationship with the coastal area in question—that many find it virtually impossible to meet them, mainly because of land losses.

In ancestral times, the foreshore-seabed continuum was regarded as *papamoana*, the continuation of *whenua* (land) into and beneath the sea, while the waters lapping on the beach are amniotic fluids nourishing the *whenua* and the *tāngata whenua* (people of the land).[53] Nevertheless, the Takutai Moana Act 2011 still refers to the "common marine and coastal area." Even as the government seeks to address injustices, the thinking remains stuck, obstinately, in a colonial framing, echoing the intransigent division between earth histories and oceanic histories initiated in Aotearoa when James Cook drafted his charts, separating land from sea with a single, continuous line.

In the Hauraki Gulf (which borders the city of Auckland), where 1.2 million hectares of ocean encompasses numerous islands and harbors, there are signs of progress toward more relational, networked thinking about land-sea exchanges and an idea of the gulf itself as having its own life and rights. Composed of Tikapa Moana and Te Moananui a Toi, the Hauraki Gulf is one of New Zealand's most valued and intensively used marine ecosystems—for food gathering, recreation, and conservation. The gulf is at risk of death by a thousand cuts—overfished, poisoned by runoff, and suffocated by sediment. Hopes for its salvation were pinned on a unique, super-collaborative marine spatial planning process—Sea Change Tai Timu Tai Pari—that produced a marine spatial plan in 2017. Those involved were tasked with making decisions "on behalf of the Gulf," in effect becoming a voice for the gulf as a living system. Māori engagement and inclusion of *mātauranga Māori* (Māori knowledge, protocols, and practice) were recognized as fundamental to the process and are reflected in the vision of the plan to "restore the health and *mauri* [life force]" of the gulf. Sadly, the plan languished, ignored and unimplemented until June 2021, when the Labour-led government published a comprehensive strategy for its implementation.[54] Action is now eagerly awaited.

Conclusion

The Hauraki Gulf illustrates in microcosm what is happening to Hinemoana, the global ocean. In the United Nations' *Second World Ocean Assessment* (2021), human impacts on the ocean were systematically assessed, using contemporary science to assess the effects of different types of maritime and land-based human activities on different taxa of sea creatures, different oceanic regions, and different types of marine habitats. The overall impression is of severe, perhaps irreversible, degradation of oceanic systems, including those related to climate change and biodiversity, putting at risk human habitation of the planet. As is the case with UNCLOS, however, the assessment's overarching perspective remains anthropocentric and utilitarian, with its talk of "ecosystem services" and the "management" of maritime resources for human purposes. There is no mention of the life force of the ocean, or of Hinemoana's kinship with people.

The question then arises as to whether humanity can effectively deal with devastating planetary changes with strategies that assume the world was created for human uses, and that people can command nature to do their bidding. When a recent virtual gathering of Nobel Prize winners addressed these dilemmas, the language was vivid, and bleak. According to Sandra Diaz, an Argentinian ecologist:

> We have incontestable evidence that the living fabric of the earth is being unraveled fast. The only reason this is happening is the present dominant model of appropriating nature. Runaway climate change, massive biodiversity loss and intolerable social and environmental inequality among people are simply the three most serious symptoms of the same root problem. They must be tackled together.[55]

Time and again, the speakers stressed the interconnected nature of these crises, and an urgent need to change patterns of thinking based on the separation of people from nature, the fragmentation of living systems, and the earth as created for human purposes.

Oceanic philosophies that presuppose the fundamental interconnection of all living beings and planetary systems might help us to imagine different ways forward for people and the planet. As Tui Atua Tamasese Ta'isi Efi has suggested, thinking about current existential challenges from the vantage point of a fish, or a bird, or a forest, or a river might provide alternatives to extractive, anthropocentric perspectives. Experiments of this kind have been happening in Aotearoa New Zealand. In 2017, for instance, in response to the Whanganui River claim to the Waitangi Tribunal, the Te Awa Tupua Whanganui River Act was passed by the New Zealand Parliament, which for the first time in the world recognized a river as a legal being, with its own life and rights. As Whanganui people say, "*Ko au te awa, ko te awa ko au—kei te mate te awa, kei te mate ahau*"—"I am the river, and the river is me. If the river is dying, then so am I."[56] In this Act, a settler nation began to re-imagine its relationship with a river, and this is now influencing other policy frameworks—*te mana o te wai*, for example, in the Resource Management Act (1991)—in which the health of waterways is given priority over human uses.[57]

This kind of thinking is also being extended to Hinemoana, the ocean. In 2019, during the 250th commemoration of the arrival of the *Endeavour* in Aotearoa, bringing the first Europeans ashore, Moananui—Te Paepae o Tangaroa Symposium was held, attended by Pacific star navigators and other Oceanic experts on fisheries and the sea. Here, new voyaging histories were shared, and ancient ways of being and knowing resurfaced. These navigators accept the sciences as enriching their own kin-based vision of the ocean, and they often carry out scientific experiments on their voyages. Thus the legacies of Enlightenment science, brought to the Pacific on board the *Endeavour*, are interwoven with Tupaia's vision of the sea as a great *marae*, in which the sea itself, the winds, the stars, and whales, are all ancestors, and as kin, people can "think like a fish."

The ocean is vast, and deep, and impenetrable. It has never been conquered by human beings. With its infinity of life forms, it holds many secrets. As the experts who attended the Moananui—Te Paepae o Tangaroa Symposium recalled, Hinemoana relentlessly eats away at the land, as in the Māori saying "te ngaungau o Hinemoana—the gnawing of the sea maiden," showing the overwhelming power of the ocean. She has never been impotent, free for Western nations to claim or dominate. Hinemoana is a reality, a living being in her own right—let's just give her a voice, they said, and a right to govern herself. With a shift away from Western hubris and human exceptionalism, the law of the sea might be rewritten to recognize an Oceanic vision, one in which the world's great ocean has its own independent life, and its own right to be healthy and flourish. From there, it is but a short step to recognizing the independent life and rights of the planet, and to putting humanity in its proper place as one planetary life form among many.

PART II

New Geo-Theologies

5

The Voices of an Eloquent Earth

Tracing the Many Directions of Colonial Geo-Theology

Jarrod Hore

In 1839, the Cambridge-trained clergyman William Clarke arrived in New South Wales to take up a chaplaincy about a day's ride inland from Sydney Cove at Campbelltown. Through the Society for the Propagation of the Gospel, Clarke was swept up from his native southeast England and deposited among the dry woodlands of the Cumberland Plain. Ministering to a wide array of colonial subjects in Campbelltown, Parramatta, Dural, and Castle Hill, Clarke regularly spent weeks on foot and horseback moving between settlements inland of Port Jackson. Clarke took his ministry seriously, first in the west and then at St. Thomas' Anglican Church in North Sydney, but the true blessing of his posting in the antipodes was the opportunity it afforded him to geologize. In the sandstone country of the Sydney Basin, and along its ancient littorals in the Hunter Valley and the Illawarra, Clarke found a site worthy of a subject that he first encountered at Cambridge in the lectures of Professor Adam Sedgwick and among the fossils, rocks, minerals, and shells of the Woodwardian collection. In Sydney, Clarke was convinced he had encountered a "new earth for geology."[1]

This realization had a series of profound meanings for an emerging scientific fraternity in New South Wales that was also well versed in the doctrines and hegemonic traditions of biblical history. In the hands of preacher-geologists like Clarke, who drew on traditions of natural theology and made them the basis of a wide-ranging earth science in the antipodes, geology was indeed the "ultimate cosmopolitan science," as James Secord argues.[2] Not only did early geology in New South Wales become one of the key domains in which questions about universal earth history and biblical tradition were raised, but it also informed those who were wrestling with the facts of colonial encounter and Indigenous difference. In this context, geology and ethnography were pursued simultaneously, if not together, and both were shadowed by cosmological questions about creation, human origins, and the operations of what Clarke referred to as Dame Nature.[3] The purpose of this chapter is to consider natural theology and earth history in a (settler) colonial context. What exactly were the relationships between colonial encounter, geological fieldwork, and biblical history? How did the Australian arcs of natural theology variously absorb Indigenous cosmologies, and how did this knowledge travel? Why were figures like Clarke and his mentor Sedgwick so invested in the concept of providence?

On one level, Clarke's thinking was part of the epochal changes taking place within Christian science in which religious differences between scientists generated debates about design, natural history, and divine providence.[4] His correspondence with thinkers such as Sedgwick, Roderick Murchison, Richard Owen, Charles Darwin, and the American James Dana provided access to the very center of geohistorical debates. At the same time, he was embedded within a local network of missionary-naturalists and lay brokers, who drew up all kinds of local physical and cultural information. New geological concepts of deep time did not just develop in step with alternative understandings of humanity, time, and place, but they were consistently exposed to them.[5] Localized understandings of geohistory underpinned the correspondence that flowed between naturalists confronting colonial encounter and their distant sources of authority and advice. Sometimes colonial geology absorbed Indigenous earth histories, practices, and cosmologies, as it did at Kurrur-Kurran on Awabakal country and at Ko-pur-ra-ba on Wonnarua country further up the Hunter Valley, but these details rarely made it back to Cambridge or London, where the "contiguity" between the "sacred and secular in nature" was read in Christian terms.[6] From this perspective, new earth histories were incredibly mutable, operating across different sites and scales and throughout flows of information that linked local missionaries with regional naturalists and metropolitan geologists. This multidirectional geological project, linked at every stage to natural theology, threw different traditions together in ways that produced a flexible and applicable kind of cosmopolitan knowledge.

Earth histories like this emerged from encounters with specific sites. They began in places like the Hunter and moved, via Clarke's antipodean network of missionaries and amateur naturalists, into colonial centers. They were transformed as they passed across Clarke's desk in North Sydney, where the pastor mediated information and put local knowledge into global circulation. In the course of tracing these new earth histories, this chapter will move through a number of scales. It will start with the local interactions of geological fieldwork, which involved both the extension of dominion over a landscape and the encounter with alternative Indigenous cosmologies in a colonial subregion. As I raise questions about disciplinary practice in the antipodes, we shift to the regional scale and the peculiar culture of comparative geo-mythology that some missionaries pursued in New South Wales. Finally, the chapter will trace flows of this local and regional information—travel, translation, correspondence, and publication—into a wider web of scientific inquiry. In all these contexts, but especially in colonial New South Wales, natural theology exerted a reliable influence on fieldwork and provided a different approach to resources and their abundance. In some ways, natural theology was turned to settler-colonial ends and Clarke and his network developed a branch of geohistorical inquiry that delivered the world to their ends and made sense of their own ambitions.

Local Nature as a Puzzle: Fieldwork and Encounter in the Hunter Valley

Within a year of his 1839 arrival in New South Wales, Clarke had "explored the whole of [the] Illawarra as far as the Shoalhaven, the Kangaroo Ground in the County of Cumberland, the Blue Mountains, and the Coal District of the Hunter."[7] This amounted to over twenty-five thousand square kilometers of territory and dramatically eclipsed the extent of many of the studies Clarke had conducted as an amateur and a student in southern England. Much of this travel was connected to his ministry, but on these trips, Clarke paid close attention to the shape and form of the country around Sydney. In an 1841 letter to Sedgwick, Clarke claimed that "the difficulties are so great in this country . . . the most industrious and extensive researches can only lead to an approximation of the whole truth. Owing to the peculiar construction of the country, its deep and impassable ravines, its enormous forests, its want of crossroads and good sections, it is only by most painful plodding, that one can make out anything satisfactory."[8] Clarke argued that he had traversed this country "again and again," and this enabled him to make some early claims about stratigraphy, the elevation of seabeds, and the geology of the Sydney Basin. The whole area, he more or less correctly surmised by way of cliffs along the Lane Cove River and at South Head, showed the superposition of vegetable fossils over mollusks and therefore had once been inundated by the sea.

In early nineteenth-century New South Wales, these kinds of expeditions generated and relied on a degree of intercultural contact. A series of recent reinterpretations of "exploration archives" by Shino Konishi, Tiffany Shellam, Penny Olsen, Lynette Russell, and others have shown that surveyors, explorers, and pastoral speculators relied on Indigenous spatial and scientific knowledge as a matter of course. Thomas Mitchell, the colony's most influential surveyor general, was almost always accompanied by an Indigenous guide on his expeditions over the Great Dividing Range and into Australia's largest river basin, the Barka-Darling. A Wiradjuri man named John Piper, for example, accompanied expeditions along the Barka-Darling River and into Australia Felix in 1836 and northwest into the interior of the continent in 1845–1846.[9] Clarke, too, often turned to Indigenous brokers throughout his career, and in the southern summer of 1839–1840, traveling with the visiting American geologist James Dana, witnessed a Tharawal corroboree while examining the geography and the coal measures of the Illawarra region.[10] In the colonial context of exploration and surveying, settlers sought out Indigenous information in a variety of ways. In some cases, it was practical in that local mediation helped settlers move through the country, and in other cases, it had direct scientific or ethnographic value. In all instances, though, these encounters or relationships were important because they informed how settlers experienced the field.

One site where the relationship between seemingly secular scientific knowledge-making and different kinds of settler-colonial encounter was laid bare was at a place called Wingen in the upper Hunter Valley, about 240 kilometers north-northwest of the Illawarra (fig. 5.1). At Wingen, the Sydney and Hawkesbury sandstones, which Clarke was in the process of tracing out in 1839 and 1840, abut older Carboniferous and Permian sedimentary rocks

FIGURE 5.1. The smoking seams of Ko-pur-ra-ba. Conrad Martens, *Burning Mountain (Mount Wingen, near Scone)*, 1874, watercolor, 61 × 91 cm. Art Gallery of New South Wales, gift of Professor T. F. Heath, 1995, accession number 142.1995.

and some more recent volcanic deposits. These strata cover extensive seams of coal dating to the Permian period, 250 million years ago. In the 1820s and 1830s, as white settlers expanded into the Hunter, Wingen became a site of interest and speculation. In 1828, a rumor surfaced in Sydney that some colonists had discovered "a volcano in the neighborhood of *Hunter's River*." The information came from a group of three colonists from Sydney who traveled into the valley and stayed on the flanks of Wingen for several months. Sensationally, they reported that when they came upon the site, "a dense volume of flame" burst from the earth and that by night "the flame can be seen distinctly rising in a sulphurous bluish column and stretching away through the atmosphere." Over time they recorded observations after digging into the side of the mountain and splitting rocks. The expedition evidently relied on a group of local Indigenous people, who "accompanied the party" and at one stage provided a traditional treatment for burns after the colonists dug a pit into the burning mountain and one of them fell in. The colonists noted that they had adopted the name "Wingen" from this group of Kamilaroi people, who regularly traveled through this area but nevertheless "gazed upon the volcano, with an expression of astonishment and dread." The colonists concluded from this that Wingen was only recently ignited. This must have appeared fair enough to the white colonists, but it is more likely that this expression of concern—the colonists recorded the Kamilaroi repeating the phrase "*deebil, deebil*"—was related to the fact that the whole party was on Wonnarua country, where the Kamilaroi were visitors too.[11]

Much of this was reported in 1828 by the chaplain Charles Wilton in his short-lived *Australian Quarterly Journal of Theology, Literature & Science*, which was geared toward the "prosperity of the colony," as construed in commercial, religious, and intellectual terms.[12] After publishing the initial notice, Wilton visited Wingen himself. He wrote a series of articles on the curiosity in various news outlets between 1828 and 1833. Although Wilton had realized it was a coal seam fire in 1829, he maintained an air of mystery and wonder in his reporting, writing that "the face of the rock, as the subterranean fire increases, is rent into several concave chasms of various widths. . . . The sides of the rock were of a white heat, like that of a live kiln, while sulphurous and steamy vapours rose from a depth below, like blasts from the forge of a Vulcan himself." After a visit in 1829, Wilton argued that the name Wingen was derived from the Kamilaroi word for fire and that the combustion observed at the site "far preceded the memory of man."[13] These kind of appeals to Aboriginal history and etymology were quite common in the colonial press, and eventually, William Clarke also became well known for the ways in which he incorporated Aboriginal place names and geo-mythologies into his writing.

Clarke even wrote, in 1846, that "Australian traditions about fire, light, the sun, and serpents" should be heard with the same respect as Greek myths.[14]

No doubt Wilton agreed, but it seems his sole reliance on earlier reports left the puzzle of the burning mountain unresolved. Perhaps if he had asked the Wonnarua people about the site, they might have been able to provide a more detailed story. The Wonnarua knew the burning mountain as Ko-pur-ra-ba, so called because they obtained oxidized red and yellow earth (ko-pur-ra) from there, which they used to make body paint.[15] Their stories link this prosaic site to a rocky outcrop across the valley known as the "Wingen maid," a woman fossilized by the great Creator, after her husband failed to return from a raid on the Kamilaroi. Contemporary versions of this geo-myth hold that the Wingen maid's burning tears rolled down the mountain and ignited the earth at Ko-pur-ra-ba.[16] Mostly likely unsatisfied with this alternative earth history, neither Wilton nor the earlier visitors ever really settled on an account of Ko-pur-ra-ba's formation, leading them into various kinds of speculation or sensation. The important point to take from this is that all these settlers preferred to understand the field as a puzzle to be solved. Partly in vain, they applied a range of different scientific, ethnographic, etymological, and even corporeal strategies to try to apprehend the burning mountain. And though the practical utility of Ko-pur-ra-ba was settled quite quickly—the first party returned to Sydney with many mineral samples from the site—figures like Wilton and then Clarke continued to present natural phenomena as objects of curiosity.

A similar dynamic played out at the other end of the valley at a place called Kurrur-Kurran, where a partially submerged forest of petrified wood became a site of interest for missionaries and geologists in the 1830s. The petrified forest is situated at the northwestern reach of Lake Macquarie, or Awaba, which is a large saltwater lake just south of where the Hunter River reaches the sea. Kurrur-Kurran first appeared in print in 1834 as part of the missionary Lancelot Threlkeld's *Australian Grammar*, which he had written with the assistance of an Awabakal man, Biraban, over the first decade of his missionary work in the lower Hunter. Biraban explained to Threlkeld that for the Awabakal, the bay of fossils was formed when "a large rock fell from the heavens" and killed a group of people who had gathered at the command of a huge goanna. According to tradition, the people had been killing lice in fires and the goanna sought to punish them for this before returning to the sky.[17] Threlkeld recorded many of these stories in the process of his missionary work and his translation of the gospels of Luke, Mark, and Matthew into Awabakal language. His missionary and ethnographic projects struggled throughout the late 1830s, though, and Threlkeld resorted to sinking a shaft into the seams of coal underneath mission land.[18] It is likely that this is what brought Kurrur-Kurran to the attention of Clarke in 1842, who by this stage was primarily occupied with questions about the age and distribution of antipodean coal.[19]

Clarke visited Kurrur-Kurran sometime in the middle of 1842 and quite quickly wrote a paper on the subject. Reading the landscape around the lake and the inlet, Clarke noticed that the stratigraphy—"subordinate beds of lignite and coal"—was consistent with the rest of the lower Hunter Valley. Where the submerged forest stood, the geologist noted the existence of a long, flat sandstone shelf, covered by alluvial deposits. He explained that, throughout this, flat "stumps and stools of fossilized trees" interrupted the surface of the water. "One can form no better notion of their aspect," Clarke wrote, "than by imagining what the appearance of the existing living forests would be if their trees were all cut down to a certain level." Based on the presence of similar petrified tree trunks in Threlkeld's shaft and some others nearby, Clarke confidently positioned these trees within the sandstone strata covering the high-quality lower Hunter lignite.[20] Clarke considered this natural phenomenon as "by far the most curious instance of the freaks of nature which have met my notice." For Clarke, Kurrur-Kurran presented a "singular picture of the past and the present," through which he could consider "the operations of most powerful though secretly evolving causes."[21] He was, in other words, encountering the puzzles of the field as part of a providential landscape over which settlers were all the time extending their intellectual and material dominion. On the local level, natural theology drove colonial geology, absorbing various geo-myths and sharpening the focus on natural resources. This meant that, in the Hunter, this form of earth science began with Indigenous political geologies (fig. 5.2).[22] Largely, though, the Wonnarua and Kamilaroi at Kopur-ra-ba and the Awabakal at Kurrur-Kurran were understood as sources of information and not masters of geohistory, a fact that allowed settler knowledges to shed these stories.

Hand in Hand Together: The Theological Contexts of Colonial Geology

Many preachers in early New South Wales shared a strong interest in the physical world. As Martin Rudwick has explained, this fact relates to the way that natural theology underpinned early nineteenth-century geological and sacred history across a range of denominations.[23] The Church of England, for example, was more or less funding Clarke's early expeditions, and although Wilton was altogether more focused on more traditional spiritual matters, his broad commitment to colonial development does not seem to have clashed

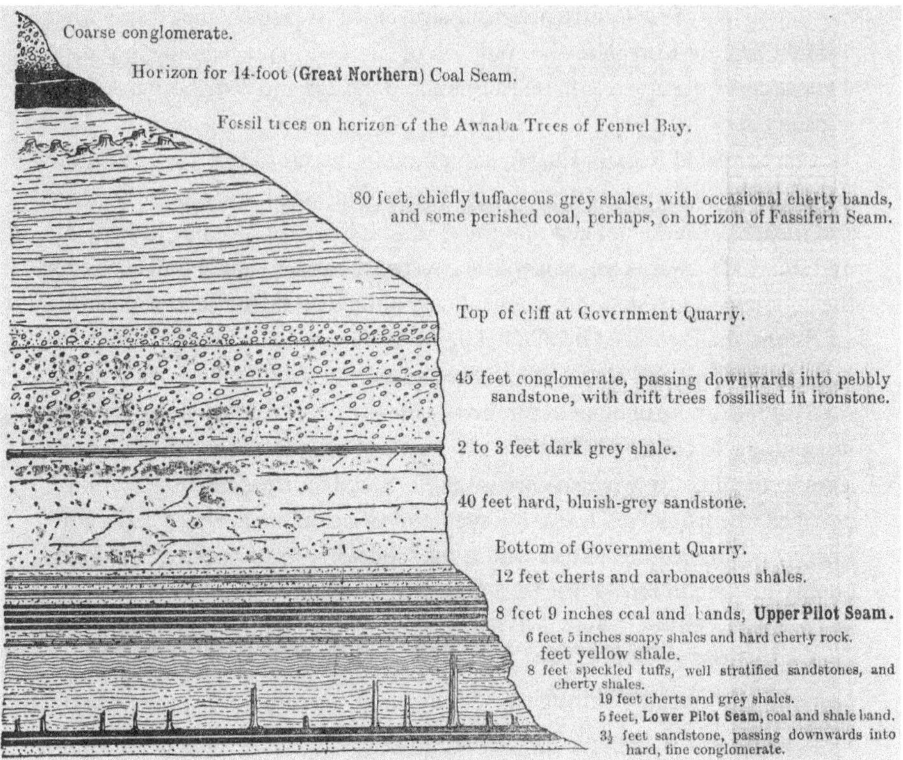

FIGURE 5.2. A cross section of the government quarry at Swansea showing the position of fossilized trees, coal seams, shales, and sandstones. The Swansea quarry was on the opposite edge of Lake Macquarie, sixteen kilometers southeast of Kurrur-Kurran. T. W. Edgeworth David, *The Geology of the Hunter River Coal Measures, New South Wales* (Sydney: W. A. Gullick, Government Printer, 1907), 11.

with his clerical responsibilities in Newcastle. Threlkeld, by contrast, was an agent of the Congregationalist London Missionary Society, which, unable to sustain the Lake Macquarie mission, forced the minister to develop local coal seams. To this we might add the late nineteenth-century Presbyterian minister and linguist-explorer William Ridley, who understood the connections between the antipodean landscape and Aboriginal people in clear terms. Drawing on the earlier work of Biraban and Threlkeld, among others, Ridley explained that sacred relationships inhered in physical places: "The religion of the aborigines in all parts of Australia includes a belief in sorcery, and a dread of numerous demons, spirits of the wood, of the river, of the mountain and the pool."[24] These men all pursued different paths into the same fundamental question about the nature of the antipodean earth. Armed with their powers of observation, put to various geo-mythological ends, and a faith in their own God-Creator, they pursued natural and spiritual knowledge simultaneously.

Wilton's *Australian Quarterly Journal of Theology, Literature & Science* was one such distillation of colonial natural theology. It's very first article set out the connection between science and religion, exhorting readers to "look through Nature up to Nature's God." Those who worked with and in the earth had a privileged role in this. Mineralogists, according to Wilton, witnessed "the mighty wonders" of the land, and geologists, "in observing the several stratifications of the Globe, and the various petrified remains of what once formed a part of animated nature, behold the exact accomplishment of Scripture." By this he referred specifically to the biblical Flood, which, from the second half of the seventeenth century, became the core concept through which natural philosophers such as Thomas Burnet aligned earth history with biblical doctrine.[25] Wilton's understanding of this alignment was notably instrumental. Writing in 1828, at the outset of New South Wales's first major investment boom, the point of all this was to pursue "active and spirited research." There were publics in both the colony and the imperial metropole who would welcome geological news from "this fifth division of the globe." Such inquiries fundamentally linked the "welfare of the Colony," "the advancement of Science," and the glory of God, through a thorough grounding of "the facts recorded in Revelation."[26]

Almost from the beginnings of the pursuit, colonial geology was distinguished, in part by practitioners such as Wilton and Clarke, by its instrumental orientation. This feature has been emphasized in histories of Australian science, which hold that Clarke in particular simply adopted the technical "precepts of the experienced British geologist" when he arrived in Australia in 1839.[27] But alongside quite new investments in certain natural resources and their exploitation, colonial geologists also took up long-standing interests in the extension of a sacred Judeo-Christian cosmology over the entire globe. Again, the intellectual fulcrum of this project was the biblical Flood. Clarke's interest in this event stemmed from his understanding—no doubt passed down from Sedgwick—that the impacts of the Flood "can be traced over the habitable globe," leaving "its records in imperishable monuments that will ever attest to its former influence."[28] Historians of science have long known about this intermingling of the practical, the sacred, and the legendary in early geology, but in the colonial context this complex earth history drew in a third set of cosmological beliefs: those of Aboriginal people.[29] As I have discussed, Clarke was drawn to these beliefs on an antiquarian level, comparing them to Greek mythology, but he also sought in them evidence of Christian revelation. In 1842, while reviewing a book on Aboriginal origins, Clarke noted that some traditions complied squarely with the biblical story of the Flood. One, sourced from the Darug people west of Sydney, involved an inundation that covered the top of the Blue Mountains, from which only two people escaped. Another, which is harder to pin down, referred to a "great flood" that destroyed all things before the earth and its animals slowly regenerated.[30]

The intermittent consilience between certain geological features, biblical stories, and Aboriginal tradition captivated many colonial naturalists, linguists, and philologists. The comparative approach that has, according to the intellectual historian James Turner, "always lain at the heart of philology," made it an appealing way to link disciplines in a colonial context, where they were typically forced into close contact anyway.[31] Many contemporary expeditions mixed geographical, geopolitical, ethnographical, linguistic, and, of course, promotional objectives as a matter of course.[32] Reviewing Phillip Parker King's narrative of the 1818–1822 coastal survey of the northwest portion of the Australian continent, Wilton's *Australian Quarterly Journal* framed the value of the publication in these same mixed terms. The reviewer was at pains to take equal notice of what King reported about his experience with Aboriginal guides Bungaree and Bundell and "the habits &c., of the Aborigines of Australia," as they were of the expedition's findings about the continent's "Natural History—Geology—Botany &c."[33] This promise of consilience was still inspiring colonists four decades later. William Ridley, who covered much of northern New South Wales and Moreton Bay as part of his missionary activity, delivered a long lecture in 1864 that addressed "the early history of the Australian race; some of their laws and customs; specimens of their language" and "a brief account of their religion" in the context of Christian missionary activity. Ridley was provocatively drawn to the same creator figure who fossilized the Wingen maid, who was known to a wide range of Aboriginal groups over much of New South Wales. This figure was described as a "Sky Father" who entered the world, made it, and exited again (in the upper Hunter, as it would happen). These creation traditions were unsurprisingly understood as "a remnant of the truth at first revealed to man—One invisible God who made all things."[34]

As both a prominent minister and as the most senior geological thinker in New South Wales, Clarke absorbed much of this information and used it to create his own hypotheses, which sometimes departed from biblical models. Anticipating Ridley's conjectures in the 1860s that the Aboriginal races made their way to the Australian continent "like a crest of foam upon the advancing wave" of the "Hindoo; Malay, and Mongolian tribes," Clarke flirted with the idea that Aboriginal geo-myths were syncretic.[35] Clarke had a close association with the explorer Ludwig Leichhardt, who, in the 1840s, had provocatively reported that some Aboriginal groups appeared to maintain practices linked to Hebrew doctrine. At around the same time, Clarke also suggested that the closest analogies to the dominant Aboriginal tradition of the creator-serpent could be found in Hindu stories like the account of Kāliya in the Bhāgavata Purāṇa and others within the Uttararāmacarita.[36] In colonial New South Wales, geology existed in a cultural context that was shot through with stories like these. Philology and comparative mythology reliably fed into colonial earth histories in much the same way that "language thinking," according to the historian of science Emily Kern, was ever present in nineteenth-century inquiries into "human diversity."[37]

In other words, it was impossible for early colonial geologists to simply adopt the practices of their forebears in Britain. The core business of colonial geology involved handling these multiple sources of information and assembling a useful vision of the earth from them. This vision of earth history took on different forms at different scales, though, and Clarke dutifully reassembled its contents when he wrote colonial geology into a wider web of science.

A Wider Web of Science: The Geological Project and the Purpose of Nature

That information about Aboriginal etymologies and cosmologies was swirling so consistently around geological questions in the colonial context makes the content of Clarke's letters to his former teacher Adam Sedgwick all the more remarkable. Writing to Sedgwick about burning mountains in 1843, Clarke offered little in the way of cultural or linguistic background. Aboriginal knowledge functioned in a practical way, guiding colonial geologists in their search for "a second Wingen" north of the Hunter and yet more coal deposits.[38] Clarke provided more detail about the seam in the upper Hunter a year later, in 1844, when he suggested that the coal had been ignited by bushfire and dwelled on the connections between the valley and the basalt soils of the nearby Liverpool Plains.[39] In the correspondence between the colonial and the imperial geologists, phenomena like Wingen were singularly useful as markers of subterranean resources. Sedgwick was on the same page. He thanked his informant in the antipodes for his details on "pseudo volcanoes" and offered to compose an abstract from Clarke's reports on the upper Hunter.[40]

These priorities also shaped how information about Kurrur-Kurran circulated through to Sedgwick in England, and they were augmented in a material sense by the movement of "natural antiquities."[41] Already we have observed how Clarke's geologizing at Lake Macquarie rested upon a set of older inquiries geared toward Aboriginal cosmologies. In 1842, he sent his paper on the petrified forest to Sedgwick and the Geological Society along with "two large specimens of the fossils." He directed Sedgwick to take his pick of these for the collections at Cambridge and leave the other to the Geological Society in London. Shortly after this, in 1845, in the midst of an inquiry into the age of the New South Wales coal seams, Clarke sent over two thousand more rocks and fossils to Sedgwick. These physical specimens were intended to form a resource for his readers in English geological journals and provide a basis for a wider investigation into stratigraphic conformities, thereby helping Clarke calibrate his hypotheses.[42] Belatedly and reluctantly, Sedgwick accepted them in 1847, when they were accessioned into the collections at Cambridge and examined for conformities. Sedgwick concluded then that the coals were fundamentally different to the English and Welsh beds, which had similar shells but different plants.

This short interchange demonstrates some of the key aspects of geological practice on the global scale in the early nineteenth century. Ideas and arguments rested predominantly on fieldwork and collection, and these activities supported hypotheses that were tested both in the field and against recorded information in the form of approximated earth archives in places like Cambridge and Somerset House, headquarters of the Geological Society. Once tested, theories about the extent of global conformities were gradually refined through correspondence, more fieldwork, and further study. For figures like Clarke, labor in the field involved considering and evaluating ancestral stories about the formation of landscapes, highly speculative assessments of human movement and evolution, and more straightforward inferences from natural history. Throughout most of this process, geologists like Clarke were publishing suggestions, arguments, and conclusions in a variety of colonial and imperial forums. In the course of this activity, the calibration of deep time became a great comparative enterprise that involved balancing material and spatial knowledge, on the one hand, with textual and hermeneutic information, on the other.[43] But while specimens and hypotheses circulated smoothly enough through the pathways of world science, other knowledges suffered from a kind of attrition. Even though Aboriginal geohistories appear to have been dismissed or erased at certain points in chains of correspondence, colonial missionaries, natural historians, and philologists continued to consider them. In the colonial world, part of the appeal of a good natural puzzle was in its attachment to Aboriginal mythology, but in the wider web of science, its appeal inhered more narrowly in its practical geological meaning. This appeal was, of course, related to certain theological concerns and folded into Christian thinking, but in Clarke and Sedgwick's correspondence, it never admitted the kind of multidirectional cosmological discussion that seemed to grip missionary-naturalists in New South Wales.

For Clarke and Sedgwick, places like Wingen induced a slightly different kind of wonder than the one experienced by Wilton and those colonists who made their way to the burning mountain. In the exchange of imperial science and theology, natural puzzles were primarily places where thinking about earth served as a way to engage with the humbling materiality of revelation. In this context, the abstractions of geological work had a higher purpose in revealing the universal scale, perfection, and designfulness of God's creation. For Clarke and Sedgwick, as for a cast of other earth-inclined natural theologists, there was a divine truth in the expanding geological horizons of the early nineteenth century.[44] Colonial geology in Australia followed James Hutton, a deistic natural theologist, and a number of other contemporary settler earth scientists, including Canada's John William Dawson and the American James Dana.[45] The approach was closely aligned to Sedgwick and Cambridge, where geology courses were also, really, courses in theistic natural theology. Sedgwick admitted, later in life, that his course always returned to the ways in which a "great, living, intellectual, and active Power must be the creative Head of the sublime and beautiful adjustments and harmonies of the universe."[46] You can see the impact of Sedgwick's teaching on Clarke's response to the natural world in New South Wales. Reflecting on the vastness of the coal reserves in 1841, Clarke gushed: "these deposits . . . have been hoisted up from the sea level to 3, 4, and 5,000 feet, what a sublime idea do we obtain of the great operations of Dame Nature."[47]

Australian historians of science have been overwhelmingly drawn to Clarke's role in the colonial reception of Darwin's *Origin of Species*, which Clarke defended on the basis of open scientific discussion while holding on to his own creationist views. Clarke's beliefs were split, according to Ann Moyal, between different geological, biological, and theological positions.[48] Viewed narrowly, this appears to be typical pragmatism, but viewed within a longer history of intricate connection between geology and theology, it strikes a different tone. Natural theology was a way of interpreting the material world "according to God's will and divine plan." This is also at the root of Clarke and Sedgwick's curious use of "Dame Nature"—an older term, then in the midst of a gradual nineteenth-century revival—to describe a generous and benevolent material world. Pushed to its logical extreme, natural theology amounted to a "divine mandate to exert dominion over nature" that lent a different edge to inquiries into earth and human history.[49] The very purpose of colonial knowledge production, in this context, was labor, toil, and extraction. To heed the scriptural call and "subdue the earth" meant knowing it, charting its features, and identifying its hidden commercial possibilities. By the time that he displayed his openness to scientific debate about natural selection, Clarke was well trained in absorbing new information into a flexible natural theological framework and deploying it to understand both God's word and God's works.

*

Natural theology and its approach to knowing the earth had a powerful colonial afterlife in the nineteenth century. As colonists made their way into a range of unfamiliar environments and cultural geographies in early nineteenth-century New South Wales—Clarke's "new earth for geology"—they encountered a series of puzzles. Some of these were framed in purely physical terms. What forces, for example, could lead to the combustion of earth under a mountain of sedimentary rocks? But they also delved into alternative cosmologies. Colonists asked, simultaneously, What is the name for this place? What happened here? And how do you know it? One of the forces driving these questions was natural theology, which opened a series of avenues through which missionary-naturalists might pursue closeness to God through a knowledge of the earth. Generally, these figures were convinced that, as expressed in the first edition of the *Australian Quarterly Journal*, "the real lover of nature indeed and the true Christian can never fail to profit by what he sees around him."[50] This mode of inquiry led some people to record and compare widely different geohistories, in the course of which missionaries and geologists were drawn into conversations about philology, cosmology, ethnography, biblical history, Aboriginal tradition, and all other kinds of speculation. These stories, which Wilton noted "necessarily lead us abroad," were part of a spectrum of actions that led from local evaluations of places and resources, through an open inquiry into all things to do with humans and places, all the way to global conversations about geological time and the nature of the earth.[51] Colonial geology, via natural theology, brought all these conversations together.

The other force that brought them together was colonial expansion. Indeed, all this thinking was tied to the practical orientation and compelling direction of colonial geological work in the early nineteenth century. The search for knowledge was also always a search for resources: intellectual and economic agendas were sutured together. Clarke and Sedgwick knew this best. A dawning awareness of geological making and unmaking suggested that the earth itself was the key to understanding the purpose of God's creation. For Clarke, there was no real limit to the possibilities of this approach.

He wrote in 1855 that "the very *facts established by geological research* . . . can never by any sound reasoner, contradict the testimony of *God's word*." Geological inquiry concerned "the very documents of Nature" that could alone illuminate "*facts* of which the Scripture is silent, but of which the earth is eloquent."[52] Clarke's declaration then, that it was no accident that good Anglo-Saxon settlers tended to establish themselves exactly where fossil fuels were buried in the earth, bound both religion and resources into a kind providential geology defined by its directionality.[53] In statements like Clarke's, we see an arrow being drawn from the deep past, through Clarke's own present, and into a racialized future dependent on technical knowledge of the earth, inductive work in the field, and a kind of firm belief in the rightness of the colonial project.

All this aligned, too, with Sedgwick's natural theology, which was almost certainly the key source of Clarke's settler-colonial assurance in New South Wales. The senior geologist had earlier been led into identical, highly racialized, conjecture. He mused, for example, in 1844, "is it not strange, that in almost all parts of the world where Dame Nature has spread out her subterranean fuel—there you are sure to find some Anglo-Saxon settlers nestling most comfortably?"[54] In nineteenth-century New South Wales, this kind of thinking sustained a specific set of investments in geological fieldwork, natural resources like coal, and the colonial earth itself. Perhaps more unexpectedly, it also generated a different approach to fieldwork that encouraged scientifically inclined ministers, missionaries, and preachers to consider their own biblical narratives alongside new earth histories, Aboriginal traditions, and other cosmologies.

6

The Spiritual Geographies of Plate Tectonics

Javanese Islam, Volcanology, and Earth's New History

Adam Bobbette

By 1975 most British and US scientists agreed that the theory of plate tectonics described earth's history in a new way. The theory claimed that the lithosphere was broken into large plates floating on a relatively liquid mantle. The plates moved in relation to each other and their contact zones were associated with volcanism, ocean trenches, rift valleys, and gravity and magnetic anomalies. The theory confirmed Alfred Wegener's controversial argument in 1913 that continents drifted, but it also purported to have found the long-sought mechanism that powered drift—convection currents in the mantle. Japan, Indonesia, and the Caribbean island festoons came to be seen as the superficial effects of plates driving into each other, melting, and exploding through volcanic eruptions; the construction of the lithosphere could be witnessed there in situ. By contrast, the continents were considered by many geologists to be more ancient earth-forming processes that were seen to have long ago come to completion. In this new narrative of earth's history, island arcs were the vanguard of earthly evolution.

Plate tectonics was described by many of its North American and European proponents as revolutionary. Martin Rudwick has suggested that geologists thought this because they had been reading Thomas Kuhn.[1] The sentiment has not entirely faded—some historians have recently narrated the history of the theory as the result of a small band of pioneer scientists working on the margins to transform the establishment.[2] Other narratives, however, have shown how many of the key geologists were not marginal at all; they were at the centers of European and US power, working in major oceanographic institutes and universities, funded by the defense industries and communication technology corporations.[3] It has also been shown that if anything was revolutionary about the theory of plate tectonics in the Kuhnian sense, it was local and confined to a small group of American geologists who, in a brief period of time, transitioned from a fervent attachment to orthodox theories of fixism to mobilism. European geologists (including Wegener), South African geologists, and especially Dutch geologists in the early twentieth century had long been engaged in vigorous debates, creating innovative geological instruments and speculating about drift, because of the landscapes they were encountering. This chapter is about that moment in the Netherlands East Indies in the early twentieth century.

The Netherlands East Indies colonial government established a volcanological survey in 1918 out of the need to protect its plantation economy from volcanic cataclysm. After a massive and catastrophic eruption at Kelud in East Java in 1919, the first permanent observatories were built on Java's most dangerous volcanoes. The financial resources and scientific attention newly oriented to volcanism meant that the survey became a center for the development of the nascent field of volcanology.[4] Scientists were taught to think in new ways about the depth of the lithosphere through their encounters with eruptions, landslides, earthquakes, ash rains, and mudslides associated with volcanism. This was coupled with a new understanding of the immensity of the underwater trenches that bordered the Indonesian archipelago. Scientists raised new questions about how volcanoes and the ocean were connected. Thinking about the thickness of geological process through the lens of volcanism, and at the scale of a region nearly the size of Western Europe, enabled geologists to take questions of drift seriously in ways that seemed impossible in other contexts.

Dutch scientists did not, though, become volcanologists in a place absent of traditions of understanding volcanism. Javanese geographers had long thought about the depth of the earth surface and the relationship between volcanoes and the sea. Volcanism was foundational for early modern polities in Central Java. The form of the polity itself was frequently expressed in the shape of a volcanic mountain, and volcanoes were seen as the original form of the earth emerging from the ocean. The court chronicles of the Central Javanese sultanates of Yogyakarta and Surakarta from the middle of the nineteenth century explain the coproduction of the dynasties with the volcanoes and the Indian Ocean in interlocked networks of exchange. The volcanoes and the Indian Ocean were conceived of as their own centers, including dynasties of deities with local, regional, and global genealogies, both ancient and recent. Each center also exchanged with the others in a spiritual geopolitics. The sultan of Yogyakarta's name, Hamengkubuwono, meant "nail of the cosmos"—because he held together this multiple, sometimes unruly, multicentered cosmos. The landscapes in between the volcanoes and the ocean were also the abodes of powerful spiritual forces located in caves, river confluences, hilltop forests, and graveyards. This Javanese spiritual geography was not unknown to colonial geologists. In fact, colonial scientific encounters with this spiritual geography, in direct and indirect ways, enabled colonial scientists to understand the landscape and volcanism in new ways.

One of the ways that Central Javanese geographical thought shaped colonial volcanology was through a broader turn to interest in Javanese culture on the part of Europeans. Dutch colonialism at the turn of the twentieth century, as in other European colonies, was defined by its liberal turn to a civilizing mission. A fascination with the pre-Islamic, Hindu, and Buddhist past of Java flourished as a part of this. The establishment of the Java Institute in Yogyakarta brought together Javanese aristocrats and elites with Dutch colonial elite ethnologists and anthropologists. The Java Institute began to publish the journal *Djåwå*, mainly in Dutch but also Javanese, which focused on the literary, philosophical, courtly, and folk traditions of Central Java. Theosophist lodges in major cities brought together orientalists, Sanskritists, Javanese elites, and members of the royal families with mining engineers, geologists, and volcanologists and encouraged the unification of religion and science. For many Theosophists, Hindu and Buddhist traditions also rejected the modern separation of theology from science, and because of this Theosophists sought essential truths in these traditions. Conceptions of the Javanese landscape also became familiar to colonial geologists through translations of the royal chronicles, called *babad*, from the Surakarta and Yogyakarta palaces. There were hundreds of *babad*, many of them written in *Kawi*, held in court libraries that liberal Dutch scholars came to recognize as contributions to world literature. A number of *babad* written between the eighteenth and nineteenth centuries had sections transliterated and translated into Javanese and Dutch with commentary by enthusiastic Dutch scholars.[5]

Dutch colonial volcanology and Central Javanese geographical thought have been conventionally treated separately. This chapter explores how they produced each other. At the same time that volcanologists were becoming interested in the depth of volcanism, the mobility of the earth's surface, and the relationship between terrestrial features, the spiritual geographies of those places were also becoming familiar. It was not that colonial volcanologists unproblematically adopted those spiritual geographies but rather that they were, in more or less direct ways, enabled by them to see the landscape in new ways. Volcanologists developed a vision of volcanic cultural determinism that considered volcanism at the center of social processes. They undertook this as they traveled along active ritual pathways to volcanic craters, supported by Javanese fieldworkers who conducted their experiments and translated conversations with locals for them.

When the theory of plate tectonics is drawn back into these formative encounters between volcanologists and Central Javanese geographical thought, its story emerges in a new light. It raises questions about what was left out of the theory of plate tectonics other than failed scientific theories or technologies. What was silenced, and by what mechanisms, of the Javanese conception of volcanism? Or, how did the Javanese tradition prefigure modern volcanology? What did it mean to transition from a conception of volcanism as the guarantor of royal sovereignty to one where modern earth sciences managed volcanism on behalf of the state? The point of this is not only to remind us that plate tectonics was not a revolution but also to reorient the intellectual geographies of the theory away from Britain and the United States—to the slopes of Javanese volcanoes.

In the Realm of the Queen of the Indian Ocean

Of particular significance in the spiritual geography of Java was the Indian Ocean. Pieter Veth, the Dutch geographer, described it in 1888, as if he was looking out from the Central Java plains, as the "vast realm of Nyai Ratoe Kidul."[6] Nyai Ratu Kidul was long recognized as the ruler of the Indian Ocean. Depending on the account, her realm encompassed the entire Indian Ocean off the southern coast of Java, a region associated with earthquakes, tsunamis, and unfriendly seas. In her region, the Indian Ocean was a fearsome area with few friendly landing places; it stood in stark contrast with the north of Java, the *pasisir*, which for centuries was a site of global traffic and where Java's main port cities emerged.

Nyai Ratu Kidul was significant not only because she ruled the Indian Ocean but also because she was married to Central Java's sultans. *Babad* from the nineteenth century explained that, in the 1550s, the Mataram Kingdom was established in part by the Sultan Panembahan Senopati enlisting Nyai Ratu Kidul's support and their marriage.[7] Senopati was said to have met with Nyai Ratu Kidul a number of times near the coast and at river confluences that drain the mainland into the Indian Ocean. She also participated in the jihad against Dutch colonialism in the 1610s and 1620s led by Sultan Agung, the third sultan of Mataram. Later, Pangéran Dipasana called on her for help with slaughtering the Europeans in the region, a project that ultimately failed, just like his rebellion. When the Mataram Kingdom established by Panembahan Senopati was divided in 1755 under negotiations with the Dutch East India Company, the kingdom splintered into Yogyakarta (with Sultan Hamengkubuwono) and Surakarta (with Susuhunan Pakubuwono), and Nyai Ratu Kidul was thereafter married to both, as well as their successors.

Dutch colonial and Javanese ethnologists in the 1920s began to understand that the marriages between the sultan of Yogyakarta, susuhunan of Surakarta, and Nyai Ratu Kidul was structured by debt. Articles in *Djåwå* documented how, in return for maintaining the security, safety, and stability of the sovereign, the sultan and susuhunan owed Nyai Ratu Kidul gifts. The Labuhan procession—literally "to throw"—was a mechanism through which those gifts were provided through an annual pilgrimage to the South Sea, where items were released on rafts into the ocean. R. Soedjana Tirtakoesoema, a translator and contributor to *Djåwå*, recorded a Labuhan procession in 1921 on the occasion of the ascension of Sultan Hamengkubuwana VIII. The offerings to Nyai Ratu Kidul consisted of sixteen types of cloth printed with patterns (some representing the sultanate of Yogyakarta), scented oil, cosmetics, incense, coins, wilted flowers, and wooden boxes with hair and nail clippings of the sultan.[8] The offering of cloth suggested that Nyai Ratu Kidul circulated between the realm of the Indian Ocean and the sultanate, donned its clothes, and could use its money in its markets. The Labuhan, in this sense, was an annual process of facilitating her capacity to circulate within the polity. Tirtakoesoema even suggested that the balance of power between the sultan and Nyai Ratu Kidul was in her favor, that the sultan was in fact her adviser.[9]

Nyai Ratu Kidul and her watery realm of waves, earthquakes, and tsunamis cannot be understood without the volcano Merapi, a mere fifty kilometers to the north. On that same Labuhan, the retinue of the sultan took the train toward Surakarta, disembarked at Kalasan, and followed the road north to the village of Ngrangkah (Umbulharjo) on the southern slope. The retinue then progressed to the edge of the vegetation line, on the uninhabited rocky slope, in view of the smoking crater, to provide the final offerings. They ate a meal of rice in the shape of a volcano (*tumpeng*) and roast chicken, recited the first chapter of the Quran, and invoked the following deities at what they called "the 'navel' of Java, Mount Merapi": Sangyang Umar, Kyai Empu Permadi, Kyai Brama Kedhali, Gusti Eyang Panembahan Prabu Jagad, Kyai Sabuk Angin, Bok Nyai Gadhung Mlathi, and Gusti Panembahan Megantara.[10]

These names were important in part because of what their honorifics signified, such as "the highest" and "greatest" but also titles and rank such as king or kyai (a Muslim scholar). Some deities were local ancestors, figures of historical importance, of Islamic descent, as well as belonging to Hindu genealogies. Kyai Empu Permadi, for instance, was a *wayang* character of the genealogy of Batara Guru, a variation on characters from the *Mahabharata*. In this instance, the character from the Hindu epic had metamorphosed into a kyai. In the spiritual-material topography of the Labuhan, the ontology of Merapi was historical and genealogical, it reflected the social hierarchies of the sultanate and combined the spiritual geographies of Hinduism, Buddhism, and Islam. Merapi in 1921 was constituted by these material and spiritual networks, simultaneously local and cosmic.

If Merapi was the Navel of Java, the Labuhan also made clear that it could not be understood independently from the Indian Ocean. Observers often noted how *babad* and other stories brought together the sultanate with the Indian Ocean and volcano in shared origins. They also noted how deities in both places traveled between the ocean and volcano. Theodore Pigeaud, a translator, philologist, and admirer of *babad* literature, showed repeatedly in the 1920s how Javanese literature grappled with the common origins of the sultanates, volcano, and Indian Ocean. In his commentary on the *S. Baron Sakèndèr babad* (likely written after the middle of the eighteenth century),[11] Pigeaud noted that in order to become the lord of Mataram, the *babad* established that the prince must initiate relations with both the volcano and the Indian Ocean. When Senopati went to meet Nyai Ratu Kidul, the *babad* explained, his assistant, Juru Martani, went to the mountains. Sultan Agung, likewise, traveled to the mountains before descending to meet Nyai Ratu Kidul.[12] For Pigeaud, *S. Baron Sakènḍèr*, and *babad* in general, were works of Javanese history that did not recognize the difference between myth and history, or between nature and social history—nature was social.

If the Navel of Java, Merapi, could not be understood apart from the realm of Nyai Ratu Kidul, this suggested questions about what connected them. The rivers that drained the volcanic slope, including its debris, into the Indian Ocean were important conduits for spiritual traffic. Lucien Adam, the Dutch assistant-resident to the sultan of Yogyakarta, who was a keen ethnologist of courtly custom while he undertook his diplomatic duties, recorded historical and legendary toponyms in the region. He recounted an episode that acknowledged the connection between the crater of Merapi and its rivers. A skirmish between a giant curious snake and Kyai Ageng Mangir at the crater resulted in the snake having its tongue cut off and turning into a *keris* dagger at a site that after was called Kendit, where the Labuhan would provide offerings and the meal of rice and chicken was taken. The snake traveled to the crater by way of the Bedog River, which feeds the Progo River, which in turn drains into the Indian Ocean.[13]

It is clear from translations of the *babad* and ethnologies of court life and its landscapes in the 1920s just how significant the Queen of the Indian Ocean and the pantheon on Merapi were together. The very polity itself originated in and through their cooperation and interaction; the crater of the volcano was historical, in exchange with local and cosmic deities, and deities could disrupt the kingdom and test the limits of the sultan's power through eruptions, earthquakes, and tsunamis. In order to maintain his power, the sultan was therefore in a relationship of debt to the deities, of providing them offerings and gifts in return for their protection, meaning the relative quiet from natural disasters. If disaster struck, it was potentially an expression of dissatisfaction of the deities with operations in the human realm. A natural disaster was by no means the expression of an anonymous natural event unfolding, or a natural expression of the structure of the earth surface—it was a human political event. What lends credence to this understanding is the nature of the offerings themselves—clothes, food, money—which suggested volcano and ocean deities participated in human affairs.

Geologists in the Realm of Nyai Ratu Kidul

When the Labuhan was recorded in Yogyakarta in the 1920s and 1930s, geologists had recently begun to connect the Indian Ocean with volcanoes. Rogier Verbeek had undertaken the first complete geological survey of Java between 1886 and 1896 with the hope of identifying profitable ores and mineral deposits, but he soon realized that volcanoes were poor sources of profit. In parallel, he also surveyed antiquities for the Royal Batavian Society for Arts and Sciences. The two surveys informed each other as Verbeek came to interpret the landscape and Java's natural history as ruins. Volcanoes were conceived as the ruins of once-great mountains that exploded as they fell apart. Sea levels had risen and fallen in geological time and Java had progressed through a series of transformations from being, at one time, connected via land bridges to mainland Southeast Asia, with fauna traveling as far as India, to becoming submerged and segregated into an archipelago of volcanic cones. As Verbeek scaled the slopes of Java's volcanoes to reconstruct this history, he also cataloged nearly seven hundred ruined temples (*candi*). He described each ruin according to its geology and stratigraphy, just as he had for fossils on volcanic peaks. He also noted the date of every stone inscription and described the political geography of the eastern migration of Hindu and Buddhist temple complexes between the seventh to fifteenth centuries.[14] In Verbeek's vision, Java and its culture were ruins. Borobudur exemplified this: a massive, inexplicably abandoned, half-buried temple—that suggested a once-great culture—surrounded by volcanoes, themselves ruins of once-great mountains. This fever for antiquities, in part, also drove the ethnologists at the Java Institute and its members' interests in *babad*. They saw that beneath the cloak of Islam, the remains of Hindu and Buddhist cultures were discernible in practices such as the Labuhan. The deities in the landscape referenced the still-living evidence of that ancient time. The traditions and rituals of the sultanates, though syncretized with Islam, maintained those otherwise abandoned cultures. In other words, the sultanates, too, were ruins of Hindu and Buddhist empires. The purpose of ethnology and geology was to reveal to the Javanese the past they had forgotten. This was often an explicit attempt to undermine Javanese Islam, which many Dutch colonists understood as a corrupting influence.

Verbeek's natural history, though, had little to say about the role of Nyai Ratu Kidul's realm in the geological history of Java. Verbeek had known little about the Indian Ocean and remarked only that along the south coast of Java ship soundings had indicated "great deeps." He remarked that a rift had occurred at some unknown time, separating the East Indian archipelago from the northwest of Australia. His lack of knowledge about the Indian Ocean was carried forward by later geologists who maintained a persistently northern orientation in their thought and tended to view Java as the southeastern most extent of the Eurasian continent, connected to the mainland via the submerged Sunda Shelf. Through periodic sea level rise and fall, the Sunda Shelf connected and disconnected the archipelago from the mainland. This northern and western orientation was formed in no small measure because of the availability of ship soundings in the Java Sea compared to the paucity from the Indian Ocean. The surplus of studies focused on the north was the result of attempts to link the Netherlands East Indies to Malaya via underwater telegraph cables. The position of Nyai Ratu Kidul's realm in the geological history of Java and the East Indies came to play a more decisive role in the 1920s, when geologists developed, by their own independent means, stories that also insisted that Java's volcanoes could not be understood separately from the Indian Ocean. Moreover, they also came to understand that the realm of the Indian Ocean and volcanoes exchanged with each other.

In 1923, 1925, and 1926, the civil engineer-turned-geodesist Felix Vening Meinesz undertook the monumental task of mapping gravity in the ocean floor. He developed a novel four-pendulum device to measure gravity anomalies from within a submarine. Gravity anomalies were indications of the distribution of mass on the ocean floor and could indicate depressions or ridges underwater. Broadly, the seafloor had been understood to be relatively flat and tectonically stable. Pendulums had not been successful on ships before because of the interference of wave action. Vening Meinesz's gravimeter was the first successful device to measure gravity at sea, and he established unprecedented numbers of gravity stations and measurements off the south coast of Java. He found there not only a "rift," as Verbeek had imagined, but a vast underwater trench nearly 8,000 kilometers long and roughly 160 kilometers wide, running parallel to Sumatra, Java, and the eastern part of the archipelago. What Vening Meinesz also immediately noted was that the massive underwater trench ran parallel to the spine of volcanoes that ran down the center of Sumatra and Java. The pressing question was if and how they related to each other. The discovery of the trench decisively shifted geologists' attention toward the south.

Simon Visser in the Royal Magnetic and Meteorological Observatory in Batavia had developed a new way of calculating earthquake epicenters, and between 1909 and 1926, he came to understand that earthquakes were originating in the Indian Ocean when they had previously been thought to have been land based. He also came to understand that some earthquakes were originating as deep as six hundred kilometers underground. It appeared that earthquake epicenters coincided with the trench and were perhaps products of it. Volcanism and seismicity were coming to be thought together with the depths of the Indian Ocean. Vening Meinesz and fellow geologist colleagues, including Gustaaf Molengraaff, Albert Brouwer, Reinout van Bemmelen, and Johannes Umbgrove, turned to ideas of continental drift as a possible explanation. In 1931, Vening Meinesz hypothesized that the crust was "moving toward a curved part of the fold-line."[15] In 1934, Umbgrove explicitly invoked continental drift as a possible source of the fold.[16] In 1936, Philip Kuenen developed a petroleum jelly and water model in a laboratory that horizontally compressed the material like he imagined the earth surface to be forming (figs. 6.1 and 6.2). The model successfully reproduced the fold, and though the nature of the mechanism—the force driving the movement—was not defined, it nevertheless suggested that horizontal compression from some force was a possible origin of the trench.[17] B. G. Escher imagined that "currents" and "vortices" under the crust were responsible (figs. 6.3 and 6.4). These ideas were forcefully reorienting the geological imaginary developed by Verbeek by shifting the center of gravity of Java away from the north to the south. The effect was no less far reaching on volcanoes as they came to be understood in relation to the trench and no longer as disconnected, collapsing ruins, as Verbeek had thought. As Molengraaff insisted, "a genetic connection must exist" between the volcanoes and the trench, and "[it] has to be sought in one and the same crustal movements."[18] As Vening Meinesz formulated it in 1931 to the Royal Geographical Society in London, the stresses in the surface created by the trench "bring about fissures in the crust, or at least a decrease of pressure, and this may give rise to the formation of volcanoes."[19] Javanese volcanoes were coming to be understood in terms of their relationship to processes of folding in the Indian Ocean. Looking at volcanoes, for geologists, came also to mean looking at the Indian Ocean.

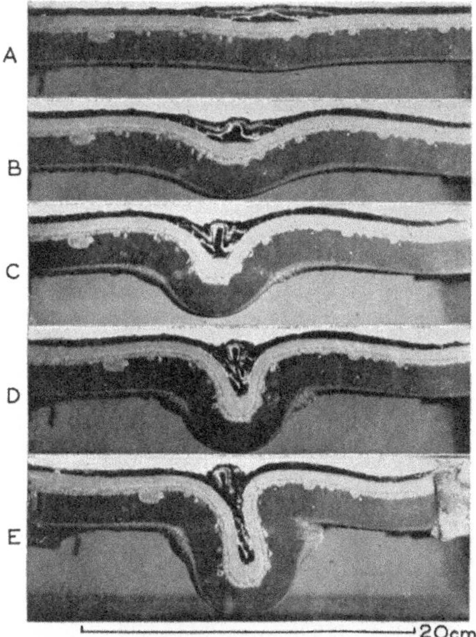

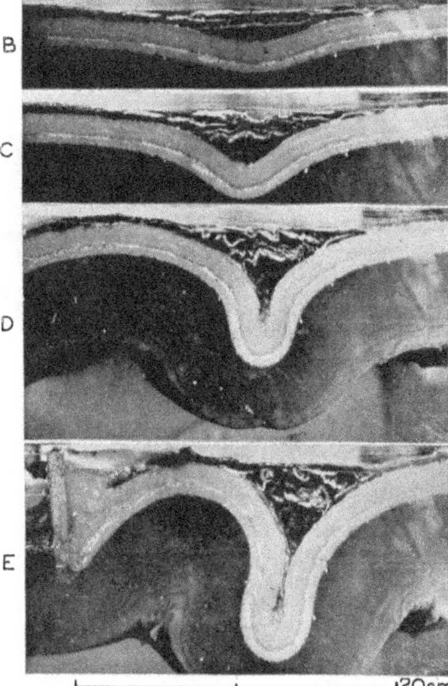

FIGURES 6.1 AND 6.2. Kuenen's petroleum jelly and water models illustrating how the surface of the earth might be folding. P. H. Kuenen, "The Negative Isostatic Anomalies in the East Indies (with Experiments)," *Leidsche Geologische Mededeelingen* 8, no. 2 (1936): 186–87. © Naturalis Biodiversity Center.

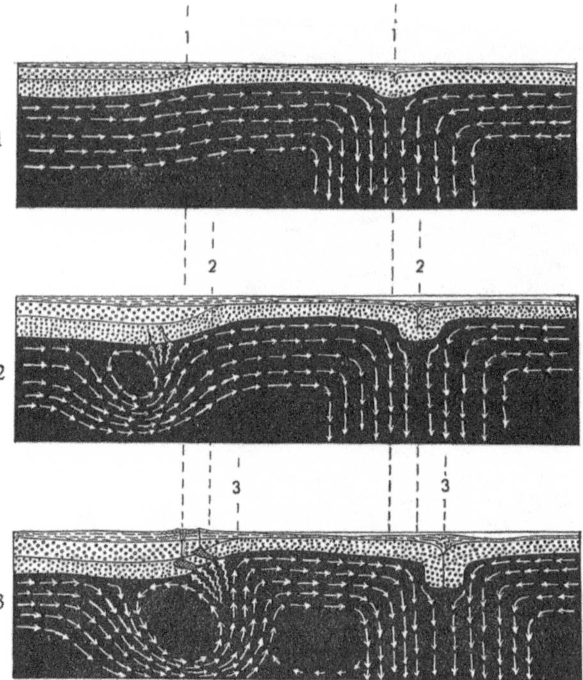

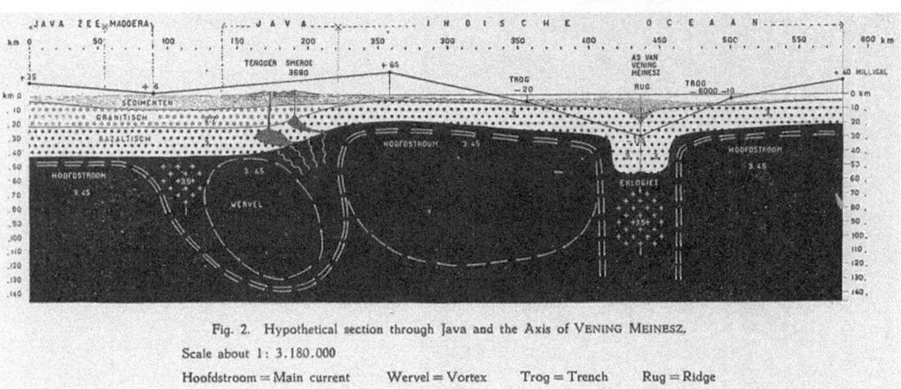

Fig. 2. Hypothetical section through Java and the Axis of VENING MEINESZ.
Scale about 1: 3.180.000
Hoofdstroom = Main current Wervel = Vortex Trog = Trench Rug = Ridge

FIGURES 6.3 AND 6.4. Illustrations of Escher's theory that convection currents and vortices under the earth's crust were responsible for creating trenches. B. G. Escher, "On the Relation between the Volcanic Activity in the Netherlands East Indies and the Belt of Negative Gravity Anomalies Discovered by Vening Meinesz," *Proceedings of Koninklijke Nederlandse Akademie van Wetenschappen* 36 (1933): 677–85. Courtesy of Royal Netherlands Academy of Arts and Sciences.

Geologists on Volcanoes

In 1921, Dirk van Hinloopen Labberton, a Sanskritist, Javanese teacher, and secretary of the Theosophical Society in Batavia, explained that in Java, in 1403 of the Saka calendar, "the firmament was the ocean."[20] He was referring to an eruption of Kelud volcano in East Java. This was one of many translations he made of stone inscriptions, *babad*, and other Javanese sources that referred to the watery nature of volcanoes. He pointed to the many descriptions of *lahar* mudslides that flowed from craters and to metaphors of gushing and spilling. He was also keen to explain his new interpretation of an eruption in 1006, in which Java was supposedly inundated by "floods of disasters." He had recently translated inscriptions that described a massive sequence of natural disasters that apparently destroyed Java "as if by tidal wave." The inscriptions, he explained, said that "all of Java looked like a sea at that time. The Sanskrit 'arnawa' means: effervescent water current, tidal wave and ocean, including '*bandjir*' [flood]."[21] This eruption was responsible, he claimed, for the destruction of Hindu Buddhist civilizations in Central Java and their migration east. Many of the ruins that Verbeek had found were the result of this cataclysm.

Reinout van Bemmelen was aware of Hinloopen Labberton's interpretations, since they were published in the *Natuurkundig Tijdschrift voor Nederlandsch-Indië*, where volcanologists also published their field reports. Hinloopen Labberton's essay was printed beside another article that for the first time announced the foundation of a Commission for Volcanology to be established to permanently monitor Java's volcanoes.[22] Van Bemmelen was a member of the Commission and later became its director. Nearly two decades after Hinloopen Labberton proposed the catastrophic destruction of Javanese culture by way of volcanoes, Van Bemmelen surveyed the peak of Merapi and noticed for the first time that the cone was inside a much larger ancient crater. He drew on Hinloopen Labberton when he wrote the following:

> The island of Java looked like the sea, the "kraton" [palace] was burned into ashes, and covered by the mountain. The stone inscription speaks of the Maha Pralaya of Java, which means the end of an epoch of the country by a natural calamity. The present active cone of Merapi rises above the one that was destroyed in 1006 A.D.[23]

Like Verbeek, Van Bemmelen understood a volcano as the ruin of a once mightier mountain. What he contributed to this view was to link the Javanese ruins in the landscape to particular volcanic events recorded in those antiquities. The language of floods and waves was not interpreted as literal water (such as a tsunami) but as a volcanic catastrophe. Volcanoes were understood, for the first time among geologists, as agents that determined Javanese social and cultural history. The fall of the Hindu and Buddhist empires, he argued, made way for the rise of Islamic empires and later the flourishing of the Central Javanese sultanates. Van Bemmelen's understanding of Merapi was developed into an expansive philosophy of volcanic determinism. As he would later put it, framing it in terms of orogeny, "*mountain building provides the very basis of our existence on earth*."[24]

It was not unusual that Hinloopen Labberton was a Theosophist. Theosophist lodges in Bandung, Batavia, and Yogyakarta were sites where scientists and Javanese elites from the royal families of Yogyakarta and Surakarta crossed paths. They were hubs for critique of the colonial system, and members of the sultanate from Surakarta and Yogyakarta presented ideas to members of the Dutch elite about the power of the Hindu Buddhist tradition to combat the corrupting influences of modernization. Interpretations of the shadow theater plays based on the Mahabharata and Ramayana were published by members of the royal family in the *Theosophist Magazine*. The father of Indonesia's first president, Sukarno, was a Theosophist; the first vice president, Muhammad Hatta, was offered a scholarship to study in the Netherlands by a Theosophist. The lodges were hubs where scientists encountered the principles of the Javanese traditions upheld in the Central Javanese sultanates. Umbgrove even complained privately that Theosophy was "flourishing" in Bandung; he wrote, "their 'lodge' has many members, including our boss who is better informed about it than he is about geology."[25] He was referring to A. C. de Jongh, a mining engineer, who had contributed an article to the *Theosophist Magazine* in 1914 that sought to lend modern scientific credibility to Annie Bessant and Charles Leadbeater's "occult chemistry."[26] Many Theosophists, including Hinloopen Labberton, explicitly encouraged a renaissance of pre-Islamic Hindu and Buddhist traditions of Central Java associated with the *kraton* (palace). This orientalist enthusiasm would have been one source for volcanologist's encounters with Central Javanese spiritual geographies.

While geologists were encountering the mystical tradition of the Central Javanese sultanates in the lodges, they were also encountering it in the landscape. In 1920, Georges L. L. Kemmerling, the director of the volcanological survey, was undertaking fieldwork on Merapi and reported that he took "an old ritual path" up to the southern side of the summit in order to take samples.[27] That "old ritual path" was almost certainly the route of the Labuhan, and it is very likely that the place at which he took samples was the same as where the Labuhan described by Tirtakoesoema the following year was reported to have taken samples of sulfur back to the sultan. The first observatory was a hut at Maron located to the west of the Labuhan path and the path to Maron converged with the Labuhan path. Observers in the hut would have used the Labuhan path during their observations. The observers' fieldwork was enabled by the infrastructure of the Labuhan. On that same expedition, Kemmerling remarked, "After all, it seemed very possible that the imbalance occurring in the Indian Ocean creates an effect in Merapi."[28] He also noted Visser's new recordings of earthquake depths in the Indian Ocean. Not only was Kemmerling following the paths constructed for the Labuhan to take his samples—he was also coming to recognize the causal links between the volcano and the Indian Ocean, the fundamental basis of the Labuhan.

The Labuhan path was not the only way that the spiritual infrastructure of Merapi enabled modern volcano scientists to do their work. The northern flank of the volcano was also frequently visited by scientists to monitor the status of the crater, and their presence there was recorded in photographs and drawings. They spent time camping on a plateau called Pasar Bubrah, which is scattered with large boulders and debris from previous eruptions. Pasar Bubrah translates to "Scattered Market" or "Ruined Market" and referred to the spirit market where deities from inside the volcano went to hawk their goods. Scientists recorded spiritual toponyms in the drawings and maps, such as Reinout van Bemmelen's depiction of Merapi from 1943 where he indicates with arrows the location of Pasar Bubrah and Masdjidanlama, the mosque where deities prayed (fig. 6.5). The scientists' Javanese assistants (*mantris*),

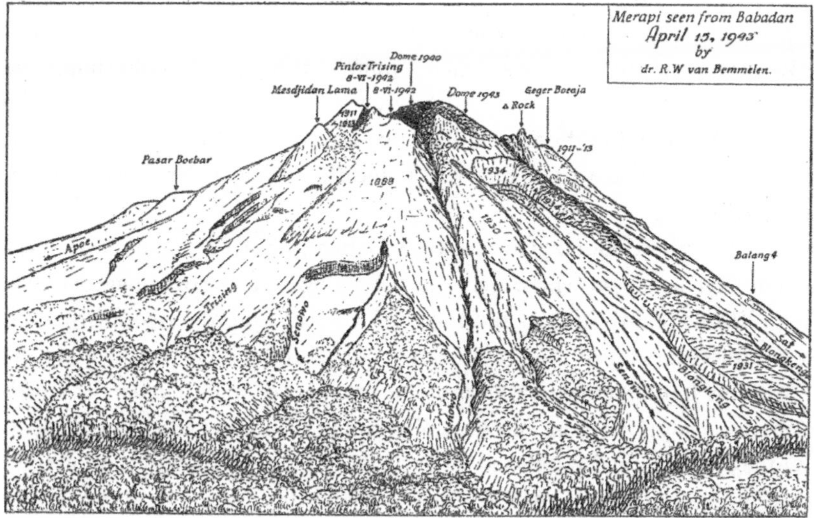

FIGURE 6.5. Reinout van Bemmelen, "Merapi Seen from the Observation Post Babadan (April 15, 1943)," *The Geology of Indonesia* ([The Hague]: Government Printing Office, 1949), 209.

who undertook experiments, carried scientists' equipment, and translated for them, would have understood the toponyms and very likely discussed their meaning with scientists. Spiritual topographies also became places for lookouts and observation posts, such as at Plawangan, which was understood as a nail in the volcano.

A New Old History of the Earth

In 1928, Vening Meinesz was invited by the Carnegie Institution of Washington and the US Navy to lead a gravity expedition on the USS *S-21* submarine in the Gulf of Mexico and Caribbean Sea because of the successes of the gravimeter in the East Indies. He was assigned two assistants, Elmer B. Collins, from the US Navy Hydrographic Office, and F. E. Wright, from the Carnegie Institution. He instructed them in the use of the gravimeter. The purpose was to study the edge of the North American continent and understand if its mass was in equilibrium with the ocean floor. They found several anomalies that suggested "departures from the normal state or balanced condition of the earth's crust," and indicated sources of seismicity and volcanism.[29] These results departed from the conventional view that the ocean floors and continents rested together in an equilibrium state. In 1932, Vening Meinesz returned to North America for a joint expedition with Princeton University and the US Navy on the USS *Chewink* to conduct more measurements at sites that suggested strong gravity belts like in the East Indies. His assistants were Harry Hess, from Princeton University, and T. T. Brown, from the US Naval Research Laboratory, Washington, DC. Hess concluded that the narrow strip of strong negative anomalies pointed to a lack of equilibrium.[30] He also suggested that the belt was due to a massive folding of the crust along the lines of that found in the East Indies and also associated with seismicity and volcanism. In other words, they had found the realm of Nyai Ratu Kidul in the Caribbean.

In 1933, the Geodetic Committee of Japan imported a Vening Meinesz gravimeter to Japan, where they discovered again a mirror image of Nyai Ratu Kidul's realm. In 1948 and 1949, J. L. Worzel and Maurice Ewing used a modified version of the gravimeter on US Navy submarines to make almost nine hundred observations off the Pacific and Atlantic coasts of North America, Hawai'i, Guam, and Australia, and in the Bering and Chukchi Seas. By 1965, Worzel had published an account of nearly four thousand gravity measurement stations at sea between 1936 and 1959.[31] The British were likewise invested in gravity measurements, and by the 1960s, Vening Meinesz's apparatus was in use by most Western states with an interest in seafloor exploration and geodesy propelled by Cold War interest in submarine warfare and communications infrastructure.[32] Vening Meinesz's 1958 book, written with Veikko A. Heiskanen, *The Earth and its Gravity Field*, was seen by students at the Lamont Observatory at Columbia University as an essential resource for gravity measurement and theory.[33] The relationship between deep ocean trenches, volcanism, and seismicity was discovered in island arc systems around the globe and came to be seen as engines of planetary evolution.

In 1962, Hess published his seminal paper, "History of Ocean Basins," in which he argued that the seafloor was mobile, spreading from mid-ocean ridges into the continents. This phenomenon could be witnessed in island arcs, like the south coast of Java, where the ocean floor was pushed into continents and produced volcanoes, earthquakes, and new land. Hess used Vening Meinesz's and Johannes Umbgrove's theories to construct his vision of earthly evolution. Like Meinesz had before, Hess imagined the solid earth soon after its formation, with no differentiation between surface and core until "the great catastrophe," an event of "convective overturn" that began to differentiate the core from the surface. This differentiation led to the formation of a single protocontinent, which was then torn apart by convective currents and took the shape of modern continents. Mantle convection continued to push the crust away from the mid-ocean ridges and into land. The great catastrophe was the foundational moment that created the modern earth; it was a story of a fall from unity to differentiation and original catastrophe; it was also an expression of optimism at the earth's capacity to generate itself anew. Island arcs were places of creative destruction, windows into the primordial processes of our planet's formation and its capacity for radical novelty. Hess's paper was pivotal for the adoption of the theory of plate tectonics among US geologists because it convinced them that convection currents were the mechanism that drove the system.

Vening Meinesz's shift of focus from the north to the south of the Indonesian archipelago revealed a system that became central to the plate-tectonic narrative. Geologists in the United States began to call it revolutionary. Bruce Heezen and Marie Tharp published their stunning representations of the world's ocean floors without water in the 1960s and 1970s and depicted the earth from the perspective of its massive ocean ridges and cavernous depressions where the ocean floor made contact with continental shelves. Their maps and diagrams were published by the Geological Society of America (fig. 6.6) but also in more popular venues, such as *National Geographic*, and introduced readers to the plate-tectonic vision of the earth. The realm of Nyai Ratu Kidul was subsumed into a larger narrative of planetary processes and history.

The narrative that island arcs were the result of plate subduction carried

FIGURE 6.6. Bruce C. Heezen and Marie Tharp, "Physiographic Diagram of the Indian Ocean, the Red Sea, the South China Sea, the Sulu Sea and the Celebes Sea" (New York: The Geological Society of America, 1964). With permission from Fiona Schiano-Yacopino.

forward the central insight of the Labuhan. For the sultanates of Yogyakarta and Surakarta, the realm of Nyai Ratu Kidul could not be thought to be separate from Merapi. Earthquakes coming from the ocean were associated with rumbles in the volcanoes, and this was understood in terms of exchange between the two realms, they did not act independently. The movement of Nyai Ratu Kidul from the ocean to the pantheon of deities in the crater signified this. Nyai Ratu Kidul's realm was oceanic, but she was simultaneously chthonic, mud and ash, salt and fresh water. When US geologist Warren Hamilton argued in 1973 that the Indian Ocean was subducting beneath Java and causing its volcanism, he surreptitiously imported the language of the sultanates; to look at Merapi was to look at the Indian Ocean—they had to be understood together.[34]

The complementarity of the two visions has continued to be born out more recently. Merapi erupted in 2010, in one of its most violent eruptions since 1930. The gatekeeper (*juru kunci*) Penewu Surakso Hargo, or more familiarly Maridjan, who was responsible for providing offerings from the sultan, died in the eruption. Maridjan lived in Umbulharjo and was likely descended from the same family of gatekeepers that conducted the Labuhan procession that Tirtakoesomoe reported on in 1931. In 2010, Maridjan had refused to leave the village despite a request from government scientists. He had also refused in 2006, claiming that the volcano would not harm his village (and was proved correct). In 2010, the head of the volcanological survey, Surono, spoke angrily to the media about Maridjan's "voodoo" beliefs. The conflict was pitched in terms of modern science versus superstition and progress versus tradition. Surono thought that it was dangerous for hundreds of thousands of Merapi residents to respect the forecasts of a mystic more than him—a seismologist trained in France in the most up-to-date theories and techniques. During the eruption, which lasted nearly four months, close to two hundred thousand people became refugees. After the gatekeeper died, calls went out for Surono to be appointed as the new gatekeeper. Seemingly, science had triumphed over mysticism.

After the eruption, Surono published a paper as the lead author in the *Journal of Volcanology and Geothermal Research*, coauthored with seventeen Indonesian, French, British, US, Norwegian, and Singaporean scientists. They argued that the eruption was the result of the transfer of magma through the "volcano's subsurface plumbing." That plumbing system connected the trench in the Indian Ocean to the volcano. It was also understood that the Opak Fault played a key role; the fault runs between the trench and the volcano and may have acted as a conduit. The Opak Fault is named after the Opak River, which lies approximately above the fault and begins near the crater and drains sediment into the Indian Ocean. It is also fed by the river beside Umbulharjo, the village where the Labuhan takes place. Nyai Ratu Kidul was known to use the Opak River as a thoroughfare to the crater, and the confluence of the Opak River and the Indian Ocean was long used as a place to meditate to meet the goddess.[35]

In 2016, Surono attended the Labuhan ceremony in Umbulharjo. He was invited as a guest of honor and greeted warmly by the new gatekeeper, Asih, the son of Maridjan. Surono joined a small group of local elites and dignitaries from the sultanate to watch the evening *wayang* performance before the offerings would be carried up "the old ritual path" that Kemmerling had taken in 1920 to Kendit, where the snake had its tongue cut out by Kyai Mangir. In an interview in *Tempo* magazine at around the same time, Surono defended the Labuhan, arguing that it resisted the modern volcanologists' imaginary that they could dominate the unpredictability of nature through the use of modern technology. The Labuhan, he argued, also kept alive the notion that offerings need to be given to the volcano, and that the relationship between people and the volcano was not about mastery.

What Surono was negotiating was the central dilemma between Javanese geographical thought and modern volcano science. Modern volcano science in Java was encouraged and enabled by the sultanates of Central Java to connect volcanism to the Indian Ocean. The spiritual geography of the sultanate gently led modern scientists to develop a narrative of the constitutive relation between the ocean and volcanism. For the sultanates, their polities were mutually constituted with the ocean and volcano; their histories were materialized in those landscapes. The ritual of the Labuhan enacted these histories. For Surono, the enactment of those histories had become more urgent than ever.

7

Geo-Spiritualities of the Flood

Political Geologies of the Great Deluge on the Mountains of Anatolia

Zeynep Oguz

In the summer of 2017, I was with Ahmet and Cengiz near the Garzan oil field in the Batman Province of Turkey's southeastern region. Ahmet was a Kurdish petroleum worker at the state-owned Turkish Petroleum Company. Cengiz, a Turkish geological engineer, was employed at the exploration department of Turkish Petroleum's Ankara headquarters. We were surrounded by the massive dolomite limestone cliffs covered by red, clayey sandstones that are characteristic of the Garzan basin. As usual, Cengiz was quizzing me about the chronostratigraphy of these formations. (*The Jurassic or the Cretaceous period? Which epoch—Miocene or Eocene?*) When he was satisfied with my answers, he pointed at a visible, vertical fault line on one of the tall, beige-red limestone cliffs and started speculating about why the fault could be displaced like that: an earthquake, erosion, or another kind of movement? As I was trying to come up with a seismic scenario, Ahmet jumped in: "I don't know about an earthquake, but what I see is also a sign of a flooding event. There are vertical marks of deposition on those sediments." "What flood?" asked Cengiz, skeptically. "Noah's Flood, obviously," Ahmet replied. "You know, Prophet Noah's ship landed a few miles from here." "Mount Ararat is hundreds of miles away from here," Cengiz said. "No, it's not Ararat. It's *Cûdî*." Ahmet was referring to the 2,144-meter tall Mount Judi located in the Şırnak Province in Turkey's southeast. He went on: "These lands, cliffs, valleys . . . *The Mountain*—they are sacred for us."

For centuries, the mountains of contemporary Turkey's Eastern and Southeastern Anatolia region—or Western Armenia and Northern Kurdistan, respectively—have occupied a central place in religious and spiritual cosmologies about the mythical deluge. Two mountains, in particular—Mount Judi and Mount Ararat—figure as the final resting place of Noah's ark after the Great Flood in creationist Judeo-Christian and Muslim beliefs. They have been the leading two—and often competing—locations worldwide for those searching for the remains of Noah's ark, as detailed in the Bible's book of Genesis and the Quran. Narratives of a deluge, however, are not solely sourced by biblical and Quranic verses; they are also found in Anatolian and Mesopotamian folk stories that date back to Sumerian and Assyrian myths. Further, Mount Judi for Kurds and Mount Ararat for Armenians hold a unique cultural and historical significance. As Ahmet put it, *they are sacred for us*—the "us" here being not Muslims or Christians but the Kurdish people, whose political-geographical imaginary of Kurdistan is constituted by its mountains, and especially by *Cûdî Dağı*.

Conversations around the Anthropocene have been leading to a renewed attention to the place of religion and spirituality in a moment of cataclysmic planetary change.[1] As the historian of religions Mary Evelyn Tucker points out, the religion-ecology nexus "is now poised to be a key participant in the dialogue involving the Anthropocene and environmental humanities."[2] Narratives of the Great Flood have been spaces where religion, science, and myth merge into each other.[3] Historian of science Lydia Barnett demonstrates that flood narratives have also been central to the idea of a "global humanity" and the emergence of a planetary consciousness during early modernity. Through the narrative of the Great Flood, philosophers and historians made sense of the "spiritual decline of humanity and the physical deterioration of the natural world" as part of the same tragic history of catastrophic and potentially irreversible ruin.[4] In this chapter, rather than exploring various narratives of the Flood or the alleged structure or inhabitants of the ark, I examine imaginaries and practices around Mount Judi and Mount Ararat as the two most popular mountains among the manifold beliefs about the final resting place of Noah's ark in Turkey and beyond. Attending to the often-neglected geological materiality of Flood narratives, I aim to unpack the spiritual-political histories and cosmologies that the two mountains are embedded in.

As mountains that simultaneously gesture at mythical histories of planetary environmental change, human political relations, and the forces of nonhuman spiritual entities, Judi and Ararat can be taken as "geo-spiritual formations."[5] Pitched against each other as resting places of Noah's ark, they are also intertwined with territorial imaginaries, colonial and nationalist projects, insurgent movements and counterinsurgency measures, and the making of extractive frontiers in the Anatolian Anthropocene. The mountains where it is believed that Noah's ark came to rest after an environmental-moral catastrophe, in this sense, are geo-spiritual formations that also bear witness to ongoing environmental, political, and cultural catastrophes in Turkey, a deeply troubled post-imperial nation-state built on the denial of genocide, the systematic dispossession of the Kurdish people, and the ecological destruction of Kurdistan.

Mount Judi

The biblical account of Noah in the book of Genesis serves as the predominant flood story in Western culture; it shares many affinities with the ancient Mesopotamian flood accounts like the Akkadian epic of *Atra-Hasis*, the Sumerian story of Utnapishtim, found in the Epic of Gilgamesh. These ancient Mesopotamian epics and the story of Noah share several common elements: God, or the gods, calls forth a flood to remove humans from the earth. In both, one chosen individual is called upon to build a large boat that carries all the animals to be spared from the deluge. After a torrent of several days of floods, the ark comes to rest on a tall mountain.[6] The biblical and Quranic accounts of the flood also parallel each other, with one significant disagreement over which mountain the ark finally rests on.

The proponents of the Mount Judi thesis argue that Mount Ararat has been mistakenly assumed to be the final resting place of the ark due to a misinterpretation of the Genesis narrative in the Bible. The Bible references Mount Ararat in the story of Noah: "And the ark rested in the seventh month, and on the seventeenth day of the month, upon the mountains of Ararat."[7] The Quran names the mountain upon which Noah's ark landed as "Mount Judi" and not Mount Ararat: "A voice cried out: 'Earth, swallow up your waters. Heaven, cease your rain.' The floods abated and His will was done. The ark came to rest upon al-Judi, and a voice declared: 'Gone are the evil-doers.'"[8] While some contemporary evangelical Christian explorers that hike to Mount Ararat's peak every year claim that the "mountains of Ararat" refers to Mount Ararat in eastern Turkey, advocates of the Judi thesis argue that the Hebrew text speaks of "mountains" in the *plural*. "Ararat" in the phrase "mountains of Ararat,"[9] they argue, refers not to Mount Ararat but the entire kingdom of Ararat, the Hebrew name for Urartu. Thus the kingdom of Ararat, they add, encompasses Mount Judi.[10] The creationist archaeologist Andrew Collins wrote in his 2014 book that there was once an ancient monastery on the summit of Mount Judi; this monastery was destroyed by lightning in 776 CE. Collins states that, after this time, the belief that the mountain was the "Place of Descent" declined, and its position was taken by Mount Massis, which was called *Ağrı Dağı* by the Turkish people and Mount Ararat by the Armenian people.[11]

Toward the end of the twentieth century, Judi would receive significant attention from British and German explorers. During World War I, Gertrude Bell explored from Aleppo along the banks of the Euphrates River down through Mesopotamia to Baghdad, where she turned around and came back up the Tigris River in Turkey. Bell reached Mount Judi in 1909. In *Amurath to Amu-*

rath, published in 1911, Bell described the shrine and the festival held there every year in honor of Prophet Noah:

> The Babylonians, and after them the Nestorians and the Moslems, held that the Ark of Noah, when the waters subsided, grounded not upon the mountain of Ararat, but upon Judi Dagh. To that school of thought I also belong, for I have made the pilgrimage and seen what I have seen.... We climbed for two hours and a half through oak woods and along the upper slopes of the hills under a precipitous crest.... As I walked through the woods I was overmastered by the desire for the snow patches that lay upon the peaks and Sefinet Nebi Nuh, the ship of the Prophet Noah, was there to serve as an excuse.... And so we came to Noah's Ark, which had run aground in a bed of scarlet tulips. There was once a famous Nestorian monastery, the Cloister of the Ark, upon the summit of Mount Judi, but it was destroyed by lightning in the year of Christ 766. Upon its ruins, said Kas Mattai, the Moslems had erected a shrine, and this too has fallen; but Christian, Moslem and Jew still visit the mount upon a certain day in the summer and offer their oblations to the Prophet Noah. That which they actually see is a number of roofless chambers upon the extreme summit of the hill. They are roughly built of unsquared stones, piled together without mortar, and from wall to wall are laid tree trunks and boughs, so disposed that they may support a roofing of cloths, which is thrown over them at the time of the annual festival.[12]

Another explorer who was intrigued by Mount Judi was German geologist Friedrich Bender, president of the Federal Institute of Agricultural Research in Germany. In the early 1950s, Bender worked as a geologist in Southeastern Anatolia and had close contact with local Kurdish people who told him stories about the resting place of the ark and Quranic stories about Noah. Both Bell and Bender wrote about a Nestorian Christian monastery located on the mountain, five hundred meters away from where the ark was believed to be located.

Gertrude Bell's accounts were influenced by both the folk beliefs she collected from the Kurdish and Arab residents of the region and from medieval texts by Persian and Arab geographers such as Ibn Khordadbeh and Istakhri, whose travelogues Bell encountered through the works of the Dutch orientalist Michael Jan de Goeje. Mehmet, a local Kurdish guide from Cizre who I spoke to in 2018, told me a similar folk story about the shrine on Judi that was transmitted over generations. He said that not that very far from the Nestorian monastery that once stood on Mount Judi, there used to be a shrine dedicated to a Muslim dervish called Lawkê Xerîp. He was referring to the same shrine whose remains Gertrude Bell photographed in 1909. Xerîp, according to the Kurdish folk story, had decided to go on a pilgrimage to the ark. Mehmet went on:

> His journey on the slopes of or *Cûdî* took seven years. He finally got very close to the ark, but the Devil appeared to him in the form of a human and told him that he had to walk for another seven years to reach Prophet Noah's ark's resting place. Believing the Devil, Xerîp begged Allah to accept his pilgrimage and take his life right there. His wish was granted and Xerîp died there.

Mehmet told me that this story was transmitted among Kurdish people of the area for years and eventually a shrine was built there to honor Lawkê Xerîp and his pilgrimage. Every year during the seventh month of the Islamic calendar, Mehmet said, people from nearby villages visited the shrine to pay their respects to the Prophet Noah and make their wishes. "We are the descendants of Prophet Noah," Mehmet added. Ethnohistorical accounts draw connections between the history of the Kurdish people and Mount Judi, although not necessarily with the figure of Noah. According to Kurdish studies scholars Samar Abbas and Mehrdad Izady, the contemporary name "Mount Cudi" preserves the ancient Guti name, referring to the Guti people who inhabited modern Kurdistan over four thousand years ago.[13]

The annual festival on the shrine of Lawkê Xerîp that both Mehmet and Bell speak of, however, has not taken place for decades, as Mount Judi was closed to public access in the 1980s following the armed conflict between the Kurdistan Workers' Party (*Partiya Karkerên Kurdistan*, or PKK) and the Turkish Armed Forces. On the backdrop of their preexisting spiritual and religious significance, it was with the beginning of the war that the meaning of the mountains of Kurdistan in general, and Mount Judi in particular, started to take on new layers of meaning for the Kurdish people, the Kurdish freedom movement, and in return, the Turkish state.

Mountains, Forest Fires, and Martyrs

The significance of the narratives about the final resting place of Noah's ark should be understood in the broader context of the important relationship between mountains and Kurds, as well as the mountains' significance to the preservation of Kurdish people's autonomy and identity. "The Kurds have no friends, but the mountains" is a well-known proverb in Turkey and greater Kurdistan that crystallizes the central cultural and historical importance of mountains to the Kurdish people and how they constitute Kurdistan, which consists of mountain chains enclosing a series of interior basins, encompassing the international boundaries of Iran, Iraq, Syria, and Turkey. Scholars have argued that the rugged and mountainous terrain of the region has "both rendered the Kurds extremely resilient to systemic changes to larger states in their environment, and also provided hindrance to the materialization of a unified Kurdish political will."[14] The topography of the region has historically kept Kurdish tribes apart from each other, delaying the development of a collective or national identity and rendering the Kurdish people vulnerable to the Turkish, Persian, and Arab nationalist and assimilationist projects. However, the same mountain ranges in the region that have separated Kurdish tribes from each other have also protected them from being fully taken over or assimilated by Turkish, Persian, or Arab states.[15]

Compared to other mountains, however, Mount Judi has a particular significance for both Kurdish cultural imaginaries and the Kurdish freedom movement in Turkey. The latter emerged as a response to the century-long repression of Kurdish cultural and political rights, first under the Ottoman Empire in the late nineteenth and early twentieth centuries and then by the newly founded Republic of Turkey from 1923 and onward. By this time, former Ottoman Empire territories, including Greece and Bulgaria, had gained independence. The Armenian genocide in 1915 resulted in the killing of more than 1.5 million Christian Armenians from Anatolia. Following a series of forced population exchanges with Greece and Bulgaria, the new Republic of Turkey came to be composed of a Turkish-speaking majority and a large Kurdish minority. The founders of the republic hoped that Muslim populations, including the Kurdish people, would be assimilated into a Turkish national identity. But the Kurds were the majority in large parts of southeastern Turkey, and they resisted assimilation, constituting what the Turkish state then referred to as the "the Eastern question," and more recently, "the Kurdish issue."

The Turkish state implemented various strategies to solve the so-called Kurdish issue, resorting to political oppression, forced resettlement, and massacres when cultural assimilation attempts did not work. A Kurdish resistance movement materialized in the late 1970s among educated, urbanized Kurdish intellectuals and political activists, and it culminated in the formation of the PKK in 1978. The PKK, a Marxist-Leninist armed organization, started a guerrilla war against the Turkish Armed Forces in 1986, and since then, it has been fighting for self-determination and independence, and recently for expanded political and cultural rights and autonomy in Turkey. The over-thirty-year war has resulted in the death of forty-five thousand people and displaced three million Kurds from their homelands.

Following the start of the guerrilla-led Kurdish armed insurgency of the

PKK in 1986, public access was closed off to Mount Judi, alongside other areas extending from the Syrian-Turkish border to the Armenian border. As the Turkish state declared a state of emergency in the region in 1987 and implemented martial law, Mount Judi became an entity of immense physical and symbolic importance for both sides, as it served as a crucial PKK base for almost thirty years, providing concealment and shelter. The aim of the Turkish state in imposing these measures was manifold: protecting potential civilians or state officials from being targeted or abducted by the PKK, limiting the mobility of the Kurdish population, blocking aid to Kurdish guerrillas by Kurdish villagers in the region, intensifying the military presence of the Turkish state in the region and control of infrastructures, and forcing Kurdish residents to migrate to western parts of Turkey.

Mount Judi, considered to be the resting place of Noah's ark, as well as the mountain where the Kurdish people originated, is now also a sacred place for political reasons for the Kurdish people. Mount Judi is where Kurdish guerrillas have lived, trained, eaten, and fought for an emancipated future to come. It is the mountain that has sheltered guerrillas against the Turkish Army. Most recently, it was transformed into a burial ground for the dead, during the start of peace negotiations between the Turkish government and the PKK. In April 2013, the PKK announced that it was moving its forces in Turkey to northern Iraq. Yet, when the negotiations collapsed, the ceasefire was terminated, and the armed conflict resumed in the summer of 2015.

During the ceasefire, the PKK built seventeen cemeteries in the region for dead Kurdish guerrilla fighters, revered as martyrs, many of whom had died fighting the Turkish Army in Turkey and the Islamic State in Syria and Iraqi Kurdistan. The guerrillas were buried with funeral ceremonies largely uninterrupted by the Turkish Army during the ceasefire period. In 2015, following the resumption of the military operations against the PKK, the Turkish government ordered the demolition of the cemeteries, accusing the PKK of using them as meeting places or for their arsenal. Between 2015 and 2017, the Turkish state bombarded and destroyed eleven cemeteries in the Kurdish cities of Ağrı, Muş, Van, Bitlis, Mardin, Hakkari, Kars, Dersim, Siirt, and Diyarbakır.[16] Destroying symbols they considered to be sacrilege, Turkish forces demolished tombstones and erected Turkish flags in the cemeteries. In mid-September 2015, for example, a military operation in the province of Muş targeted a PKK cemetery in the area. Walls and gravestones inscribed with PKK symbols were demolished. Members of the pro-Kurdish Peoples' Democratic Party claimed the destroyed structures included a mosque and a *cemevi*, an Alevi house of worship.[17] A cemetery on Mount Judi was among those destroyed by the Turkish Army.

In 2020, a series of uncontrolled forest fires raged through Mount Judi and the surrounding regions. Every year, and especially since the Turkish government's unilateral decision to break off peace talks with the PKK in 2015, devastating wildfires blazed in northern Kurdistan, consuming thousands of hectares of forest and cultivated land, reducing biodiversity, and eliminating wildlife. Since the first half of the 1990s, the Turkish state has been using forest fires as a counterinsurgency strategy, as forced village evacuations were often followed by forest burning.[18] Again, the Turkish state's motivations behind forest burnings were manifold: the PKK fighters often hid in the forests close to villages, which allowed them proximity to food and other resources and offered protection.[19] The strategy of forest destruction aimed to not only force guerrillas to live in remoter areas but also to destroy the Kurdish population's overall livelihood, which relied on agriculture and animal husbandry provided by the forest environment.[20]

During the series of forest fires that struck towns around Mount Judi in 2020, the sentiment of the Kurdish population affected by the disaster seemed to confirm the historical legacy of the Turkish state's use of ecocide as a counterinsurgency strategy. Alternative news sources that interviewed locals reported that many viewed the forest fires as events far from "naturally occurring." Kurdish villagers used the term "application" or "execution" when they referred to the fires, implying that there was a malevolent agency behind the socioecological catastrophe.[21]

The forest fires on Judi therefore can be seen as a continuation of the counterinsurgency and "de-Kurdification" strategies of the Turkish Army—strategies that have been centered around the weaponization of ecological and geological environments—for example, flooding caves by building dams and burning entire villages—since the start of the war.[22] I have written elsewhere about how limestone caves in Turkey's upper Tigris valley have become central to the Turkish state's developmentalist and counterinsurgency projects at the same time.[23] Terraforming the cavernous topography has been a crucial task for the Turkish state's territorialization and counterinsurgency efforts.

The recent destruction of the guerrilla martyrs' cemetery and the fires on Mount Judi constitute a similar practice of territorial control and counterinsurgency warfare, but in this case, they also involve acts of profanation and spiritual violence. The cemetery has become a locus of contention between the Kurds and the Turkish state in their most recent phase of the struggle.[24] Political scientist Banu Bargu suggests that in bombing cemeteries, "the desecration of the dead becomes a new site of articulating identity, of producing the ethnic, spiritual supremacy of the Turkish nation."[25] This is because, as anthropologist Hişyar Özsoy puts it, "the Kurds resurrect their dead through a moral and symbolic economy of martyrdom as highly affective forces that powerfully shape public, political and daily life," thereby "promoting Kurdish national identity and struggle as a sacred communion of the dead and the living."[26] The cemetery on Mount Judi, thus, became a geological and ecological site of political and spiritual contestation.

The monastery and the shrine on Mount Judi also came under attention when, in the spring of 2018, special operation teams from the Turkish Army reached the previously inaccessible areas of Mount Judi. This included the martyrs' cemetery located near the shrine of Lawkê Xerîp. Turkish soldiers destroyed the tombstones, opened the graves, removed the bodies, and flattened the surface to return the shrine "back to its original condition."[27] Mount Judi thus emerged as a pivotal ground upon which counterinsurgency warfare was being waged through competing notions of spirituality and sacrilege. Yet it also became a ground for asserting the Turkish state's power through the positive logic of developmentalist and extractivist regimes, where the destruction of insurgent spaces and peoples could be utilized to increase oil and tourism revenue.

In 2018, for example, newspapers heralded that Turkey was "finally set to drill for oil in area freed of terror."[28] One news report claimed that following "successful operations" near a highland against the PKK on the outskirts of Mount Judi, Turkish Petroleum had finally begun exploration. Garnished with colorful photos of the newly erected drill rig, group pictures of soldiers, the governor of Şırnak, and Turkish Petroleum employees, the article quoted the president of Turkish Petroleum saying that this had been a project they had been "developing for a decade but were unable to implement due to terrorism."[29] The governor also confirmed these statements, remarking that the area had been "a high-terrorist-activity zone and was almost impossible to reach. There used to be PKK camps here. Now, we—engineers, workers, the governor, police—can easily come here."

If the hydrocarbons locked in Mount Judi were opened to an extractivist model of reterritorialization as resource frontiers, the shrine of Lawkê Xerîp would present a tourism opportunity. On May 25, 2021, "The Second International Symposium of the Flood, Noah's Ark, and Mount Cudi" was held at the University of Şırnak. I watched the live streaming on YouTube from my then-apartment in Chicago.[30] The speakers—including the governor and the state-appointed mayor of Şırnak—stressed the importance of reviving the traditional festival held on the shrine every July. Others discussed Mount Judi and Şırnak's potentials for religious tourism. The short talk from the head of

Şırnak's chamber of commerce made a striking comparison: "We used to talk about the number of dead bodies in these lands, now we are finally talking about the number of tourists."[31]

Mount Ararat

The second contender for the final resting place of Noah's ark is Mount Ararat, the highest peak in Turkey, rising in isolation above the surrounding plains and valleys and providing a panoramic view of the Armenian, Iranian, and Turkish highlands. A dormant volcano with no crater, whose most recent eruption was in the last ten thousand years, Mount Ararat has long been a primary figure in evangelical Christian narratives of the Great Flood. Yet it started to receive renewed attention in recent years, with the (re)discovery of a ship-like formation on it, known as the "Durupınar site." I listened to the story of the Durupınar site at a conference on "Mount Ararat and Noah's Ark" in 2018. The irregular formation was first sighted on May 19, 1948, by a Kurdish shepherd named Reşit Sarıhan, as he was grazing his goats on the steep hills of the mountain. Ten years later, the same formation was identified by an aerial cartographer, Captain İlhan Durupınar from the Turkish Air Force. While flying above Ararat for a joint mission between Turkey and NATO in October 1959, Durupınar had encountered a strange shape below him: the 150-meter formation looked exactly like a boat. Thinking that the shape could not have been formed by nature, the captain photographed it and informed the Turkish government about his discovery. In September 1960, *Life* published pictures of the site with the headline, "Noah's Ark? Boatlike Form Is Seen near Ararat."[32] The claims about the irregular formation, however, were quickly refuted by scientists. Before the *Life* coverage, a group from the US-based Archaeological Research Foundation surveyed the site, concluding that the formation did not have any archaeological remains and it "was a freak of nature and not man-made."[33] From 1977 through to the late 1980s, self-proclaimed archaeologist and explorer Ron Wyatt promoted the site as the remains of Noah's ark and organized several expeditions to the site. Despite these attempts, however, archaeologists and geologists continued to discredit the ark hunter's claims. In the 1990s, Turkish geologists and geomorphologists explained the formation of the shape in geological terms. One of them was Murat Avcı, a geologist and aerial surveyor who worked between 1957 and 1966 in the Turkish Air Forces as a jet-engine mechanic, and from 1966 until 1973 for the Mineral Research and Exploration Institute of Turkey as a field geologist. For Avcı, the shape came into being through the formation of an earthflow through earthquakes, glaciation, and the subsequent glacial meltdown processes, which naturally transformed the slab into a ship-like feature. In 1994, a US Merchant Marine officer, David Fasold, climbed to the site with geologist Lorence Collins. In a cowritten paper, Fasold and Collins argued that the boat-shaped formation was a natural geological formation that merely resembled a boat.[34]

Hoaxes, Myths, and "Terrorists"

The scientific debunking of the Durupınar site did not put an end to expeditions to Ararat. On the contrary, there has been a renewed interest in the Durupınar site over the past decades, and visitors are still welcomed by a road sign that directs them to Mount Ararat. In 2010, several American news outlets started to report that remains from Noah's ark may have been found on Turkey's Mount Ararat. A Hong Kong–based evangelical group, Noah's Ark Ministries International (NAMI), had announced that they had discovered a wooden structure in an ice cave. They were almost certain that the preserved remains belonged to Noah's ark. The group claimed that they had discovered several compartments of the ark, some with wooden beams, and suggested that the compartments were used to house animals. They even asserted that the structure was carbon dated to yield an age of 4,800 years. Professor Randall Price, an evangelical Christian, ark hunter, and former member of NAMI, and Dr. John Morris, president of the Institute for Creation Research based in Dallas, Texas, who was a consultant to the NAMI team, later revealed that NAMI's claims were a hoax. It turned out that what the group claimed to be a petrified wooden structure was in fact a combination of volcanic rock and modern-day wood. Price argued that a group of Kurdish workers might have planted large wood beams taken from an old structure in the Black Sea area at the Mount Ararat site.

In Turkey, especially after 9/11, there has been a noticeable effort to provide an ecumenical reading of the biblical and Quranic story of Noah that might compete with the prevalent idea of a clash of civilizations between East and West. English-speaking Christian authors invite their readers to investigate the biblical and Quranic stories of the patriarchs and prophets, insisting that, despite theological differences, the Quran can provide some helpful information to understand the silences of the biblical text and vice versa. The notorious Turkish-Islamic creationist and sex-cult leader Adnan Oktar (also known as Harun Yahya) claims that if the existence of the ark is proven, it will foretell the second coming of Jesus Christ, this time as a Muslim. For Oktar, with the return of Jesus, Muslims and Christians will finally come together and save humanity from the "dangers of Darwinism and atheism."[35]

Despite the hoaxes and dubious evidence surrounding Mount Ararat expeditions, several tour companies continue to take thousands of people to the summit of Ararat every year. While researching less high-profile attempts to locate the remains of the ark on the slopes of the mountain, I came across the website of Amy Beam, an American expedition organizer and guide, who, in 2007, started Mount Ararat Trek (https://www.mountararattrek.com), specializing in taking travelers to Mount Ararat summits. Beam has devoted a major part of her website to forty-seven warnings against climbing Mount Ararat. After recommending the essential items to bring on the expedition (adhesive bandages, pain relievers, and oxygen tanks) and listing some of the downsides of mountain climbing, such as the lack of electricity and bathrooms and the risk of heat exhaustion, heatstroke, frostbite, altitude sickness, and hypothermia, Beam's warnings take a different turn. She cautions potential climbers that they will not be seeing Noah's ark on Ararat, adding that the biblical archaeologists', evangelical explorers', and other travelers' claims of having found the remains of the ark have proven to be hoaxes.

Finally, Beam mentions the Kurdish issue—the military control, the presence of the Turkish military and the PKK in the area, and the kidnappings and other incidents that might take place during climbs to the summit. Beam's warnings were not misplaced. On July 8, 2008, three German climbers were kidnapped from Mount Ararat base camp by the PKK in retaliation for the German government shutting down the Kurdish ROJ television station, which broadcast from Germany and beamed into Turkey. The three climbers were part of a group of mountaineers from Bavaria taking part in an organized expedition to climb to the 5,165-meter summit. On July 8, PKK guerrillas walked into the climbers' camp at 3,200 meters, took three members of the group, and disappeared, releasing the captives after two weeks.

Just as Ararat's popularity seemed to be fizzling out due to security concerns and fraudulent tour companies, in 2019, a small group of creationist archaeologists and geophysicists scanned the structure of the supposed ark using ground-penetrating radar and resistivity imaging technology and released three-dimensional subsurface images of the ship-shaped formation at the Durupınar site (fig. 7.1). They claimed the three-dimensional shape, which has the appearance of an "immense ship," did not include features that could be "explained by natural rock erosion."[36] They noted that the internal dimensions of the length, width, and height of the ship-like formation were identical to the dimensions of Noah's ark found in the Bible. John Larsen, a New Zealand–based geophysicist, concluded:

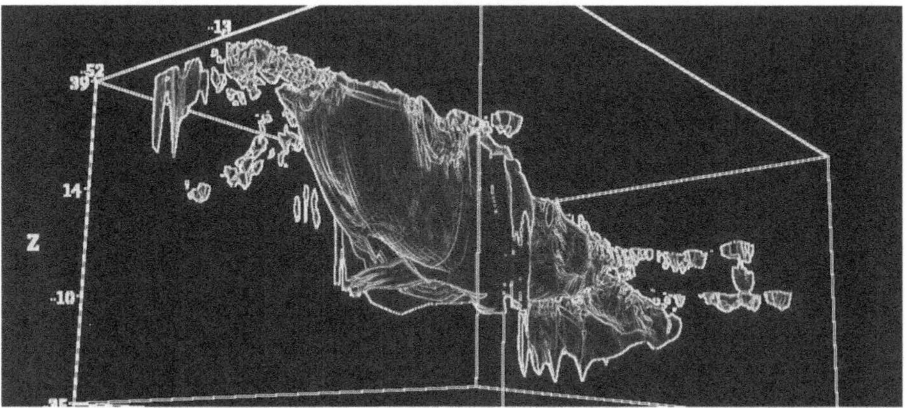

FIGURE 7.1. Resistivity scan of ship-shaped formation at Durupınar site. John Larsen, "3D Image Looking Upwards at the Front Left Section of the Bow," from Jake Wilson, "Noah's Ark: Selection of Images," 2020. With permission from Jake Wilson.

> Each of these pieces of evidence, point to this object as being man-made, and of ancient origin. In the history of the earth, the only ship of this size, which has ever been built from timber, and matches every feature which has been found on this site, is the Ark which God commanded Noah to build.[37]

The release of the three-dimensional subsurface images of the Durupınar formation sparked a new wave of tours and expeditions to the area. Today, the Turkish Ministry of Tourism is looking for ways to promote the site and turn it into a global center of creationist tourism. In this, as a government official remarked at the "International Symposium of Mount Ararat and Noah's Ark" in October 2021, "Ararat has the potential to become Turkey's preeminent national symbol."[38] Yet, capitalizing on Mount Ararat is not only important for tourism but for Armenia as well, due to Ararat's iconographic significance.

Mount Ararat has been a central figure in the making of the cultural history of Armenia, contemporary Armenian culture, and the legacy of the Armenian genocide in 1915. Ararat has been the principal national symbol of Armenia and considered a sacred mountain by Armenians as a locus of Armenian nationalism and irredentism. The making of Mount Ararat into a sacred place, like Judi, has a history. Ararat came to represent the destruction of the native Armenian population of eastern Turkey (Western Armenia) in the national consciousness of Armenians, especially in the aftermath of the Armenian genocide of 1915, which is officially denied by the Turkish state. Today, in most Armenian homes in the modern diaspora, there are pictures of Mount Ararat, a bittersweet reminder of the homeland and of future national-territorial aspirations of "going back to the homeland." As the highest mountain in the region, Ararat is visible from many locations in Armenia. It figures in poetry, sculptures, paintings, postage stamps, and songs. The omnipresent image of Ararat rising above Yerevan and its outskirts constantly reminds Armenians not only of Noah's Flood but also of another, more recent catastrophe, and its ongoing effects in the present. Ararat's omnipresence blends into the ongoingness of the denied Armenian genocide of 1915 and the remaining Armenians' exodus from their ancestral homelands.

Other Catastrophes

The entanglements of cosmologies around the Great Flood and geological formations in Eastern and Southeastern Anatolia allow earth historians and anthropologists to rethink how territorial imaginaries, violence, the past, and the future figure in cosmological and geological imaginaries. Competing narratives about the final resting place of the ark are deeply linked to geopolitical and territorial imaginaries of and in Turkey, Armenia, and Kurdistan, evocative of the ways in which geo-spiritual formations of the Anthropocene are always already grounded in planetary *and* regional histories of political action, imagination, and violence.

Kurdish accounts draw historical links between the sacred figure of Prophet Noah, Mount Judi as the final resting place of his ark, and Kurdish ethnic identity. As Mehmet's words indicate, the Kurdish people are regarded as the "Children of Noah," descendants of those that fostered the birth of a new civilization after a planetary environmental-moral catastrophe. The sacred status of Mount Judi was reconfigured through the colonization, assimilation, and displacement of the Kurds by the Ottoman Empire, first, and then the subsequent nation-states of Turkey, Iran, Syria, and Iraq. Mount Judi provided protection, comradeship, and shelter to the Kurdish insurgents—and a sacred resting place to Kurdish martyrs. The spiritual-political significance of Mount Judi turned it into a target for the Turkish state, which carried out its symbolic and material counterinsurgency project through forest fires and the destruction and desecration of martyrs' cemeteries. The final step for the Turkish state was the reterritorialization of Mount Judi into a resource and religious-tourism frontier. Similarly, the Turkish Ministry of Tourism has been promoting both mountains by promoting Noah's Flood and ark-related events. In addition to the potential for transforming Mount Ararat into a hallmark of religious tourism, promoting Ararat also serves to violently counter the principal icon of the Armenian nationalist imaginary.

Yet these mountains also render such totalizing attempts of violent erasure futile. Competing narratives about the aftermath of a catastrophe, Mount

Judi, and Mount Ararat conjure up historical catastrophes, or other kinds of ongoing violence: the Armenian genocide and its systematic denial by the Turkish state and most of the Turkish people; massacres that took place after Kurdish rebellions; tens of thousands of killed guerrillas; and, finally, the guerrillas' desecrated graves. When some catastrophes are erased, repressed, or rendered unspeakable by the perpetrators and their benefactors, the mountains where a mythical catastrophe took place become geological formations that capture modes of resisting the ongoing violence of erasure.

A few months after our conversation, Mehmet wanted to introduce me to his distant cousin Ali, telling me that Ali would be a better fit to tell me about "the significance of Mount Judi for us." I telephoned Ali one afternoon, and we met up for tea. I learned that members of his extended family, including a cousin from his father's side, "were on the mountains," meaning that they had joined the ranks of the PKK in the past decades. In passing, Ali told me his paternal uncle's son, Memo, had been killed by the Turkish military four years ago. They couldn't retrieve Memo's body, but Ali believed that he was buried in the graveyard on "Cûdî Dağı, on sacred land. On ancestral land." I asked Ali how he mourns for his loss now that the guerrilla martyrs' cemetery has been destroyed by the Turkish Army. He told me that the army had removed the bodies, and they did not know where Memo was. "It's okay," he said. "Prophet Noah's ark is also nowhere to be found. They keep looking. They won't find shit. But *we know* it was there once upon a time. You know? Waters, bodies, ships—they all vanish at some point, but the mountain will be there—it will remember."

PART III

New Elemental Histories

8

"Glass Worke"

Precious Minerals and the Archives of Early Modern Earth Sciences

Claire Conklin Sabel

"Was the artisan drunk?" Phra Narai, king of Ayutthaya (r. 1656 to 1688), was reported to have asked upon receiving the sword he had commissioned from the English East India Company (EIC) in December 1680. It seemed that "the Honourable Company had noe kindness for him else they would have gott it made better," the king remarked. The sword had been presented by factor Richard Burnaby on behalf of the EIC in hopes of cultivating Narai's favor; but it had not succeeded. Burnaby stood aside, humiliated, as Narai ordered the unsatisfactory article to be smashed into pieces and disposed of. This incident marked the culmination of a series of failures by the EIC to deliver to Narai the evidence of resources and knowledge that he demanded from English representatives to his kingdom.[1]

Just a few months before, Burnaby had submitted a memorandum to Narai testifying that, as evidence of the EIC's "devotion to His Majestie's service," they had "taken all imaginable care in provision of as many of those curiosityes His Majestie gave in comand as the shortness of time would possibly admit." Burnaby assured Narai that, for those goods that could be delivered, "the Honourable Company imployed their best endeavours for provision of those rarityes." Despite these efforts, several of Narai's requests could not be completed as ordered. "The best artists in England, & consequently wee may without vain glory say in all Europe," Burnaby apologized, "could not frame such representations" in pure quartz or rock crystal, as Narai had specified.[2] The main obstacle was that there were no pieces of rock crystal in England large enough to carve into the "representations" that Narai requested, including the sword handle that was to be enameled and decorated with precious stones.[3] In addition to lacking the raw materials, the EIC had failed on a second account: they had not been able—or perhaps willing—to procure any of the "ingineers, gunners, refiners, goldsmiths" desired by Narai to relocate to Ayutthaya and serve his court.[4]

Narai made many such requests over the course of his thirty-two-year reign. Diplomatic exchanges in late seventeenth-century Siam have received considerable attention, both because they involved spectacular displays of material culture and because they are exceptionally well-documented.[5] This chapter considers one facet of this activity to highlight the ways that new patterns of long-distance maritime trade in the seventeenth century could be instrumentalized to inquire about mineral resources in both Europe and Southeast Asia. By requesting artisans and artisanal objects, "representations in . . . *glass worke*," from foreign representatives, Narai was seeking evidence that the English could outperform his own and other regional polities in their ability to obtain and manipulate the precious minerals that were central to commercial and political prestige in the early modern world. Seeking to establish diplomatic relations with the court at Ayutthaya, EIC merchants and their investors in London were similarly hoping for access to natural and crafted commodities that were not found at home. But what emerged from these exchanges were not only negotiations over goods and services. Both parties evaluated each other's knowledge of sourcing and working precious minerals as a means of assessing their respective claims to earthly, territorial, and spiritual power.

I argue that this episode suggests that precious minerals, as particularly value-laden natural objects, have broader historical and methodological virtues for the study of the earth sciences in the early modern period. Gemstones illustrate the role of long-distance commerce in moving objects, information, and expertise at a *global* scale, creating the conditions necessary for the globe itself to be empirically investigated as a material entity. But they also demonstrate that these investigations were not the exclusive domain of natural philosophers: as prized commodities, precious stones were among the choice materials of diplomatic exchanges and religious relics, which both attracted extensive documentation and commentary. Gems produced records of wealth and power expressed in mineral terms, forms of earthly knowledge of interest to rulers, merchants, and naturalists alike. This chapter follows the archival traces left by Phra Narai's pursuit of crystal, to demonstrate that by embracing gems' complex geopolitical and cosmological contexts, historians of science can unearth new sources for the varieties of early modern knowledge about the earth.

I take the category of "earth sciences" to be capacious and inclusive of a wide range of practices and concerns that are not limited to questions about the antiquity and origin of the earth. In doing so, I follow recent scholars who have demonstrated the reliance of early modern European sciences, such as paleontology and mineralogy, on the practical labor of diverse practitioners.[6]

Many learned Europeans' theories of the earth's formation did take on a distinctive form of "Mosaic physics" in the late Renaissance, which prioritized biblical chronology.[7] But exclusive focus on the project of reconciling Christian religious authority with pious natural philosophy nevertheless obscures the existence of a wider range of applied approaches to mapping and understanding the earth, many of which furnished data, specimens, and insights for elite theorists, merchants, and rulers alike. Much earthly investigation was motivated by the practical concerns of healing, divination, agricultural improvement, navigating unfamiliar territory, and especially locating and extracting mineral wealth, both within and beyond Europe.[8] Historians of South and Southeast Asia, meanwhile, have tended to emphasize the pragmatic aspects of colonial geographical thought in exercising territorial control and affecting conversion in the eighteenth and nineteenth centuries. Articulated in new scientific institutions of surveying, mapping, and representing the globe, these practices displaced or obscured existing ways of knowing and using the earth.[9] This chapter examines such activities at a moment when European powers were still relatively peripheral to regional commerce. Exploring the political economy of mineral knowledge in commercial relations between Western Europe and mainland Southeast Asia allows us to look beyond normative divisions of "pure" and "applied" sciences and "religious" and "scientific" earthly knowledge in the seventeenth century. Gemstones served common functions in distinct cosmologies, whose valuation of precious minerals both served a range of extractive projects and illuminated different ways of knowing.

John Tresch has usefully outlined a role for history of science as "the comparative study of materialized cosmologies—ideas of the order of nature that are enacted, embodied, elaborated and contested in concrete settings, institutions, representations and practices."[10] It is just such a comparative history that this chapter proposes gemstones can offer. Taking the court at Ayutthaya as a case study, the chapter will survey the wealth of earthly knowledge mediated by mineral commodities in the commercial networks that converged on this major entrepôt. I first sketch the common role that precious stones played in manifesting political legitimacy, spiritual prowess, and virtuous knowledge across much of Eurasia—the common cosmological conditions that positioned gems as signifiers of territorial and terrestrial power. Next, I describe the political and economic context in which these mineralogical networks took shape in the seventeenth-century Indian Ocean world, in which Ayutthaya was a significant player. And last, I examine the kinds of information and claims that precious minerals mediated in the exchange between the EIC and Phra Narai's court at Ayutthaya, when the EIC sought to negotiate new trading privileges and to monopolize tin exports from the Malay peninsula in the 1680s. Understanding the mineralogical dimensions of these exchanges locates a wealth of historical evidence in the commercial archives of entities like the EIC. These archives can further our understanding of diverse cosmologies of the earth and their interactions, especially in times and places where the textual record disproportionately represents European perspectives.

Gems as Cosmological Objects

Precious minerals, above all gemstones, were a common feature of early modern knowledge traditions across Eurasia. Stones' powers derived not only from their remarkable physical qualities and rarity but especially from their difficulty to obtain, typically from hard-to-reach places. Over the course of the sixteenth and seventeenth centuries, unprecedented global travel between growing maritime empires, long-distance commercial networks, and globalizing infrastructure for knowledge production gave gems a new role in mediating information about the earth. But unlike precious metals that were melted into transactional currency, gems remained most valuable when embedded in distinct cosmological matrixes: astrology, mineral medicine, natural history, spiritual authority, and political order. In diverse worldviews, gemstones represented relationships between heaven and earth, human and divine, wisdom and wealth. As exceptional, translucent objects emerging from the dark recesses of the subterranean world, they could act as a material metaphor for power, wisdom, enlightenment, and purity.[11]

Precious stones were the ideal vehicle for information about distant places because of their ubiquitous value, giving them prestige and power as they traversed different terrains. This is borne out in many premodern textual traditions, which accord gems significance precisely because of their unique origins. Gems' rarity was not only associated with their physical qualities and limited quantities but crucially from their origination *elsewhere*, whether elsewhere on the terrestrial globe or in the extraterrestrial cosmos. Possession of these rare and difficult-to-find objects thus imbued their owners with powers associated with their divine origin. Phyllis Granoff argues that in Indic religions, jewels' magical properties were seen as a "particularly fitting material for religious visions, which themselves seem to come from a different world," stressing that jeweled visions represent a pan-Indic belief in the material power of precious stones.[12]

Across South, and later Southeast and East, Asia, jewels were the metaphor par excellence of Buddhist virtue and practice, epitomized in the "Triratna," or "Three Gems," of Buddhism: the Buddha, the dharma (doctrine), and sangha (community), with an analogous trinity in Jainism. Jewels serve a variety of metaphorical and mythical functions in Buddhism, as a source of comparison for forms of virtue and excellence and as powerful relics.[13] The Indic foundations of Buddhism, Jainism, and Hinduism share an emphasis on precious stones as originating from the bodies of divine beings.[14] Variations on a common origin story attribute gemstones to the body, organs, or tissues of a slain demon, which disintegrated and became the scattered seeds of gemstones all over the earth. These origins were typically repeated in technical gemological texts of the premodern era, corroborated by geographical locations and the associated properties—often medical—of gemstones.[15] The bodily associations of gems connected them both to human therapeutic purposes and to the *embodiment* of divine beings.[16] Other Indic sources attribute oceanic origins to gemstones, as well as recording gems that fall from the sky. Gems appear during the churning of the sea of milk, when herbs cast into the ocean turned into jewels. Mount Meru, the physical and spiritual center of this shared cosmology, has four faces made of precious minerals: gold, lapis lazuli, ruby, and crystal. The Hindu Buddhist goddess Manimekhala, whose name means "Girdle of Gems" in Pali, carries a radiant gem that is the source of lightning. She serves as a protector of the seas across Southeast Asia and is also the namesake of the Tamil epic poem, the *Maṇimēkalai*, composed in the first millennium.

Religious traditions across Eurasia therefore informed a wide variety of textual accounts of gems, from lapidary texts itemizing every kind of gemstone and their properties, to literary and liturgical sources that associated the presence of gemstones with the manifestation of the divine in the earth and heavens. In addition to their widespread use as personal adornment and medicine, precious stones were a prized material for religious artifacts—especially devotional objects and reliquaries. As Granoff has argued, looking at material practices, such as relic worship, highlights commonalities among Indic religions whose textual traditions and evolution are often treated independently by scholars. Such common material practices are indeed evident well beyond the Indian subcontinent, especially in the medium of rock crystal, which features prominently in Buddhist, Islamic, and Christian devotional objects.[17] In Christian reliquaries, rock crystal was associated with Christ's spiritual purity, while śarīra, crystal beads formed in the cremated remains of powerful Buddhist monks and nuns, are material traces of their spiritual power.[18] The material role of gemstones in maintaining a world order "stressing purity and permanence" has likewise been upheld for millennia by adorning revered religious sites and objects, especially sacred images of the Buddha, with precious materials including gemstones. The most spiritually potent jewel for many laypeople in South and Southeast Asia, the *navaratna* (nine-gems), represents celestial order in nine precious stones (*ratnas*, in Sanskrit), each with an astrological influence.[19]

Although embedded within distinct religious traditions, gemstones served common cosmological functions across premodern Eurasia, indexing spiritual power with the rarest and most precious of substances. This made them uniquely transmutable between different worldviews, a feature epitomized in the appropriation of religious relics from one tradition within another, such as the well-known examples of magnificent rock-crystal ewers from medieval Fatimid Cairo being prized as Christian reliquaries. Such plunder only increased as imperial conquest and long-distance maritime travel furthered the violent extraction and possession of gemstones and jeweled objects from local control, expediting their global trajectories.

Gems' boundary-crossing natures were spectacularly demonstrated in the case of emeralds. Emeralds were mined in ancient Egypt, but there were only minor deposits of emeralds known in medieval Eurasia, found in modern-day Pakistan, Afghanistan, and Austria. By contrast, American emerald deposits had long been exploited by Indigenous Muisca and Muzo miners for a variety of sacred practices. After arrival of the Spanish and the founding of the New Kingdom of Granada in 1538, Colombian emeralds violently seized from Indigenous ritual practice and mining sites were exported in unprecedented quantities. Emeralds traveled via Spanish galleons to Manila, and from there to imperial centers in Asia, especially to the Muslim courts of the Safavid, Ottoman, and Mughal Empires, where green was venerated as the color of heaven.[20] Ancient traditions were the foundation of new early modern geopolitical configurations, expressed in dazzling displays of emergent empires (fig. 8.1).

The Indian Ocean Gem Trade in the Seventeenth Century

The ancient material, textual, and cosmological associations of gems persisted into the seventeenth century but were significantly influenced by the new trade routes and religious networks of the early modern period. Thus, long-standing gemological traditions and texts began to reflect the altered conditions of the marketplace, prioritizing expertise in gems' physical properties over stories of their extraterrestrial origins.[21] This found expression in the growth of patronage of goldsmiths and jewelers, in the production of religious, courtly, and personal artifacts made from precious minerals, and in a surge of documentation of the gem trade. Travel accounts written by jewel dealers, such as Jean-Baptiste Tavernier and Jean Chardin, were among the

FIGURE 8.1. The Mughal Emerald, 1695–96, dated 1107 AH, 217.80 carats, 5.2 × 4.0 × 4.0 cm. Rectangular-cut carved emerald, the obverse engraved with Shi'a invocations in Naskh script, side drilled for attachments. Museum of Islamic Art, Doha, Qatar. Photo © Christie's Images / Bridgeman Images.

most popular Asian travelogues in seventeenth-century Europe, in part because they told of spectacular wealth and luxury at other Eurasian courts.[22] One the earliest autobiographies composed in South Asia was written in 1641 by Banasaridas, the son of a Jain jeweler and gem-merchant.[23]

The Kingdom of Ayutthaya was a prominent example of the efflorescence of commercial centers in the Indian Ocean world in this period (fig. 8.2).

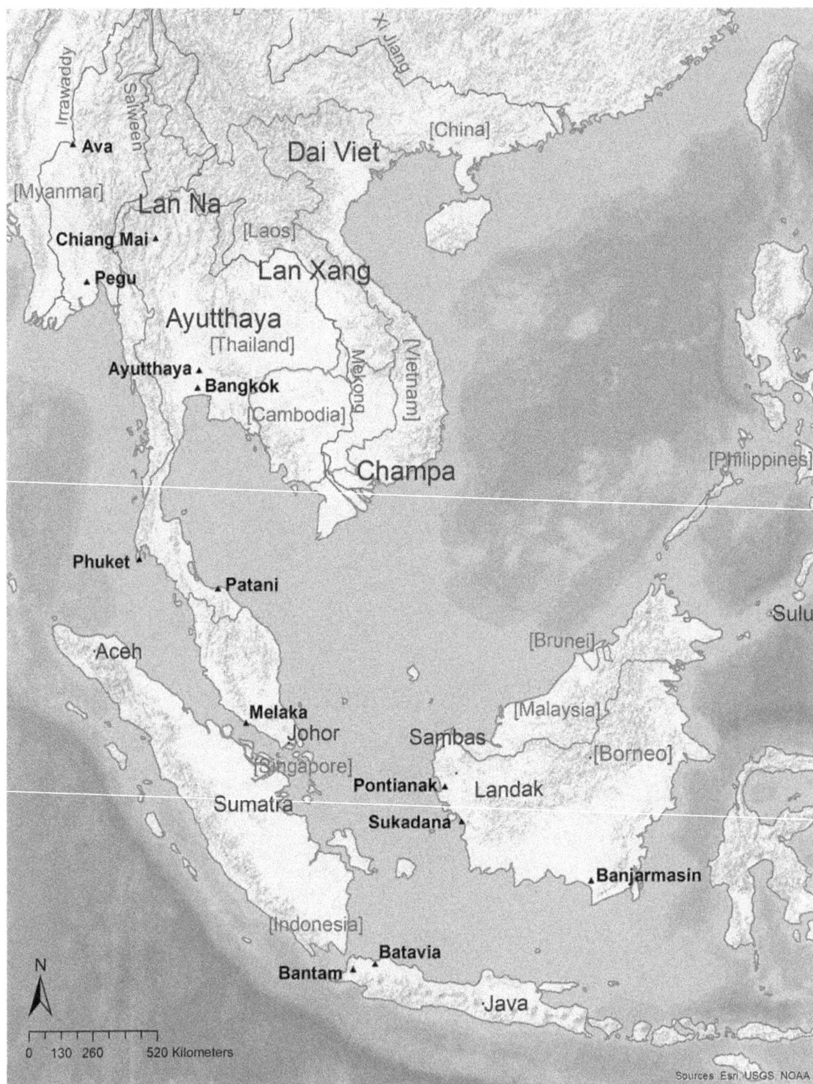

FIGURE 8.2. Polities of early modern Southeast Asia. Map based on "Mainland Southeast Asia before Modern Boundaries," in Thongchai Winichakul, *Siam Mapped* (Honolulu: University of Hawai'i Press, 1994). Created by Girmaye Misgna, Mapping & Geospatial Data Librarian, University of Pennsylvania Libraries.

Increased mobility, urbanization, population growth, economic expansion, and state centralization led to unprecedented wealth, consumption of luxury goods, and patronage of the arts.[24] After almost a century of warfare, with the 1605 ascent of King Ekathotsarot (r. 1605–1610), Ayutthaya emerged as a major commercial center and a pivotal entrepôt between the wider Indian

Ocean world and maritime and mainland East Asia.[25] The seventeenth century in Ayutthaya marked a decisive shift from a polity dependent on land-based warfare to one focused on maritime trade. Siam (as it was called by Europeans) became an important crossroad of Eurasian commerce, attracting merchants from Japan, China, and the Philippines to the east; the Ottoman, Safavid, and Mughal Empires to the west; and regional merchants from mainland and insular Southeast Asia.[26] The monarchy in turn invested significantly in both internal trade and their own legitimacy, producing huge displays of wealth that attracted significant commentary from foreign observers.[27] Following Portuguese and Dutch ships across the Bay of Bengal, English merchants first arrived by ship (fittingly named *The Globe*) to Ayutthaya in 1612, seeking precious stones, among other valuable commodities.[28]

Shifting from territorial offense to maritime commerce transformed the monarchy at Ayutthaya, which became the most powerful player in the Siamese economy.[29] This intersected in spectacular ways with the tradition of sacred kingship in Siam.[30] As in much of Southeast Asia in the period, Hindu and Buddhist traditions were blended in courtly culture at Ayutthaya, from the appointment of Brahmanical court astrologers to annual pilgrimages to the Buddha footprint in Saraburi. Kings of Ayutthaya, and Phra Narai especially, embraced a Hindu-inflected Buddhism, while remaining open to Christian and Muslim interlocutors.[31] Narai was a patron of Persian merchants and artisans, as well as more open than his predecessors to Christian missionaries. In addition to the Persianate and European influences in the court, Chinese, Japanese, Arab, Malay, Greek, and Armenian merchants found favorable positions or relationships with the court.[32]

A court poem composed in Ayutthaya (c. 1680s) epitomized these trends by shifting the locus of its praise from the traditional figure of the monarch to the matter of the city itself, endowing it with qualities invoking both Buddhist and Hindu gemological traditions:

> The city is excellent. It has everything.
> It deserves more praise than heavens high, made and beautified by Brahma.
> A city of delights. A royal city with the nine gem attributes.
> The world's finest, built by Lord Rama himself.
>
> The fame of Ayutthaya rings down from sky to earth
> Just scan your eyes across the world—it is the single celestial flower.
> Thousands and millions have found no single, tiny blemish.
> The Three Gems illuminate the world and all the heavens.[33]

The jewel-like qualities of the city were much commented upon by seventeenth-century visitors, who took Ayutthaya's temples and treasuries to be indicators of fabulous mineral wealth: in 1621, Dutch merchant Cornelis van Nijenrode reported that Ayutthaya surpassed "any place in the Indies (except for China) in terms of populace, elephants, gold, gemstones, shipping, commerce, trade and [agricultural] fertility."[34] Increased consumption of and celebration of jewels over the course of the seventeenth century can also be seen in depictions of Mughal emperors wearing jeweled objects,[35] and in the rise of *Schatzkammers* in early modern Europe, which prioritized mineral objects as the most expensive kind of treasure.[36]

But supplying the increased demand for rare minerals depended on access to stones. Colombian emeralds were a rare example of colonial conquest of actively mined gem deposits before the eighteenth century. More often, diplomacy was essential to brokering the commercial and missionary relationships that provided access to gems. Precious stones, as mutually prized objects, served as a premium currency of diplomatic exchange.[37] While gems and jeweled objects did not import new cosmologies, they both signaled and participated in changing political, commercial, and religious realities. Those working gems grafted older traditions onto new world views, such as the adaptation of rock-crystal carving of devotional objects from Buddhist to Catholic imagery in sixteenth century Sri Lanka.[38] At the court of Ayutthaya, Narai engaged EIC merchants to procure precious stones from South Asian sources for his new royal regalia, benefiting from the commercial roles that European merchants could play without necessarily transforming them into brokers of Christian ideologies.[39] Ibn Muhammad Ibrahim, the Persian author of a detailed report from the Safavid embassy to Ayutthaya in 1682, observed that Narai "has endeavored to acquire fine furnishings . . . in general, he is striving to raise his name before the world and establish contact with powerful, world-ruling sultans . . . [He has] sent delegates to the potentates of India as well as to the rulers of many other lands and seeks to understand Siam's position in the world."[40]

Nevertheless, merchants and diplomats brought their own cosmologies with them. French Jesuits equipped by the Académie des Sciences arrived in considerable numbers to late seventeenth-century Ayutthaya, hoping to use their mathematical models of heavenly bodies as an instrument of conversion.[41] While Narai was certainly interested in European astrological techniques and encouraged construction of Jesuit observatories at Wat San Paulo in Lopburi and at Ayutthaya, in practice he relied on the authority of Brahmanic astrologers at his court. Their prophetic knowledge, built on the Buddhist Traiphum (literally, "three worlds") cosmological system, accounted for the movements of the heavens and the passing of the seasons, informing many aspects of political and spiritual life.[42] Expertise in astronomy and astrology were considered a royal virtue in Ayutthaya, and kings were major patrons of astral sciences.[43] In addition to a particular interest in astronomy, Narai was an active patron of the arts and devoted considerable energy to decoration of his palace at Ayutthaya. Thus, many of the gifts brought to Narai by European visitors included works of sophisticated artisanship, such as scientific instruments and wrought precious materials.[44] These were also the items Narai prioritized in his requests from embassies, in addition to artisans who could bring their expertise to his court. The second Siamese embassy to France in 1684 included young Siamese boys who were to be trained as artisans in France and then return to Ayutthaya.[45] Siamese ambassadors visiting France in 1686 carried a list of objects to acquire, including fifty-four hats, eighteen garments, eighty items of fabric, twenty-six items of goldsmiths' work (including globes and other scientific instruments), and forty-five thousand items of glass.[46]

Rather than designating European observational practices and instruments as "scientific," sequestering Siamese cosmologies and technical knowledge in the domain of religion, we might better understand the diplomatic exchanges that took place at Ayutthaya as demonstrations of expertise and techniques, embodied in both people and objects. Just as astronomical observations could be techniques of conversion, the pursuit of minerals like rock crystal (and its artificial forms as glass) was a means of obtaining knowledge of the cosmological, courtly, and artisanal traditions of rival commercial powers.

Rock Crystal and Earthly Knowledge at Ayutthaya

The wide range of European merchants and missionaries who spent time at Ayutthaya produced extensive reports on their time there, documenting both the commercial and cosmological dimensions of the kingdom.[47] The destruction of Thai textual sources during the Burmese sack of Ayutthaya in 1767 produced a distinctive archival asymmetry for the period, prompting creative reading of the preponderance of European sources with extant Thai and other accounts.[48] The documentary record that does survive reflects the imbrication of religion, politics, commerce, and science, especially through elaborately planned diplomatic visits that reached their apex during the reign of King Narai.

A rare glimpse of Narai's international ambitions can be seen in a surviving fragment of instructions to the Siamese ambassadors to Portugal, a mission which was thwarted by shipwreck in 1684.[49] The extant text is a translation, copied by French missionaries. It shows the court at Ayutthaya anticipated a range of questions about its kingdom and prescribed somewhat general, and occasionally misleading, answers. The instructions include tantalizing but incomplete accounts of the kinds of practical knowledge exchanged between foreign kingdoms; for example, instructions began as follows: concerning commerce, "if you are asked what trade goods and what unusual items there are in the kingdom"; for geographical considerations, "if you are asked what are the tributary kingdoms of Siam and what are the kingdoms adjacent to it"; and about court procedures, "if you are asked how many sons, and princes of the blood the king has." Regarding the mineral resources of Siam, the ambassadors were instructed to describe what could be found in the mountains, but the named minerals are no longer legible in the text. The instructions also note, in fascinating detail, that precious stones were brought from Manila (possibly an allusion to stones coming from the Americas) and that "a lot of merchandise is found there [Ayutthaya] which comes from outside [the country]," something observed by many European visitors.[50]

Narai's particular interest in artisanship is also a theme of European merchants' and missionaries' records. As Bhawan Ruangsilp notes, the court of Ayutthaya had two options for acquiring foreign knowledge: "information gathering by its men travelling abroad, and the employment of foreign experts."[51] Europeans reported various foreign artisans working for Narai, who seems to have found mineral knowledge particularly valuable. The Dutch East India Company (Verenigde Oostindische Compagnie, or VOC) noted a goldsmith at the royal palace in Lopburi and a pyrotechnics expert who advised on cannons and gunpowder.[52] The French missionary La Loubère reported that Monsieur Vincent, a French physician who had traveled east from Persia and had some knowledge of chemistry, had been involved in advising Narai's mining concerns.[53] These included a wide range of minerals from gold and copper to antimony, but Vincent had also reportedly "discovered a mine of crystal."[54] The VOC capital at Batavia (Jakarta) offered an opportunity to seek new artisanal skills more locally, even if recruitment was not always successful. A mason was dispatched back to Batavia when it became clear that he was a small-scale stone cutter, rather than someone who could advise on quarrying in the mountains as Narai desired.[55] Such exchanges were not unidirectional. La Loubère noted that "Mr Vincent inform'd me that he has seen in the hands of the Talapoins [Buddhist monks], who secretly busie themselves in these researches, some samples or pieces of Saphires and Diamnds that came out of the Mine."[56]

In contrast to the French and the Dutch, who established a more stable presence in Siam over the course of the seventeenth century, the English struggled to find sure footing. The first EIC factory at Ayutthaya lasted from

1612 to 1623 and did not reopen until 1661, coinciding with the recent ascent of Phra Narai in 1656. It finally closed in 1685. English aspirations for Ayutthaya persisted because the EIC hoped to find a viable alternative to the Dutch presence in maritime Southeast Asia, a connection to Japanese and Chinese maritime commerce, and a new market for goods from the Indian subcontinent. Moreover, both the EIC and private traders were interested in accessing the mineral wealth of mainland Southeast Asia, which had been consistently observed by foreign visitors to be rich in precious metals and gems.[57] EIC agents came to focus on the tin deposits in the region as a potentially lucrative commodity to resell around the Indian Ocean.[58] Tin mines were active across the Malay peninsula and its nearby islands, including Phuket, and many were under control of Malay states that were nominally vassals of Ayutthaya. Narai offered the EIC privileged access to these deposits just as he was dealing with uprisings at Patani, on the other side of the Malay peninsula, and likely desired additional military assistance in quelling the rebels.[59] Yet despite receiving several coveted tin monopolies, EIC agents struggled to convince Narai of the value of their own wares, and they repeatedly failed to secure a commitment from Narai to import a minimum annual amount of English goods to Ayutthaya.

Letters sent back from Ayutthaya to EIC factories at regional headquarters in West Java and Gujarat recorded Narai and his deputies continuing to press for material evidence of English commodities, by requesting sophisticated goods from England made from carved crystal, based on wooden molds or "forms" that Narai had provided, which may have been either devotional objects or decorative vessels. These were at least of a plausible scope, as opposed to the miniature furniture that Narai requested VOC merchants supply him in full-size in glass.[60] When the instructions reached the EIC headquarters in London, the council there confirmed receipt of the order. Agents in London reported in October 1677 that "wooden samples of severall cups" that Narai had sent "are exactly made in two severall sorts of mettall, the one glass the other christalline, and some few other things added." However, this was a compromised fulfillment of the order, as "there is no christall of such big peeces to be had as to make vessells or bowles of 10 inces over, therefore they are made in the cristalline mettall." This was also true for Narai's sword hilts, which were done "in christall glass & philosopher's stone" and additional gifts, including "prospective glasses" were noted to be "exquisitely made and cost much because of their worke, being well enammelled."[61]

What did Narai have in mind with these requests? Evidently the EIC found it significant to distinguish between "christall of such big peeces" and "christalline metall"—the latter likely an early reference to glass manufactured with lead, a novel imitation of carved rock crystal.[62] This suggests that Narai had been quite explicit in requesting objects made of rock crystal. Such items were similarly prized in the *Kunstkammers* of Europe, where a hard-stone carving tradition had flourished since the sixteenth century, and previous European embassies had sent crystal gifts. Narai was perhaps more likely familiar with foreign rock-crystal objects of Mughal, Safavid, or Qing production, worked from rock-crystal deposits in South Asia, the Middle East, and East Asia.[63] The chronicles of Ayutthaya suggest that Narai was particularly partial to precious stones: a section on foreigners' gifts to Narai singled out his receipt of "crystal plates" embellished with gold and emeralds, gem-set rings the size of tamarinds and lotus buds, and sword hilts decorated with diamonds, pearls, and "red emeralds" from dignities from Machilipatnam, Aceh, and beyond.[64] Though more common than those of other gemstones, rock-crystal deposits were still relatively rare—large flawless pieces were even rarer, immensely costly, and difficult to carve. Such crystal objects had been produced in Ayutthaya from the fourteenth century, as well as in the kingdom of Lanna to the north (fig. 8.3).[65] They represented extraordinary value both in labor and in natural resources, but the EIC sources suggested that both Ayutthaya and the EIC accorded further significance to these requests.

EIC factors sent Narai's specifications from Ayutthaya to their superiors at Banten (Bantam) in West Java in 1675. In reviewing the orders to be passed on to headquarters in London, EIC merchants took note of additional geopolitical context that had accompanied the request for rock crystal. After acknowledging Narai's requests for crystal work, EIC agent Henry Dacres gave an example of another incident involving molds (casts, or models) of precious stones. This involved Sukadana, an independent polity in western Borneo, which controlled a share of the island's diamond mines—the only known source of diamonds in Asia outside of India.[66]

> Amongst severall of the things mentioned in the invoice are some intended for you to make presents of to the King [Narai] and to the Barcalon [phra khlang]. Wee understand that aboute 8 or 10 yeares since the king of Succradana [Sukadana], sending some counterfitt moulds of great diamonds unto the King of Siam, was thereat inticed to send a prow unto the king of Sucradana for procureing the diamonds demonstrated by the said counterfitt moulds. As soon as the vessell arrived there the Sucardanians dispoled them of the cargo, killed the noquoda [captain], and to this day keepeth his men as slaves. The son of the said noquoda, named Lelup, hath fallen into some of our hands, which wee have now sent upon the jounk that hee may bee restored to the King, which wee are told by the Siam manderine will be acceptable unto him.[67]

FIGURE 8.3. Seated Buddha, circa 1400–1500, Thailand, rock crystal with gold and rubies, 10.8 × 4.4 × 3.8 cm, Asian Art Museum of San Francisco, gift from Doris Duke Charitable Foundation's Southeast Asian Art Collection, 2006.27.38. Photograph © Asian Art Museum of San Francisco.

This precedent was evidently a salient piece of geopolitical intelligence that demonstrated both desire for precious stones and the ways they could mediate regional conflict. In this case, Narai's pursuit of diamonds had led to sabotage, and later to an opportunity for English agents to regain favor. Requests for crystals could therefore be weighted with political significance, and claims to possession of unparalleled mineral resources could be costly if untrue. The common reference to wooden molds in this anecdote, and in the requested crystal objects, suggests a procedure for material investigation into the available mineral resources of other territories. Such models both mediated and assessed claims to access, extract, and skillfully work significant mineral deposits.[68] Siam was not the only regional power with aspirations for Borneo diamonds. In 1698, the sultan of Landak, also in western Borneo, asked for military assistance from the Bantam sultanate and the VOC to attack its rival Sukadana and take over their diamond fields. As a result of a successful conquest, both Sukadana and Landak became vassals of Bantam for almost a century.[69]

Precious stones also carried further earthly intelligence beyond factories, workshops, and courts. The Irish natural philosopher Robert Boyle, who in 1672 had published his *Essay about the Origines and Virtues of Gems*, maintained a particular interest in gemstones as a means of understanding the mechanics of the mineral kingdom and saw South and Southeast Asia as the source of the best gemological material.[70] Boyle was doubly interested in East Indian trade as both an investor and adviser to the EIC, whose networks also advanced the missionary work he supported. While undertaking his mineralogical research, Boyle personally financed the study of Malay—perceived by Europeans to be the dominant lingua franca of East Indian trade—among English orientalists, and he commissioned a translation of the gospels into Malay to be distributed by English agents.[71] Here, again, spiritual and material empirical inquiry were inextricably linked through commerce.

A member of a French embassy to Ayutthaya visited Boyle in the mid-1680s. Boyle was particularly interested in the eyewitness testimony of travelers who had visited Southeast Asia, who could provide specimens and first-hand accounts of mineralogical conditions that he had only read about in travel accounts. In an extensive conversation with his French informant, Boyle noted a wide range of phenomena: a riverine source of native gold in Cochinchin (modern-day Vietnam), the distinctive flow of tides around the "capital of Siam" (the city of Ayutthaya), and the "Great Vertue in ye stone yt is said to be found in an Indian serpent," as well as the medical properties of a kind of petrified crab, "brought over, as he told me, as a Present for a great

Minister of state, a couple wch were taken in ye act of Generation, & being petrify'd retain their former union or connexion."[72] In fact, every aspect of this interview touched on geographical and *mineralogical* conditions of mainland Southeast Asia, above all expressed in its distinctive objects—for example, a stone from a two-headed snake and a strangely patterned lump of gold. Removed from their original context, these objects, and the descriptions of them, traveled along diplomatic, commercial, and missionary routes, conveying partial information about the topography and mineral resources of the region. They were then introduced into a new cosmology of European missionary science—in which natural philosophy was embedded—which was nevertheless dependent on the same commercial sources of information and materials.

By the late 1680s, however, both French proselytization and Narai's efforts to strengthen Siam's position by patronizing foreign merchants backfired spectacularly. The year 1688 saw revolution in Siam as well as in England. One of Narai's officials usurped the throne, executing Narai's heirs and closest allies and expelling all Jesuits, to take control of the monarchy's lucrative monopoly of trade.[73] Despite distinct differences in his style of rule and his preferences for foreign luxuries, the new king, Phra Phetracha (r. 1688–1703), still welcomed jewels as gifts.[74] While Ayutthaya maintained strong trading ties with other parts of Asia into the eighteenth century, the kingdom was definitively defeated, and its capital sacked, by Burmese forces in 1765–67. When the new Rattanakosin Kingdom (1782–1932) emerged in its wake some fifteen years later, new copies of the chronicles of Ayutthaya were commissioned, but the culmination of almost a decade of Siamese-Burmese wars left few surviving Thai sources from this rich and turbulent period.[75]

Conclusion

The coeval emergence of new forms of commerce and science in the seventeenth century influenced many forms of knowledge about the earth and its resources that depended on the interaction of complex cosmologies, political relationships, and commercial exchange along new routes of maritime trade and proselytization. As the example of Narai's reign at Ayutthaya shows, precious stones' unique natures made them the object of many different desires, articulated within cosmological systems that shared an interest in the same valuable objects even if they attributed them different significance. Novel diplomatic, missionary, and commercial partnerships across seventeenth-century Eurasia disseminated and collated information about the geographical origins of minerals through objects, artisans, letters, and rumors. Through these networks, participating powers could gain a better understanding of the distributions of minerals and the territorial powers that controlled them.

These networks would not exist without the many different forms of labor and knowledge required to extract stones and metals from the ground, and then to barter, refine, and work them to the point that they could become a currency of diplomatic exchange. Even within the more plentiful EIC sources, we still know little about the artisans who produced rock-crystal objects or labored in the diamond mines of inland Borneo: more work remains to be done on the meaning of gems for those who worked most closely with them. However, commercial sources documenting the gem trade and other precious minerals have much to reveal about the forms of knowledge represented in objects and their makers, through which understanding of the earth and its matter was conveyed.

Moreover, the case of gemstones offers a long-term perspective on the interdependence of the earth sciences, extractive industries, and the state. In recent years, geologists have proposed gemstones as new markers of plate-tectonic boundaries, turning precious stones into key indicators of the definitive theory of the modern earth sciences.[76] This chapter has suggested that, for Narai, rock crystal could do the work of marking other kinds of boundaries—of territorial power, of mineral wealth, and of divine authority. Rather than looking for definitive ruptures with the past ways of knowing the earth and its histories, historicizing the earth sciences as a set of evolving and interrelated techniques of testing and monitoring geopolitical power, territorial advantage, and strategic resources allows us to understand them within a long tradition of competing cosmologies—each seeking to understand, and to assert, a place in the world.

9

"The Agent of the Most Dire of Calamities"

Ice, Waste, and Frozen Futures

Alexis Rider

As we enter the second decade of the twenty-first century, the pace at which the cryosphere is changing is truly staggering. Icebergs the size of Manhattan are calving from the Ross Ice Shelf, Himalayan glaciers are predicted to have entirely disappeared within decades, Arctic permafrost is increasingly impermanent, and the species and systems that rely on polar temperatures are evermore discordant. In the era of the Anthropocene, the cryosphere is emblematic of how deeply enmeshed we already are in environmental crisis: as it melts, ice is enacting a new earth. The environment we know is history.

The paradoxical nature of the cryosphere in the Anthropocene is glaring: as ice has begun to disappear, it has become increasingly visible—the symbolic example of species-level impact on the globe.[1] In the humanities and social sciences, this visibility has manifested in a proliferation of "cryo-histories": works that turn attention to the frozen parts of the earth and their inhabitants, both human and nonhuman, who for too long have been dismissed by the West as somehow outside of time, history, and knowledge-making.[2] Such cryo-histories often focus on the important role of the cryosphere in the development of the earth sciences in the twentieth century, when the poles became sites of climatological import as ice coring revealed them to be a "treasure trove" in which the events of the past were "separately and safely filed for future reference."[3] Less sustained attention, however, has been given to the role of ice and the cryosphere in earlier periods—in particular, when the scientific proposition that the earth was an interconnected environmental system that operated on an immense temporal timescale was gaining definition and form.

In this cryo-history, I turn to the familiar story of the development of the earth sciences in the nineteenth century and refract it through ice. In so doing, I ask: Can the study of the earth's frozen states—that is, its elemental freezing and thawing through Quaternary time—reveal "new" earth histories?[4] By "new," I mean two interconnected things. The first is a history that emphasizes the importance of the cryosphere in shaping social conceptions of environmental, "deep," time—both past and future. By looking to early theories of the ice ages, and the uptake of these theories in popular publications, we can see how anxieties over environmental change on long temporal scales were framed as the clash of natural rhythms with human progress. In the nineteenth century, this clash, shaped by the burgeoning science of thermodynamics, positioned future ice ages—and the existent ice sheets of the polar regions—as threats to the British Empire, filled with agency that exceeded human control. That ice sheets would return, unbidden, revealed the fragility of the epistemological position on which Western notions of mastery of the deep time of the earth—manifested in the burning of fossil fuels—was built.[5] Ice thus came to represent the limits both of energy and engines: no amount of control could avert the inevitability of a cold, inhospitable future. So, while I am turning to a time (the nineteenth century) and space (the British Isles) that have been areas of central attention for earlier histories of the earth sciences, I want to ask: How did cultural responses to thermodynamics—energy and entropy, or more simply cold and heat—impact the essential anthropocentric moment where science, empire, industry, and deep time met?

My second "new" earth history asks whether attention to the hard stuff of ice can reveal something new about our present cryo-historical crisis. It is well documented that the emergence of the earth sciences is tied closely to the acceleration of industrialization and resource extraction in the eighteenth and nineteenth centuries: both share a common interest in planetary deep time, and together these fields shaped, and continue to shape, scientific interrogation into planetary environmental change.[6] Told with attention to acceleration, extraction, and heat, this story is a familiar one—it is a story of the Anthropocene.[7] By instead turning to a history of cold, and a story of formidable icy danger, I hope to show a different moment of human relations with deep time, one in which ice temporalities unsettle stories of purported progress and reveal the limits of species-level thinking.

Cosmic Ice

The theory of the ice ages—the proposition, popularized by Louis Agassiz, that great ice sheets had once covered the now-temperate lands of Northern Europe and the Americas—is often cast as the last piece of a geological puzzle solved by naturalists in the nineteenth century. While the excavation into the earth and the discovery of fossils had provoked geological theories during the seventeenth century, an icy explanation for a surface scattered with erratic rocks and diluvium was not taken seriously in scientific circles until the 1840s. By the 1860s, most European naturalists were converted to Agassiz's continental glacial hypothesis, and, as they became attuned to the signs of past ice, the landscape of Britain—and particularly Scotland—rapidly transformed: huge boulders and fertile topsoil were understood as evidence of the action of ice.[8] In 1863, after several years of field research and scouring of the available literature, Archibald Geikie presented his paper "On the Phenomena of the Glacial Drift of Scotland" to the Geological Society of Glasgow. In it, he asserted that "how much soever geologists may differ in their views of the mode in which the ice acted . . . they are agreed that ice, in some form, has had a chief share in the production of [glacial drift] deposits." Geikie is clear, however: he sees land-ice as the "true theory" and is advocating for a recognition of what he calls "glacial agency."[9]

Though ice as a powerful and land-shaping geomorphological agent was accepted, what had *caused* ice sheets to amass and extend so far south from the poles remained unknown. For geologists interested in using material evidence and fieldwork to establish a sequential stratigraphic order for the deep history of the earth, the principles behind the climatic changes were of less importance. However, for Scotsman James Croll, steeped in a northern British milieu focused on the rapidly growing science of energy, the question of what caused the ice ages was deeply intriguing. "Little did I suspect," he wrote in his autobiography, "that [developing an ice age theory] would become a path so entangled that fully twenty years would elapse before I could get out of it."[10]

Born to a farming family in 1821, Croll was only enrolled briefly in school. According to him, he was "rather a dull scholar," and learned more from *Penny Magazine* than formal education.[11] Unable to attend university, Croll spent his life cycling through a series of jobs: he was a millwright, a joiner, a teashop and temperance hotel owner, a traveling salesman, and a journalist.[12] In 1858—at the age of 37—Croll applied to be the caretaker at Anderson College's natural history museum in Glasgow, the Andersonian, a position he later saw as the fateful outcome of all his previous failures. The work was not very demanding: the museum was only open four hours a day. But importantly for Croll, it gave him access to two scientific libraries containing over seven thousand books. It also put him in contact, albeit peripherally, with the Scottish scientific community, including William Thomson (later Lord Kelvin), the son-in-law of the president of the university, who was at the time busy proselytizing thermodynamics.

Drawn in by the literature available at the Andersonian, Croll began voraciously exploring pressing questions in the natural sciences. In Scotland, two fields dominated: geology and physics, and Croll was explicitly more interested in the latter. The reason, he explained, was that geology "appeared so full of details and so deficient in rational principles, being so much . . . of observation and experiment."[13] It is therefore striking that, in 1867 at the behest of Geikie, Croll left the Andersonian to be an assistant geologist for the British Geological Survey. Tasked with office management, never fieldwork, it was there that Croll published his most comprehensive work, a culmination of his research on the cause of the glacial epoch from the previous two decades, titled *Climate and Time in Their Geological Relations: A Theory of Secular Changes of the Earth's Climate* (1875). Rather than an account of what geological material could be gathered as evidence of glaciation, Croll was interested in what could have caused such extensive global freezing. As he would spend the rest of his life arguing, he believed the glacial epoch could best be explained by identifying "some great, fixed, and continuously operating cosmical law."[14]

The suggestion that cosmical influences could be the key to understanding the cause of the glacial epoch was nothing new. Naturalists had proposed that the temperature of the earth was affected by passing through hotter and colder parts of space, by the planet shifting on its axis, by a differing distribution of land and water, or by the changing obliquity of the earth's orbit.[15] In Croll's mind though, none but the last of these hypotheses could be true, as "every one of them is irreconcilable with the idea of a regular succession of colder and warmer cycles," and were therefore incompatible with mounting fossil evidence of a more temperate Arctic in the past.[16] In order to explain this fluctuation, the great cosmic law Croll sought to uncover must be cyclical: a feature he found in the changing eccentricity of the earth's orbit. This, combined with the precession of the equinoxes and the variation of the tilt of the earth on its axis, Croll was sure, was the determinate force of the glacial epoch. In his introduction to *Climate and Time*, Croll argues that if the glacial epoch could be attributed to this variation in eccentricity, the dates of ice ages could be calculated. This would enable something unique that geologists had long been searching for: "a means of determining geological time in absolute measure."[17]

To trace the eccentricity of the planet millions of years into the past and future, Croll drew on the calculations of astronomers Urbain Le Verrier and Joseph-Louis Lagrange. He was not the first to do so. Other naturalists, including John Herschel twenty years earlier, had explored the possibility of eccentricity as the generator of climate shifts. Like the naturalists before him, Croll saw that "it was quite natural, and, in fact, proper to conclude that there was nothing in the mere increase of eccentricity that could produce a glacial epoch"—the change in temperature would simply be too minute.[18] But as is laid out explicitly in a subsection of the introduction to *Climate and Time* titled "Important Consideration Overlooked," Croll argues that

> although the glacial epoch could not result directly from an increase of eccentricity, it might nevertheless do so indirectly. Although an increase of eccentricity could have no direct tendency to lower the temperature and cover our country with ice, yet it might bring into operation physical agents which would produce this effect.[19]

Moving beyond the direct connection between a high eccentricity and a glacial climate, Croll turned his attention to these "physical agents," arguing that the circulation of ocean currents, extent of ice coverage, formation of cloud and fog, and changing sea levels would be produced by changing eccentricity, thereby driving a constant increase in ice coverage, which would in turn lead to a glacial epoch. Croll was thus identifying a key aspect of climate science and earth sciences more broadly: feedback mechanisms that, through the interaction of environmental processes, increase change. Through ice, Croll thus connected geology to climate, and climate to a global system, in novel and productive ways. Ice sheets hadn't simply built up—all the natural and interconnected rhythms of the planet had collaborated to produce them. Like a well-oiled machine, the earth, under cosmic influence, could make ice. And as Croll calculated (fig. 9.1), the ice-making machine had, and would again, spring into action every fifty thousand years.

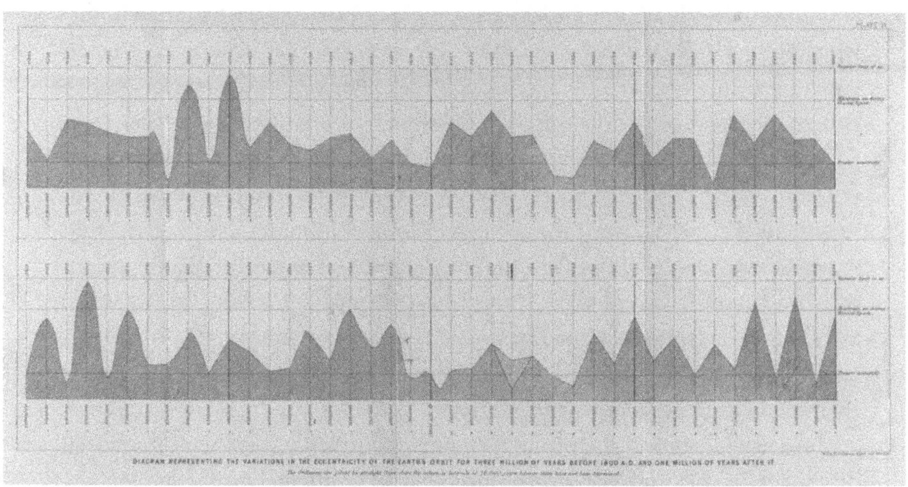

FIGURE 9.1. "Diagram Representing the Variations in the Eccentricity of the Earth's Orbit for Three Million Years before 1800 A.D. and One Million Years after It," in James Croll, *Climate and Time in Their Geological Relations: A Theory of Secular Changes of the Earth's Climate* (London: Daldy, Isbister & Co., 1875).

Hot and Cold, Energy and Waste

Croll's work on the glacial epoch, and the Victorian response to a prospective return of the ice (described below), cannot be separated from the broader discussions about energy that dominated the northern British scientific community during the latter half of the nineteenth century.[20] "Discovered" in 1840s, the science of energy—primarily the study of heat and work—was being deployed to explain both the functioning of the earth as a system and to articulate and justify the extraction of resources and labor across a rapidly industrializing world. As Cara New Daggett has argued, the science of energy therefore played a central role in defining temporal and spatial relations from the mid-nineteenth century: industry, powered by the steam engine, was "fed by fossil fuels . . . converting planetary deep time into industrial quick time, as the lives of hundreds of plants and animals become a thrust of pistons."[21] On one hand, questions of how energy was converted and produced underpinned debates over theories of the age of the earth at a geological scale: Thomson famously undermined geological estimations of the planet's age by arguing that, as the earth was an energy source that was slowly dissipating heat, it was far younger than geologists claimed.[22] On the other hand, energy and productivity became a central analytic of industrialization, and the metric with which to measure peoples and places. That is, it underpinned and justified the extraction of fuels and labor of people by insisting that an efficient use of energy was the central goal of modern industrialization. The factories and ports of Scotland, filled with steam engines, powered with human bodies, working to produce goods, were thus the material manifestation of an abstract idea: energy ever-flowing through fuel, people, and commerce. And, of course, energy production—used to measure both material and people—was a fundamental metric of colonization and empire, working both as a *measure* for labor and a *metaphor* for civility. As Kathryn Yusoff has argued, the two seemingly distinct scales of the geological and the human were inherently enmeshed, with claims about "productivity" used as a means of placing groups within a naturalized geological scale. Those deemed less productive were seen as tied to the deep geological past, "frozen in time" and unable to harness energy without the coercion of their European counterparts, who positioned themselves as the embodiment progress.[23]

Running alongside the productive power of energy was the ominous and disconcerting second law of thermodynamics: entropy, the assertion that changes in energy were irreversible, and energy tended to be more dispersed and disorderly over time.[24] If a concentration of energy was productive, its entropic dispersal was exemplary of waste. In fact, a focus of both physicists and engineers was to reduce the waste of steam engines: to keep energy productive and stave off entropic diffusion. Again, entropy was vividly discussed at both an industrial and planetary scale. For the former, the loss of energy drove the search for increasing efficiency in machines and systems; for the latter, the loss of energy suggested a world in endless and inevitable decay, a globe doomed to a "Heat Death" that would one day render it uninhabitable to humans. As Daggett explains, the contradiction inherent to the science of energy—on the one hand, it was the articulation of human progress and power; on the other, the affirmation of the lack of human control and decay—shaped European responses to deep time, causing them to ricochet "between a worldview that assumed a balanced Earth, uniquely fit for human pursuits, and one that recognized the Earth as indifferent, and possibly antagonistic, toward the survival of the human species."[25] While entropy as a thermodynamic concept was complex, the future it seemed to predict, particularly to the general public, was one centered entirely on the cold. For example, in Camille Flammarion's best-selling *Astronomie Populaire*, a plate depicts the end of the world, where, "surprised by the cold, the last human family has been touched by the finger of Death, and soon its bones will be buried under the shroud of eternal ice" (fig. 9.2).[26] Coldness was therefore increasingly understood as something metaphysically abhorrent: less an alternate climatological state than the marker of the dark, inevitable, and life-ending nature of an energy-driven world.

Tracing the impact of the Heat Death hypothesis, Gillian Beer argues that the "conversation among articulate Victorians about . . . the prospects for life on Earth in a cooling solar system worked, as half-formulated anxieties will, to generate much imaginative thought and production."[27] These same anxieties and imaginaries manifested in relation to the glacial epoch. Already primed with the prospect of an entropic world dissipating energy, Croll's theory of growing and shrinking ice sheets spoke to Victorians of a more immediate and threatening cold. But unlike the Heat Death hypothesis, which was presented as an event that was both a temporal extreme and planetary totality, the ice ages were slated to return in the not-so-distant future and to impact specific regions of the planet. And, crucially, Croll and other interlocutors of his ice cosmology identified evidence of the ice age in two existent regions: the North and South Poles. As we will see, the ice ages were thus understood as a material manifestation of the deep past and future, and existent polar ice sheets were cast as the threatening wastelands of the present. Ice—past, present, and future—was therefore anathema to the productivity

FIGURE 9.2. Depiction of the last human family, frozen at the end of the world. Plate from Camille Flammarion, *Astronomie Populaire: Description Générale du Ciel* (Paris: C. Marpon and E. Flammarion, 1881), 101.

that was being celebrated as essential to human (read, Western) progress. It was a force that, no matter what, would engulf places like the United Kingdom in the inevitability of deep time.

Polar Imaginaries

The poles, for Croll, were essential scientific evidence.[28] First, they could reveal the validity of the feedback mechanisms he had identified, as the ice that amassed there clearly persisted through warm summer seasons. Second, materials found in the poles, particularly fossilized flora and fauna, would help reinforce the theory of glacial and interglacial periods—a central tenet of Croll's theory.[29] But most importantly, the poles offered a unique insight into the frozen past: in the melted, interglacial present, the landscape of Scotland and England provided only residual information about where ice had been. At the poles, in contrast, the mechanism of the glacial epoch—powerful quantities of ice itself—could be found. But Croll was not alone in his interest in the polar regions: in fact, his turn to the planet's remaining ice sheets echoed an obsession that was percolating at a national level.

The nineteenth century saw myriad and increasing attempts to explore the Arctic seas, reach the North Pole, and navigate potential passages from Europe to the Americas in the name of trade and national pride. Many voyages met with disaster, and few succeeded in their intended goal, but as Arctic exploration became a national enterprise, the polar regions increasingly gripped scientific and popular imagination. These famous Arctic voyages—such as the Franklin expedition, which left England in 1845 and disappeared without a trace—"kindled the fires of the Romantic imagination," driving an intense fascination with the region among the Victorian public.[30] This interest extended to the South Pole, which, despite the race to Antarctica between Britain, the United States, and France in the late 1830s, remained even more mysterious.[31] As Michael Bravo has argued, the allure of the polar regions was often abstract: "The desire to stand at the North Pole captured the desire to master our geographical existence, to find a solution to our painful predicament of being earthbound."[32] Similarly, Russell Potter shows that, while trade routes were motivating factors for exploring the frozen north, the uselessness of the Pole itself "was given from the start—but its symbolic value, as the axis of cartography, the point round which all geographic knowledge whirled, was very great indeed."[33] It is no surprise, then, that speculative versions of the poles so utterly captured the Victorian imagination: the unforgiving landscape, cast as empty, inhuman, and beautifully terrifying was presented by writers, artists, and explorers as the sublime antithesis of civilization.[34] As Potter has shown, the Arctic rapidly permeated British cultural production and entertainment: representations of the Arctic in panoramas and magic lantern shows, as well as descriptions of the poles written by explorers in books and periodicals, helped disperse a conception of the frozen poles to the broader public.[35] I would suggest that their materialization of the globe's frozen future added to this imaginative allure. By turning to the far North and South as evidence of his planetary theory, Croll was therefore turning to a space, rife with visual power, that had already thoroughly captured the Victorian public's imagination.

Despite his self-fashioning as a dry reporter of fact—he once lamented to a friend, "I can feel as a poet, but cannot write as a poet"—Croll was equally enamored with the sublimity and Romanticism through which the West gazed at the poles.[36] The glacial theorist never ventured to the frozen North or South—he never expressed any desire to go, and even if he had wanted to, his health and finances would have prevented it. Instead, he drew on accounts of polar voyages to the Arctic—particularly Greenland—and Antarctica written by explorers who were writing for a public audience and thus attune to the allure and marketability of the polar sublime.[37] To take one example, in *Climate and Time*, Croll quotes Isaac Israel Hayes's *The Open Polar Sea: A Narrative of a Voyage of Discovery towards the North Pole, in the Schooner United States* (1867), transcribing:

> Our station at the camp was as sublime as it was dangerous . . . There was neither hill, mountain, nor gorge, anywhere in view. We had completely sunk the strip of land between the Mer de Glace and the sea, and no object met the eye but our feeble tent, which bent to the storm. Fitful clouds swept over the face of the full-orbed moon, which, descending towards the horizon, glimmered through the drifting snow that scudded over the icy plain—to the eye in undulating lines of downy softness, to the flesh in showers of piercing darts.[38]

The empty and inhospitable landscape of Greenland is conjured with grand poetic flair and sublime imagery—the tent, a speck of humanity, bracing against a life-threatening expanse of ice and snow that is as beautiful as it is dangerous. For Croll, embedded within this vivid description was evidence of the shape of the ice cap—a smooth surface through which no hill, mountain, or gorge was visible—that bolstered a crucial aspect of his glacial theory: the way ice sheets dispersed from center to periphery like pancake batter.[39] Through narratives of the Arctic, infused with the sublime, Croll was thus able to draw from the present evidence about the past.

At the same time, Croll drew parallels from the past about the present, seeing his theory of the ice ages as a powerful means of understanding the physical reality of the polar regions—a landscape so few Europeans had seen. "We, in this country, have long been familiar with Greenland; but till very lately no one ever entertained the idea that that continent was buried under one continuous mass of ice, with scarcely a mountain top rising above the icy mantle," he wrote in *Climate and Time*. "And had it not been that the geological phenomena of the glacial epoch have for so many years accustomed our minds to such an extraordinary condition of things . . . the Greenland ice would probably have been regarded as the extravagant picture of a wild imagination."[40] For Croll, the glacial epoch was a useful imaginative exercise, one that prepared the mind to grasp the existing ice sheets of the interglacial nineteenth century. The poles thus served as a kind of evidentiary temporal metonym for Croll. As the title of his 1879 paper, "On the Thickness of the Antarctic Ice Sheet and Its Relation to the Glacial Epoch," suggests, the correlation was direct. "If," he wrote, "conclusions in reference to the thickness of the Antarctic ice be true, they must hold equally true for the ice of the Glacial Epoch."[41] At the poles, one could gain access to a distant and disorienting past, while at the same time the distant past was a means of understanding the frozen present.

So, as theories of the ice age developed and were taken up by the more general public, the notion of a frozen past—and possible frozen future—became entangled with the already prolific representations of the poles, making the Arctic and Antarctic metaphor and metonym of the frozen eras of the earth's history and future. For a public already in the throes of a polar obsession, the temporal and spatial dimensions at work in discussions of the glacial epoch and of civilization quickly blurred. The Arctic (and to a lesser extent the Antarctic) were leveraged as existing examples of regions held in the clutches of an ice age—barren and inhospitable places characterized as "blank areas covered with eternal snow and ice."[42] And as the poles were aligned with the deep past of the earth, the colonial assumption that the regions were outside of time—or at least running significantly late in conceptions of social and cultural advancement—was reinforced. Thus, the future threat of the creeping ice was correlated with the encroachment of the uncivilized periphery on the center of civilization, and the frozen tundra cast as a present-day example of the underbelly of thermodynamics: a waste(d) land—a place where energy was impossible to harness.

Interglacial Beings

As journals and newspapers published lectures and public-facing articles by naturalists about the frozen past of Europe, describing how "glaciers, frozen rivers and lakes, and floating icebergs had converted most of Britain, and the whole of Northern Europe, into a waste of ice and snow," public interest in the glacial era grew.[43] And the increasing acceptance of the astronomical cause of the glacial epoch necessarily brought the future of the earth into discussions. The implication for Croll's theory as well as this overlay of the deep past on the polar regions were not lost on those consuming this new iteration of earth history: if cosmic causes did drive large-scale glaciation, they would one day realign, and ice would expand even further from its center of dispersal, creeping down to now ice-free land. Victorians were thus positioned as what Gillen D'Arcy Wood calls "interglacial beings," existing in a fortuitous moment of global melt.[44] At the same time, the poles—sites seemingly so remote—were increasingly cast as a natural threat, with the ice they harbored a formidable enemy lurking at the edge of what Victorians saw as the vital centers of civilization. As the author of a newspaper article titled "Is Another 'Ice Age' Approaching?" wrote in 1891, "Altogether, our future is not cheering."[45]

Popular accounts of the returning ice age explicitly framed the threat in terms of territory and empire. In *The Cause of an Ice Age*, for example, Irish astronomer Sir Robert Ball explained to his readers, "We must imagine that the cap of ice and snow which is normal at the Pole . . . refuses to remain confined within that Arctic circle . . . it creeps downward and invades the temperate latitudes."[46] At the same time, the ice, "ponderous in mass and irresistible in power, forms a *destructive engine* of almost illimitable capacity; by its influence fragments of the living rock are ripped from their bed, crushed to pieces, reduced to mud."[47] A glacial epoch is thus filled with devastating agency, as a nearly anthropomorphized material—or an out-of-control engine—relentlessly invades. Ball concludes with a frightening description of the future. "Slumbering in the Arctic regions," he writes, "lies at this moment the agent of the most dire of calamities."[48] And, Ball makes clear, this "agent" won't slumber for long. "It is . . . a consequence of the Astronomical theory of Ice Ages that they must return in the future," he writes. One day, "the ice-sheets will again return and desolate those regions which now contain the most civilized nations of the earth."[49]

Ball's image of a future where all the trappings of civilization—cultural and material—will be destroyed by ice was far from unique. In the mid-1880s, J. Horner, a frequent contributor on science to the six-penny monthly journal *Our Corner: A Monthly Magazine of Fiction, Poetry, Politics, Science, Art, Literature*, wrote a reflection for the journal entitled "Time and Change." The piece concludes:

> A thought arises in our minds—will another glacial period ever supersede our present climate? Why not? Argument and analogy favor the probability. . . . What more reasonable, then, than the inference that the climate will again grow arctic in character, and that glaciers will scour away and almost utterly efface all the triumphs of our proud civilization in these islands? The idea that we are living in an inter-glacial epoch seems strange and startling, but so far from bearing the stamp of improbability, it is, reasoning from the data of the past, as certain to come about as the rising of the morrow's sun.[50]

Like Ball, Horner saw ice as a threat to civilization and drew explicit attention to this newfound strangeness of being an interglacial. Similarly, in 1893, *Chambers's Journal*—which, like *Our Corner*, published across the arts and sciences—ran a piece titled "Is an Ice Age Periodic?" Its anonymous author writes that, in the future, a glacial epoch will return, and "the natural features of the land would be torn and scarred, the population driven out or destroyed, and the puny works of man ground and pulverised into effacement by the enormous abrading and crushing force of the moving masses of ice."[51] In a more tongue-in-cheek vein, *Punch* published a "glacial diary" in February 1881, written in an imagined future when the glacial period has returned to London. The fictional diary-keeper hoards ginger lozenges to use as currency; visitors to Madame Tussauds, suddenly frozen solid, are added to the collection; parliamentary meetings are moved to hot baths; and *Punch* headquarters relocated to the interior of Mount Vesuvius. The deep past and the present-day Arctic are fully entangled in the *Punch* satire: the diary-keeper notes that pterodactyl hunting has begun in London, while the accompanying illustration shows a crowd fleeing at the arrival of iconic (and very much contemporary) Arctic megafauna (fig. 9.3).

For all these authors, and their broad audience, the ice was therefore a creeping, inevitable threat to what they considered the definition of civilization. As Robert MacFarlane suggests, the glacial epoch was to the nineteenth century what nuclear winter was to the twentieth: the ominous threat of an uninhabitable Snowball Earth.[52] The distinction, however, is that a nuclear winter is decidedly anthropogenic: produced by the actions of humans operating at a geological (and nuclear) scale. For Victorians, the inverse was true: an extensively glaciated future earth was entirely beyond their control. Despite the sense that industrialization was separating humans from nature—bursting the limits of time through access to vast quantities of ancient energy—the return of the ice ages was the insistence of a rootedness within deep time that no amount of industrial intervention could overwhelm, that no amount of progress could counteract. Furthermore, while Snowball

FIGURE 9.3. "A Scare! What Appeared to Our Terrified Artist Last Week (Jan. 24) to Be the Approach of the 'Glacial Period,' and the Beginning of the End," *Punch*, February 5, 1881. Image courtesy Punch Cartoon Library / TopFoto.

Earth may be a useful analogy, what worried Victorian readers and writers was the devastation not of the globe but rather their specific "civilized" locale, and the social structures—naturalized via Western scientific assertions about racial hierarchies and progress—that undergirded their sense of civilization.

Thus, Victorian interlocutors predicted that the return of the glacial epoch would bring with it the invasion of purportedly primitive, unproductive peoples. After the advent of a next glacial age, one author asked if "Macaulay's savage from southern climes shall, or shall not, at some future time stand on London Bridge and contemplate the ruins of a fallen greatness."[53] The oft-repeated literary trope of "Macaulay's savage," or "Macaulay's New Zealander"—first evoked by Whig essayist and politician Thomas Babington Macaulay—was used by a range of authors to explore the potential collapse of the British Empire driven by myriad causes, not just glaciation.[54] Echoing this fear, in a reflection published a decade later in Boston, Macaulay's southern "savage" gazes over a London ravaged by ice. "Is it in the womb of Time and the decree of Fate that the ranks of poor humanity are to be scourged and decimated by another glacial deluge?" John C. Elliot asks. His answer is unequivocally yes: "London and New York, Berlin and St. Petersburg will be no more . . . Sunny France will be a Siberia knowing no summer, and Paris a frozen solitude of snow-filled streets."[55] Elliot goes on to suggest that the United States must "carve out refuge for her people in South America against the time when they will be driven out of the Northern continent by the irresistible advance of the all-effacing ice-sheet." He concludes that in the distant future the "proverbial New Zealander of Macaulay [will] actually sail away to view in reminiscent mood the place where once had stood Old London Bridge."[56]

The overlaying of a natural force—ice—with fears over the migration of peoples deemed less productive/civilized requires reflection. It is particularly striking that in these imagined futures, the influx of glacial ice does not bring with it Indigenous Arctic peoples but rather a "savage" from the South. The slippage highlights precisely what is at stake for these authors as they contemplated the glacial epoch: the British Empire and the assertion of a distinctive position for white Europeans (and their American counterparts) within "natural" deep time. That Aotearoa New Zealand would be immune from the ice age suggests either a belief that only the Northern Hemisphere would freeze or, more likely, that somehow these "savage" populations were particularly able to coexist with ice. By conflating tropics and poles, populations of people—defined within a race-based classificatory system upheld by Europeans—are being positioned along a temporal scale that can exceed the geographical as the marker of "uncivilized." And ice, the material manifestation of entropy, is the natural vehicle for this generalized trope of "savage": both share the resounding quality of being the antithesis to progress and the manifestation of an uncivil past.

Conclusion: Difference and Power in Ice

The fascination with the return of ice, articulated in the language of empire—center and edge, civilized and savage, human and inhuman—therefore reveals the entanglement of the geological and geographical that Croll and others embraced when using the poles as metaphor for the past, and vice versa. It is also an articulation of the way progress and productivity, taken from thermodynamics, had permeated social ideals. Furthermore, the fear that invading ice would bring with it "savages" from the South reveals exactly how certain peoples were placed within deep time: seen to embody the unproductive, wasteful nature of ice itself. Ice was a threat to environment and empire, able to highlight, through its very naturalness, the fragile social hierarchies on which Britain depended. Emplaced in the immense temporal scale of the geos, the Victorian scientists and public who contemplated the future return of ice responded with fear that moved from the natural world into the social, shaping claims not just about their climatological future but also their cultural displacement within it.

There is an interesting paradox at work here, one that is pertinent to the current framing (and debates) around the Anthropocene. Even in such an early moment of self-aware species-level thinking—the recognition that ice operates on a scale that exceeds human time and that over such an expansive timespan a nonhuman material can change the contours of vast swaths of the planet to the discomfort of the human species—difference was paramount. Which is to say, Victorians leveraged the time of ice to articulate the purported stratification of the species at precisely the moment when ice was showing their sense of difference to be irrelevant. The claim that today species-level thinking is problematic when applied to the environmental crisis is well trodden: culpability and action that respond to the reality of violent and exploitative histories is essential. But here we see how difference has been, equally problematically, baked into deep time since its reification as a way of making sense of the natural world. The observation that the assertion of difference and the oppression of difference is wielded by those in power is not new, but thinking with ice helps us see it in action in the nascent earth sciences.

Of course, today, as the planet warms at a pace that far exceeds any natural rhythms, heat is the most apparent—and urgent—metric of deep time and seems the ultimate arbiter of our planetary future. By being attentive to historical moments when human and deep time clashed and shaped claims about humans and the earth both, we can perhaps gain insight into our current cryo-historical moment, where anxiety around ice is likewise laden with questions of power, exclusion, and assertions of what a civilization looks like. Ice, as it becomes increasingly visible even as it is rapidly disappearing, is therefore a means through which to articulate new histories and advocate for new futures.

10

Hydropolitics for a New Nation

Hydrological Origins and Limits for the Australian Interior

Ruth A. Morgan

In a paper communicated on his behalf to the Royal Society of New South Wales in July 1917, South African geologist Alexander du Toit presented a lengthy analysis of Australia's Great Artesian Basin.[1] Having visited the continent on the eve of the Great War for a meeting of the British Association for the Advancement of Science, he had been intrigued by a vast underground system: "probably no other geological problem is so many-sided . . . bristling as it does with puzzles of all kinds," he marveled.[2] In sharing this paper, du Toit was intervening in a debate as to the origins and limits of the basin's waters that dated to the turn of the century. One side held that the subterranean waters were recharged by rainfall; its chief protagonist, New South Wales government geologist Edward Fisher Pittman, refuted du Toit in what would be his last published comments on the matter.[3] Pittman's antagonist, John Walter Gregory, chair of geology at the University of Glasgow, continued to promulgate until his death in 1932 his position that the basin's waters were entirely finite—long after the young nation's geologists, led by Pittman, had reached a consensus that the subterranean waters of the continent's arid interior were replenishable.[4]

That the continent's arid interior might pose environmental limits to the population and progress of the settler nation preoccupied Australians throughout the first half of the twentieth century. Deemed more climatically amenable to whites than the tropical north, the center nevertheless was hydrologically ill-suited even to the pastoral economy. Or so the likes of Griffith Taylor argued. Others, meanwhile, dismissed such gloomy assessments of the interior's carrying capacity.[5] Focusing on the interwar expressions of these views, historians have understood them as part of wider international conversations about global population and racial degeneration that were infused with environmentally determinist thought.[6] This chapter's consideration of the prevailing hydrogeological concerns at the turn of the twentieth century to the end of the Great War contributes a prelude to these interwar debates, in which Australian anxieties about the settler nation's progress intersected with geological concerns as to the history and structure of the earth. Here, in the wake of the nation's federation in 1901 and prolonged drought, biopolitics encountered hydropolitics, in which the continent's uncertain hydrological endowment became the subject of intense scrutiny.

Since Meinig's *On the Margins of the Good Earth*, the ways that settler colonists responded to, negotiated, and understood the climate extremes wrought on southeastern Australia by the El Niño–Southern Oscillation have preoccupied environmental historians and historical geographers, with a resurgence during and since the Millennium Drought (1995–2010).[7] Such meteorological concerns have brought Australian studies of settler-colonial climes into conversation with a wider historiography that is concerned with imperial encounters with unfamiliar environmental conditions and climate phenomena. European and North American territorial and maritime empires not only accumulated vast sums of quantitative and qualitative climate data from their expanded networks but also reckoned with diverse climatic and hydrological challenges to their imperial projects.[8] These challenges were multiscalar, warranting both locally specific responses and incorporation into a growing scientific body of climatic and hydrological knowledge.[9]

This historiographical interest in the making of atmospheric territory has turned now to the subterranean, where imperialists also sought to assert their vertical territoriality through the application of Western science.[10] Through new technical understandings and methods of sensing and visualizing the ground cut in sections, geographer Bruce Braun argues, this territory was essentially "produced," by "open[ing] up new epistemological spaces which, in turn, made possible new domains for economic and political rationality."[11] Among the historical studies of the mineral riches that this verticality afforded, water has rarely figured.[12] In arid and semiarid locales, however, the location of water above and below ground was paramount to improvement and profit. As this chapter shows, these were not only local concerns but were also related to scientific efforts to determine the age of the earth and to ascertain the mechanisms of geological and climatic change. The origins of the continent's artesian waters were a matter of grave importance for the arid interior's settler occupation and development.

✶

The Great Australian Artesian Basin of du Toit's paper has since become known as the Great Artesian Basin, one of the world's largest artesian basins, which extends across 1.7 million square kilometers, or one-fifth of the continent (similar in size to the US state of Alaska). The multilayered confined aquifer system lies beneath the continent's semiarid and arid regions, lands characterized mostly by low rainfall with high tropical seasonal rainfall in the most northern reaches.[13] Rain that falls on the ranges of eastern Queensland slowly percolates through the sandstone sediments before finally discharging from springs on the basin's margins in western Queensland and South Australia (fig. 10.1).[14] Long known to Aboriginal peoples, who have occupied parts of central Australia for at least the past forty-five thousand years, these

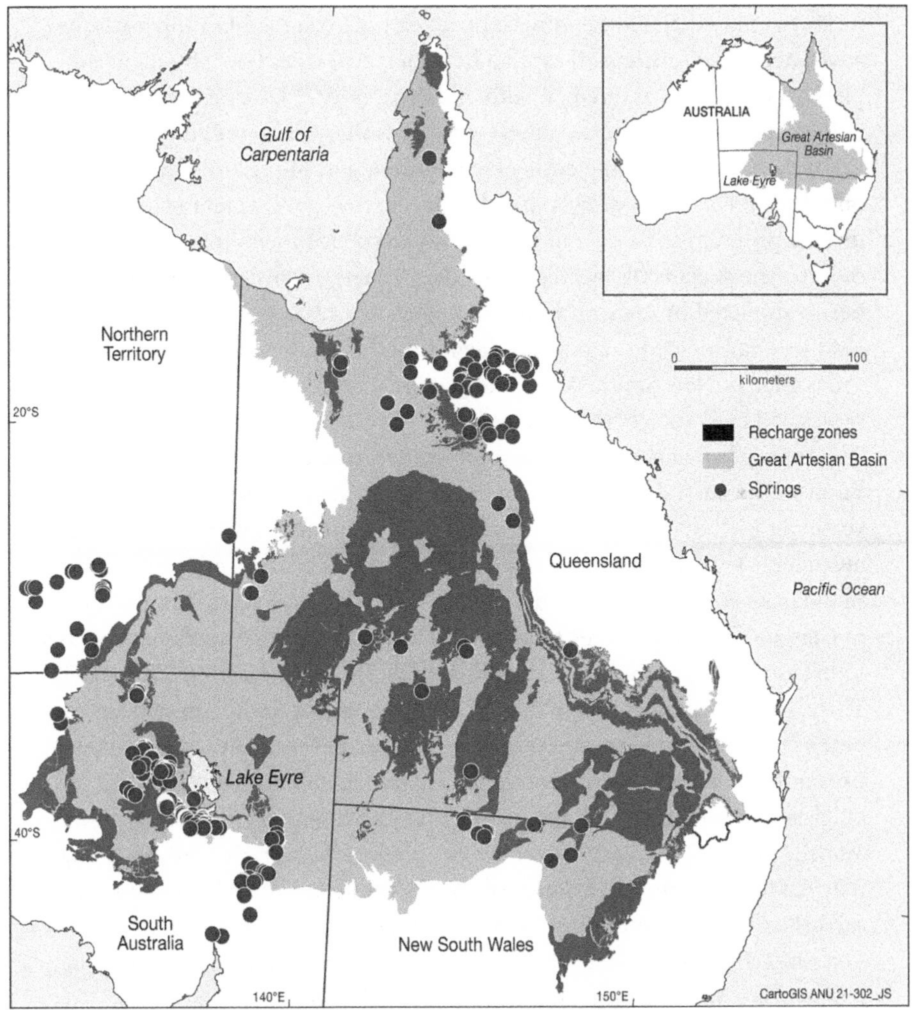

FIGURE 10.1. The Great Artesian Basin, showing recharge zones (intake beds) and springs. Map produced by CartoGIS, Scholarly Information Systems, the Australian National University.

springs became vital oases in these otherwise waterless lands for European explorers, pastoralists, and camel handlers from the mid-nineteenth century.[15]

Speculation as to the presence of groundwater in the interior had circulated since the 1820s, only growing in the wake of the successful tapping of artesian water in Paris in the 1840s, as well as in Prussia, Italy, Austria, Algeria, and the United States.[16] But it was not until the sinking of a bore hole at Killara station west of Bourke in New South Wales in 1878 that the possible extent and structure of an artesian basin began to take shape.[17] That year, engineer Thomas Rawlinson had proposed his theory that only percolation into an interior basin could account for the loss of vast quantities of river water he had observed along the semiarid fringe of the colonies. On this basis, he advocated closer inspection of what he anticipated would be "abundant supplies of water" that assured "wealth and prosperity."[18] Despite the optimism of the likes of Rawlinson and others, such as Walter Gibbons Cox and Robert Logan Jack, such were the costs of exploration and the risk of failure it was only after the drought of 1885 that the Queensland government sponsored a drilling program of its own in the colony's west.[19] Although this effort failed, others were successful: by December 1892, nineteen government bores had been completed or were in progress in Queensland alone, while private investment had yielded 524 bores by 1899, 505 of which were successful.[20] Over the border in New South Wales, meanwhile, the colony's artesian water reserves had been mostly mapped by 1900.[21]

Gregory had encountered the southwestern reaches of the Great Artesian Basin in the austral summer of late 1901. Then chair of geology at the University of Melbourne, he had embarked on an expedition with a small party of students to investigate the fossil remains that had been recently uncovered in the Lake Eyre basin in northern South Australia. Five years later and now in Glasgow, he compiled his frequent dispatches to the metropolitan press as to his journey's progress into a volume titled *The Dead Heart of Australia*.[22] There, he fleshed out his own theory as to the origins of the artesian waters of the continent's interior. These waters were not derived from rainfall as the likes of Rawlinson, Cox, meteorologist Henry Chamberlain Russell, and geologists Ralph Tate and T. Edgeworth David had supposed.[23] Rather, Gregory argued, their origins were "juvenile" or "plutonic" (after Hutton), meaning they emanated from the condensed vapors of molten rock deep within the earth. Such waters were called "juvenile" because they were surfacing for the first time. This theory, Gregory believed, explained the artesian water's high temperature and chemical composition as well as the "tidal" or geyser-like behavior of certain bores or "flowing wells."[24] These wells, he concluded, "are the modern, artificial outlets from a vast reservoir, which is almost entirely closed, and the waters discharged from it must have collected during the course of centuries, and probably of millenniums."[25]

From this position, Gregory called for the conservation of the interior's artesian water, for "it is not safe to assume that they [the wells] will flow on for ever."[26] In doing so, he chafed against the prevailing enthusiasm for bores, which by then provided not only water for stock and irrigation but also water for wool scouring, railways, western towns, steam-generated electricity, mines, and health spas.[27] The need to exploit these reserves was unquestioned, as explorer Ernest Favenc declared, "Our vast fertile downs were never destined to be idle and unproductive for months and months, dependent only on the niggard cloud o'erhead. . . . The answer to this problem is to bring to our rich alluvial surface the waters under the earth."[28] The exploitation of artesian water promised to make what had been a place of settler transience into one of permanence.

The following year, Pittman took the opportunity to refute Gregory's plutonic theory in his 1907 Clarke Memorial Lecture before the Royal Society of New South Wales.[29] The state's government geologist since 1891, Pittman had overseen the growing use of, and dependence on, artesian water.[30] Before the Royal Society, he dissected *The Dead Heart*, methodically addressing each claim of the "distinguished author." Although Pittman was careful to appear respectful of Gregory's "great reputation," he argued that the professor had no authority to speak on the matter as he had only visited the artesian area near Lake Eyre, never Queensland nor the "porous intake beds" of New South Wales. Moreover, citing the consensus among US geologists that underground water was derived from rainfall, Pittman described Gregory's plutonic theory as too "far-fetched" and "complicated."[31] In closing, Pittman turned to the declensionist tone of Gregory's title, *The Dead Heart*. "When the heart ceases to beat, decay of all the other members of the body speedily follows," he noted, inferring that Gregory's gloomy assessment of the interior's prospects extended to the entire nation. The government geologist was sure to note that "there is still a fair amount of vitality in the head and limbs of our national body, as represented by the more fertile districts nearer to the coastline."[32]

Over the course of the following decade, the pair circled each other in the meetings and proceedings of learned societies in Australia, Britain, and the United States, with occasional contributions from other members of the geological fraternity. This "Australian groundwater controversy," as it has been called, has received relatively little attention in the history of earth sciences, with historians focusing mostly on the personalities of the protagonists of a seemingly petty spat.[33] How these positions were derived from contemporary geological thought has gone largely unexplored. That similar ideas to Gregory's were entertained in the United States with considerably less fanfare suggests that the question of the origins of subterranean water in the Australian interior was of heightened significance in the early decades of the twentieth century.[34] I turn now to a genealogy of Gregory's interpretation, and then to its implications for groundwater management in Queensland and New South Wales, in order to account for the protracted and heated nature of the debate.

In his first foray into the origins of artesian water in Australia, Gregory had not advocated for their plutonic origin at all. In an article published in *The Argus* in 1903, he had professed that origin of artesian water in Victoria was "all derived" from local rainfall.[35] His about-face in *The Dead Heart*, in which he dismissed the role of rainfall in favor of the deep, plutonic origins of artesian water, followed closely the work of the Austrian geologist Eduard Suess. Suess had advanced his own theory as to the origins of springs in the intervening years between Gregory's writings for the Australian press and the publication of his book in 1906. In a 1902 lecture delivered in the spa town of Karlsbad, the subject of his first paper over fifty years earlier, Suess surveyed a selection of thermal springs and argued that they contained no trace of rain or meteoric water.[36] He elaborated further in the fourth volume of *The Face of the Earth*, translated in 1905, explaining that juvenile waters arise when the hydrogen gas issuing from the earth's interior under very high pressure and high temperatures combine with the oxygen of the atmosphere. As such, "hot springs, so far as they are juvenile, are a milder form of the volcano."[37]

In advancing this theory, Suess was challenging the tenets of the hydrological cycle, which had become widely accepted since the seventeenth century. Separate quantitative studies by French physicists Pierre Perrault and Edmé Mariotté, and English astronomer Edmund Halley, had enabled them to account for the rise of spring water by calculating the differences between rainfall and river discharge, and the evaporation of seawater and rainfall.[38] Since then, as US hydrologist Oscar Meinzer observed, "the old hypotheses became more and more shadowy until they lurked only in obscure haunts like emaciated ghosts."[39] In some European circles, however, these ghosts had been summoned in the late nineteenth century. German geologist Otto Volger, for instance, declared in 1877 before a meeting of the Society of German Engineers, the notion that rainfall was the origin of springs was "unfounded" and "fallacious."[40]

Given the Austrian geologist's work was not widely known in England at the turn of the century, it might seem unusual that his views held such influence over Gregory.[41] Yet it had been Suess who had introduced him to the prospects of East African geology in his 1891 account of the results of Hungarian explorer Count Samuel Teleki's expedition to Lake Rudolf (now Lake Turkana, Kenya).[42] An encouraging correspondence ensued, and Gregory seized the opportunity to join the Villiers expedition to British East Africa shortly afterward in late 1892.[43] When that expedition foundered, Gregory (with Suess's support) mounted his own expedition to Lake Baringo in what he would later call the Rift Valley. In honor of his Austrian mentor, who had already suggested such a "valley" extended through the African continent to the Dead Sea, Gregory named a deep basin he and his Zanzabari party encountered "Lake Suess."[44] By this point, Gregory's biographer surmises, he had become "a life-long Suess disciple."[45]

Other aspects of *The Dead Heart* also reveal the influence of Suess's work. The first chapter of Gregory's 1906 collection provides an extended account of the local legend that had drawn him to the Lake Eyre basin five years earlier. This legend pertained to the fossil remains, which belonged apparently to the *kadimakara*, an animal that lived in the sky and had climbed down to the earth in a former age when Lake Eyre was lush and verdant. It had become stranded, and thus doomed, when a bushfire burned the three gum trees that were the pillars of the sky and its only means of retreat.[46] Gregory had learned of "how the Kadimakara came down from the skies," as the chapter's title put it, from anthropologist and geologist Alfred W. Howitt, who had encountered the area around Lake Eyre first in 1859, to ascertain its pastoral potential, and again in 1861 on the relief expedition for the ill-fated transcontinental explorers Robert O'Hara Burke and William Wills. Neither experience had endeared the interior to Howitt; his endeavors nevertheless bore intellectual fruits, as he gleaned ethnographic details about the local Dieri people from Lutheran missionaries, which he subsequently shared with metropolitan audiences in Australia and Britain.[47]

Anticipating Gregory, Howitt had asked whether the Dieri narratives were derived from memories of the past when kadimakara "still lived in the marshy trails of the Lake Eyre deltas."[48] Gregory, for his part, sought "especially [to] ascertain whether they were laid down on the bed of Lake Eyre, when the lake was much larger than it is at present."[49] That central Australia had undergone a period of desiccation was a widely held view by the mid-1880s: the furrowed channels of ancient riverbeds were the seeming geological evidence of a much wetter deep past.[50] Although Gregory speculated as to the origins of the "Dieri and their allied tribes" of the Lake Eyre Basin, his interpretation of the wider significance of the kadimakara goes unsaid.[51] Read in concert with the opening chapter of the first volume of Suess's *The Face of the Earth*, however, the genesis of Gregory's interest in the kadimakara becomes more clear.

Gregory had greatly appreciated Suess's "poetical imagination" in his 1898 review of the French translation.[52] This first volume, published in Vienna thirteen years earlier, proposed an alternative history of the earth that challenged the gradualism of Charles Lyell's uniformitarian thesis, including the Scottish geologist's rejection of a biblical deluge.[53] In seeking to overturn this "geological quietism," Suess turned to a wide array of evidence, among which were the recently translated fragments of the Izdubar epic, later known as the Epic of Gilgamesh.[54] Excavated some twenty years earlier, the translation of the fragments in 1872 had prompted much excitement in Britain and North America, for they appeared to confirm the Flood and other biblical stories.[55] Suess's detailed study of the ancient text dominates the first chapter of *The Face of the Earth*, titled "The Deluge." In line 47 of this epic, "Anunnaki cause floods to rise," Suess found "conclusive proof to the geologist," that the catastrophic ancient deluge could not have been the result of rain.[56] As subterranean-dwelling spirits, the Anunnaki represented the source of the geological agency of vast quantities of water that rose "*out of the deep*."[57] Only an earthquake ("a seismic convulsion") could account for the floods of the deluge, albeit one that was not universal.[58] As Deborah Coen describes Suess, "No individual at the turn of the twentieth century had a clearer vision of geology as a global science, and earthquakes played a special role in it."[59] Gregory hoped to emulate that vision.[60]

The inspiration of *The Face of the Earth* aside, Gregory also admired Suess's social contributions. As president of the Geological Society, he addressed a ceremony for the unveiling of the "Eduard Suess Memorial Tablet" in 1929, fifteen years after the Austrian geologist's death.[61] Gregory paid tribute to Suess's "sound judgement and moral courage," sharing stories of the geologist's successful campaigns to improve the water supply of Vienna and to regulate the flow of the Danube.[62] These causes, Coen argues, were a reflection of Suess's conviction, as a veteran of the 1848 revolutions, that scientists had a duty to serve their society, which he articulated in his 1867 study, *The Ground of the City of Vienna: According to Its Matter of Formation, Composition, and Its Relationship to Civic Life*.[63] His friend Percy Boswell remembered Gregory for a similar belief in the important role of science in "social and educational development."[64] It had been this view that had led Gregory to accept the chair of geology at the University of Melbourne in 1899, where he could contribute to "one of the most instructive large-scale social developments in progress," Australia.[65]

For Gregory, then, it was his responsibility as a scientist to bring his geological expertise to bear on a question of national concern, that of the hydrological limits of the continent's interior. In the decade prior to his expedition to the Lake Eyre basin, legislative attempts to regulate bores in Queensland had been defeated, as pastoralists rejected the prospect of the state's intervention on their properties. The majority of the state's artesian bores had been drilled on leasehold land, and their waters flowed freely, mostly to waste, as they did in New South Wales.[66] Similar concerns were shared by geologists in the United States, where the nation's westward expansion since the Civil War had also relied on drilling deep wells where rivers were scarce.[67] In New South Wales, Pittman attributed recent observations of the diminishing yield of the state's bores not to waste or excess but to "the protracted drought through which we have just passed [which] has materially lessened the amount of water absorbed by the intake beds."[68] By this time, the Federation Drought (1895–1903) was well underway. One of the worst droughts recorded since British colonization in 1788, its impact was exacerbated by economic depression and environmental degradation. In New South Wales alone, the total output of wheat during 1902 fell to just 11 percent of the average for 1899 to 1904.[69] The number of sheep in Australia was halved, and the cattle population fell by 40 percent.[70] With artesian water as the "drought antidote," in the words of a government engineer, the young nation's dependence on the Great Artesian Basin was growing.[71]

Compelled to continue his intervention, Gregory resumed his theorizing on the origins of the waters of the basin in the *Geographical Journal*.[72] His confidence may have been buoyed by the favorable citation of *The Dead Heart* in British petrologist Alfred Harker's 1909 work, *The Natural History of Igneous Rocks*.[73] Determined to settle the debate, Pittman convened a meeting of government geologists from the states of Queensland, Western Australia, South Australia, and Victoria in Sydney in 1912. With no mention of Gregory, they agreed that "the ascertained facts indicate that the water is almost wholly, if not entirely, derived from rainfall; and that it percolates the porous beds under the influence of hydraulic conditions."[74] Given the diminishing flow of bores in New South Wales and Queensland, however, they recommended closer regulation of underground water resources in each state and limiting its use to pastoral pursuits and town supplies. They further argued for curtailing new irrigation ventures until their investigations could ascertain the volume of water available.[75]

A second interstate conference followed in July 1914 in Brisbane.[76] Having taken evidence from Queensland pastoralists, the gathered geologists again sought to clarify the source of artesian water, citing the general misapprehension on the matter. One witness had suggested that waters were fed from the mountains of New Guinea, or as far away as the Himalaya.[77] In a thinly veiled criticism of Gregory's continued ruminations on the topic, they blamed the "pernicious" publication of "certain recent literature" for this confusion. The geologists reaffirmed their position that artesian water originated in rainfall that percolated through beds of porous sandstone and collected in the artesian basin.[78] A month after the conference, Pittman prepared an illustrated report on the Great Australian Basin for the Sydney meeting of the British Association for the Advancement of Science, in which he again directly challenged Gregory, whose theory he dismissed as "personal opinion" that was "unsupported by anything in the shape of definite evidence."[79] Accompanying figures (fig. 10.2) and photographic plates (fig. 10.3) provided proof positive that rainfall indeed replenished the basin, identifying the very locations where this process occurred.[80] Pittman and his allies might have been more persuaded by Gregory's argument had there been clear evidence of volcanic activity in the vicinity of the Great Artesian Basin.[81]

Although Gregory, who was present at the meeting, did not comment, Pittman's report did not go unnoticed. President of the geography section, Sir Charles P. Lucas, observed in his address on the topic of irrigating deserts, "I must leave to more learned and more controversial men than I am to discuss whether [the basin's] supplies are plutonic or meteoric, and how far in this matter you are living on your capital."[82] For his part, Gregory was alive to the implications of his opposition's view, to which he attributed the declining yields and pressures of wells in Queensland and New South Wales.[83] An expectation of endless artesian water supplies, replenished by rainfall, was fostering largely unfettered extraction. The belief held by some pastoralists and geologists that, unless tapped, the artesian water would flow out to sea, only legitimized its use to avoid such waste.[84] By 1914, the flows of these artesian bores had reached their peak—some 1,229 had been drilled in Queensland by this time, and several hundred had already ceased to flow.[85]

*

Regardless of the consensus among the geologists in the employ of the Australian state governments, Gregory maintained his position on the (mostly) juvenile or plutonic origins of artesian water, albeit with some modest adjustments to his reasoning. By 1930, he had folded his assessment of the flowing wells of central Australia into an explanation of geological change at the planetary scale. In a lecture to the Institution of Mechanical Engineers, which was published in *Nature*, Gregory boldly claimed that this plutonic water was also the source of the world's oceans. This mechanism was part of what he called the "machinery of the Earth," an autogenic process that "rendered possible the evolution of man and still controls his destiny."[86] His interpretation of the hydrological limits of both the Australian interior and the earth guided his interwar interventions into national and international population

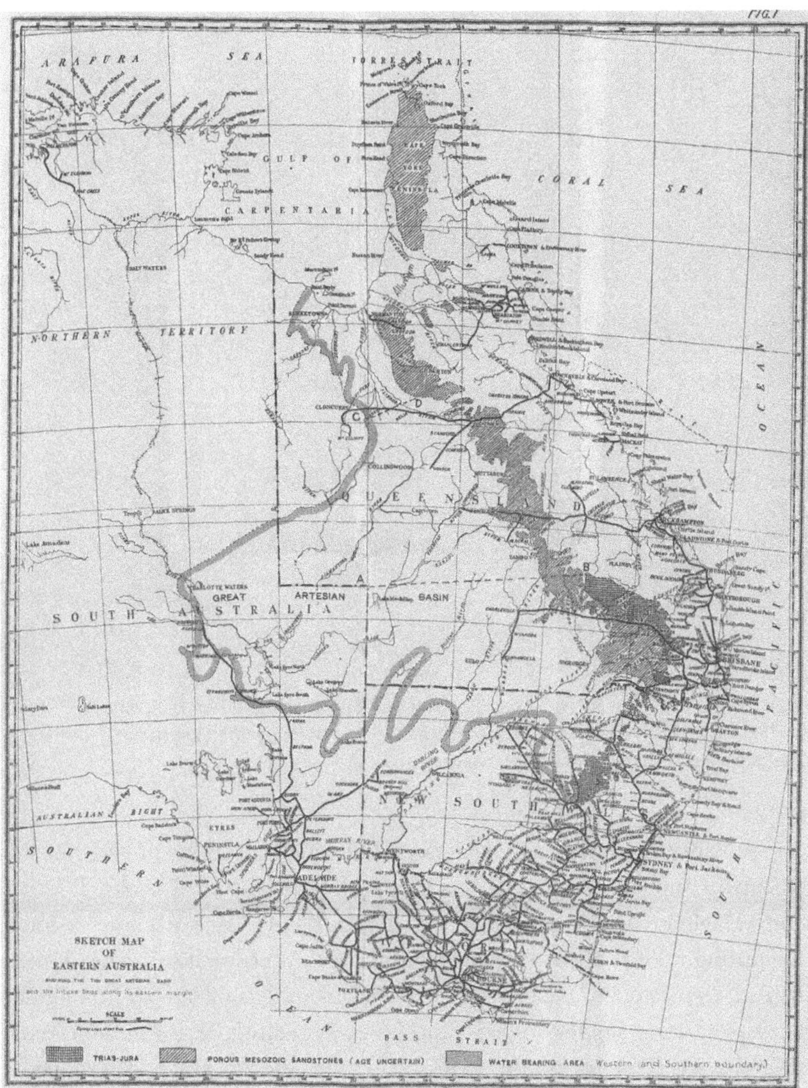

FIGURE 10.2. "Sketch map of eastern Australia, showing the Great Artesian Basin and the intake beds along its eastern margin," in Edward F. Pittman, *The Great Artesian Basin and the Source of Its Water* (Sydney: Government Printer, 1915), plate 2. National Library of Australia, nla.obj-226128581.

debates, in which he regarded the continent's arid interior as geographically unsuited to a "dense population"—reason enough to justify restricted entry to non-Europeans.[87] Presumably the one hundred million people he had earlier deemed that the continent could "maintain in comfort" would reside elsewhere.[88]

FIGURE 10.3. "Cliff of porous Trias-Jura sandstone, Blyth Creek, Roma District, Queensland," in Edward F. Pittman, *The Great Artesian Basin and the Source of Its Water* (Sydney: Government Printer, 1915), plate 3. National Library of Australia, nla.obj-226128582.

The kadimakara narrative also persisted, albeit for different ends. In the explanatory notes to accompany his 1932 geological map of Australia, T. Edgeworth David invited his readers to join him in Wells's "time machine" to journey through the continent's geological history. After the last ice age, he explained, the waters of Lake Eyre slowly shrank and salinized, and the "giant marsupials, their supplies of food and water dwindling, perish in thousands, bogged around the dwindling mud springs and the shrinking waterholes."[89] Whereas Gregory likely interpreted the kadimakara as a check on pastoral expansion, David's was a more bullish position that understood their demise as evidence only of the primitivism of the interior's pre-European inhabitants.[90] After all, David's faith in the meteoric theory of artesian water gave him immense confidence in the potential of the interior: "To-day, aided by improved transport by rail, motor car, and aeroplane, and further helped by science," he proclaimed, "[the white] man is now dotting the desert with oases around his artesian wells and dams, and is drawing the green of the wheat belt ever further inland."[91]

Just as Gregory found in the kadimakara a parable for his times, so too twenty-first century historians and geographers have turned to the ethnographic record for cultural parables to inform human resilience and adaptation in a warming world.[92] Among these parables are stories of perseverance from the Australian arid zone, the largest area of desert in the Southern Hemisphere. Far from the continent's "dead heart," these deserts are now known to have been peopled from at least forty-five thousand years ago. Such a long human history places the desert's occupants amid the transformations of the late Pleistocene, during which conditions in the continent's drylands became drier, windier, and colder as they expanded almost to the coast.[93] Archaeological evidence indicates that these peoples developed highly mobile ways of life to survive these challenging times, which took some ten thousand years to abate.

Returning to the debate over the origins of central Australia's artesian waters reveals both the fluid nature of hydrogeological thought at this time and its wider implications for the young nation. Regardless of whichever theory prevailed, the exploitation of the Great Artesian Basin would rest on a particular configuration of historical, environmental, and political circumstances, interpreted through opposing ways of reading the face of the earth and its depths. Whether finding in favor of finitude or abundance, determining the nature of the continent's "flowing wells" was a matter of consequence that would continue to shape the past and future of the arid interior.

11

Earth Time, Ice Time, Species Time

The Emergence of Glacial Chronology

Emily M. Kern

Understanding how much time had passed since the moment of the earth's creation was one of the driving questions for European geologists at the start of the nineteenth century, while comprehending how much time had passed since the origin of the species—any species, but especially the human species—possessed evolutionists, physicists, and earth scientists from 1859 through the century's end.[1] Increasingly, it was evident that the human species had had a longer and stranger career than suggested by recorded history, and on an earth that both was and was not very much like our own.[2] This was an earth where sea levels might precipitously fall and rise, one where periods of temperate and congenial climate alternated with epochs of massive and expanding sheets of ice burying lands from the Hebrides to the Gulf of Bothnia, stretching south from the Arctic to the Italian Alps and the Spanish Pyrenees. Analyzing these strata in 1839, Charles Lyell coined the term "Pleistocene," taken from the Greek roots for "most" and recent."[3] This naming captured both the strangeness and familiarity of this earthly epoch, for the Pleistocene was characterized by fossil deposits where the vast majority of represented species were known to be still living—although many of these species were only found now in much colder climes and more distant (often northerly) locales.[4] But strange-yet-familiar fossils were not the only things preserved in the geological strata of the Pleistocene age. Between and above these layers of rock shaped by ice, water, and time were relics of distant human cultures—mostly stone tools but occasionally skeletal fragments and even haunting and evocative paleolithic cave art. What remained unclear was *when*. How much time had elapsed since these prehistoric toolmakers had hunted, painted, and buried their dead in the wake of the retreating ice?

As Joe Burchfield has shown, the study of geological time in the second half of the nineteenth century was centrally concerned with determining the age of the earth, a question that also held significant implications for the plausibility of Darwinian evolution. The long, slow process of reeling in the "limitless eons of the mid-century uniformitarians," in Burchfield's words, reintroduced boundaries to the earth's deep chronology.[5] Estimates of the earth's age, whether based on thermodynamic principles, in the style of Lord Kelvin, or on the analysis of rates of denudation and sedimentation naturally impacted the estimates of the rates of species' evolution—but they also impinged equally directly, if less obviously, on hypotheses about the rates of cultural evolution in the study of human prehistory. But over much the same period where the Pleistocene was taking shape in geology, so too was the chronology of European prehistory.

In the conventional history of geochronology, this unification of human time and earth time on the same linear timeline would seem to properly belong to the period after the development of radiometric dating.[6] The same central principle applies in both cases, even if the radioisotopes of interest—uranium-238 for the age of the earth and carbon-14 for human prehistoric artifacts—decay at vastly different rates.[7] And yet, for several decades before the development of practical radiometric dating (at any scale), researchers attempted to bridge the temporal gap between human time and planetary time by using a third type of time—time as measured through the periodic advance and retreat of European glaciers during the Pleistocene era. Ice time bridged the gap between *geos* and *bios*—earth times and lifetimes. This was a revolution to match the one identified by Martin Rudwick at the beginning of the nineteenth century, which "burst the limits of time," although one conducted on scales that far outpaced the wildest estimates of early nineteenth-century religious or historical chronologies.[8] As revolutions go, it was a slow and painstaking one, gradually accreting over the decades in a manner that more than slightly resembled the creeping progression of glaciers over the European landscape in millennia past. But it was a revolution, nonetheless, that opened new horizons in the study of prehistoric cultural change and ultimately helped to completely reorient the conventional geography of human evolutionary origins.

Well before the advent of radiocarbon dating, prehistoric researchers were thinking about human prehistory on a global scale, hypothesizing the contemporaneous ages of different specimens from widely distributed locales and even occasionally and boldly assigning some preliminary absolute dates, following new theories that linked the ice ages to cyclical variations in the earth's orbit and exposure to solar radiation.[9] These were phenomena with a known periodicity that could be used to temporally bind the advance and retreat of Pleistocene ice and the human artifacts contained within the deposits the ice left behind. Considered in the geological record, the remnants of Pleistocene ice operated like an accelerated dimension of earth history, one that archived segments of (still-faster) deep human history within its strata.

Critically, these analyses of the Pleistocene ice were based on evidence specifically taken from sites in Europe and North America. Extending the evidence of the ice ages to other parts of the world (including even just other parts of the Northern Hemisphere) would prove controversial, despite the popular imagination of the ice age (and one shared by certain scientists) where the frozen landscape of the Arctic or Antarctic extended to cover the entire globe. This controversy was due to both larger debates over the causes and mechanisms of planetwide climate shifts and because of the challenging implications that these extensions presented for common understandings of the unilinear process of prehistoric cultural development and contemporary justifications of European racial and cultural superiority.

Glacial geology and human prehistory had a long period of scientific germination in the nineteenth century, but the critical period of glacial chronology and prehistoric correlation studies was bracketed by World War I and World War II. That geology and prehistory are both global and political should not be surprising.[10] However, this was also the period when Europe was being slowly but inexorably loosened from its position as a central nexus of world power and its place as the racial or cultural type-example of all human prehistory. In this chapter, I explore this moment of transition in temporal regimes by examining the critical and innovative folding together of classical stratigraphy with new research in climatology and meteorology in the interwar period. New ways of seeing the earth as a climatological system in both the present and the past in turn made visible—if partially, imperfectly, and frequently tendentiously—a new history of the global human past.

Glaciers and Time in the Nineteenth Century

Stratigraphic analysis—classification of rocks based on their position, context, type, and fossil contents—had been a defining feature of European geological study since the late eighteenth century, when Enlightened naturalists and state mining engineers set out into the mountains, ravines, and underworlds of their home territories to map mineral deposits and surface formations. But outside of the loosest estimates of rates of sedimentation and erosion, it was virtually impossible to tell how long ago a specific stratum was laid down, especially during the geologically recent Pleistocene. Another persistent element of the European Pleistocene, as previously noted, was that it was marked by evidence of movements of either ice or water, sometimes so forceful as to have moved enormous granite boulders a significant distance from their source outcroppings—although the extent, variety, causes, and implications of these movements remained a topic in high contention.

By the middle of the nineteenth century, Swiss naturalist Louis Agassiz's theory of a great ice age—which had covered the continent in a vast sheet of ice, dropping enormous boulders known as glacial erratics as it went—was increasingly accepted both in Europe and North America. Further exploration, however, led investigators to propose not just one but multiple ice ages that had successively advanced and retreated across the European landscape. Swiss geologist Adolphe von Morlot examined superpositioned glacial deposits in the Alps and preliminarily identified four distinct phases of glacial action—two advances and two retreats.[11] Morlot estimated that a modern delta near Lake Geneva containing Roman, Bronze, and Stone Age implements had accumulated over approximately ten thousand years. A more ancient delta, elevated forty-five meters above the lakeshore and ten times as large as the modern delta, was assumed to be ten times as old, or one hundred thousand years, perhaps dating to the beginning of the last period of glacial retreat. In his 1863 study *Geological Evidences of the Antiquity of Man*, Charles Lyell evaluated the glacial geology studies of Morlot and others with particular attention to their "chronological relation to the human period," noting age- and rate-estimates and the potential to correlate among the studied glaciers of Scandinavia, the Alps, and the British Isles. "But it must be confessed, that in the present state of knowledge," he wrote, "these attempts to compare the chronological relations . . . must be looked upon as very conjectural."[12]

Ice and time were linked together in glacial geology studies from the outset. But the connection was made explicit in the work of Scottish autodidact James Croll in his 1875 *Climate and Time in Their Geological Relations*.[13] As Alexis Rider discusses in chapter 9 of this volume, Croll theorized that the ice ages were caused by variations in the eccentricity of the earth and used the orbital calculations of French mathematician Urbain Le Verrier to identify the most recent eccentricity maxima to date and delimit the periods of glacial advance. He arrived at an estimate of more than eighty thousand years since the last glacial retreat—a figure that he felt accorded well with the calculation of William Thomson (later Lord Kelvin) that the total age of the earth was likely no more than one hundred million years old.[14] Croll's work was frequently cited in the last quarter of the nineteenth century, but his temporal calculations came under increasing scrutiny (as, indeed, did Thomson's) and were ultimately rejected for placing the conclusion of the last glacial period too far back in the past to be reconciled with new geological and archaeological evidence, particularly from North America.[15] Even if the chronology of ice, humankind, and the earth was relative and frequently speculative, eighty thousand years increasingly seemed to be far too remote a time to be meaningfully entertained.[16] Croll had, however, immediately understood the implications of the kinds of questions that could be answered—the age of different fossils, estimates of the rates of species' evolution—if only glacial time could be neatly parsed.[17]

Although glacial action was not unknown outside Europe—Agassiz, for one, had greatly encouraged the growth of glaciological study in North America, and extensive records exist of *pakeha* settler studies of the great glaciers on the South Island of Aotearoa New Zealand—the most detailed analyses were deeply tied to specific European locales. In his lucid and lively geological study *The Great Ice Age and Its Relation to the Antiquity of Man*, Scottish geologist James Geikie described his meticulous research into the glacial deposits and chronology of his native Scotland, before moving to summarize evidence from studies conducted on the European continent to fill out the rest of the story.[18] "Glacial Phenomena of Asia, Australia, etc., and South America" was covered in a single chapter late in his narrative, reflecting above all the paucity of detailed geological data from many parts of the world outside Europe. Yet the evidence from glacial remainders in North America and continental Asia did support the conclusion that these regions too had experienced a glacial epoch: Mount Ararat, in the Caucasus, and Mount Damavand, in what was then Persia, were both capped by snowfields and small glaciers, while further east the Himalaya gave "abundant and unmistakable evidence of a great extension of the glaciers at no very distant geological date."[19] Geikie, however, lacked detailed evidence of successive gravel deposits anywhere in Asia with which to compare the data gathered from geological and archaeological studies of central Europe, particularly the Swiss glacial lakes. In locations even more remote from the Northern Hemisphere, local observers Geikie cited doubted whether existing glaciers like those in New Zealand or at the summits of Mount Kenya or Mount Kilimanjaro in East Africa or leftover glacial moraines could be meaningfully temporally connected to the European evidence.

Between 1901 and 1909, the German and Austrian team of Albrecht Penck and Eduard Brückner published their monumental study *Die Alpen im Eiszeitalter*, or *The Alps in the Ice Age*—a product of twenty-five years of fieldwork looking at the glaciations in different parts of the Alps.[20] (The first volume was dedicated to none other than James Geikie, a longtime correspondent of Penck's, who had reached similar independent conclusions about the glacial history of Scotland.) In the first of three volumes, Penck laid out a general timeline for Pleistocene glacial history, identifying four series of glacial deposits, which were associated with four separate advances of the ice. He named them after type localities in the Alps, in alphabetical order from oldest to youngest: Günz, Mindel, Riss, and Würm. Penck also estimated—again, based on sedimentation and erosion rates—that roughly seven thousand years had elapsed since the end of the Würm, and perhaps twenty thousand years had elapsed since the Riss.[21] Penck and Brückner's work was comprehensive and detailed—an enormous boon to any researcher interested in tackling the ice-age problem.

Around 1914, another young researcher took up the question that had compelled Croll—the relationship between solar radiation variation and the ice ages—but did so with a significantly more sophisticated mathematical background. Serbian mathematician Milutin Milankovitch ultimately published his *Mathematical Theory of Thermal Phenomena Produced by Solar Radiation* in 1920, much of which had been composed while he was an Austrian prisoner of war between 1914 and 1918.[22] Milankovitch proposed that the climatic shifts between glacial and interglacial periods in the Pleistocene were due to shifts in the shape (or eccentricity) of the earth's orbit, the tilt angle of the earth's axis (obliquity), and its precession, or the degree of wobble in the axis of rotation. Milankovitch's work was useful not only as a proposed mechanism for long-term changes in the earth's climate, such as the historical ice ages, but also because this explanation supported the conclusion that these glaciations were widespread across the Northern Hemisphere and might have had planetwide effects observable in the geological record. Both Milankovitch and many of the geologists and prehistoric archaeologists who picked up his work were careful, however, to note that his theory did not explain the causes of the ice age, merely why these patterns of glacial-interglacial variation might exist. However, because these three variables were periodic and calculable, Milankovitch's work pointed to an absolute chronology for Pleistocene glacial time. This development would have critical implications not only for climatological studies but also for work in prehistoric archaeology.

Glacial Correlation and the Problem of Human Antiquity

For much of the second half of the nineteenth century, the study of past European climates and glaciology proceeded hand-in-glove with the study of European prehistory. As Charles Lyell noted in 1863, when looking for evidence of prehistoric human activity in Europe, researchers kept hitting a hard stratigraphic boundary in the form of "boulder clay," below which no artifacts seemed ever to be found.[23] This boulder clay, as the name suggests, was compressed, clay-heavy deposits studded with boulders that had been scraped up by the passage of glaciers and ice sheets over the landscape. Lyell observed this boundary did not seem to appear at the same temporal level all across Europe—in Denmark, boulder clay appeared in the strata right below the relics of the "recent" prehistoric periods of the Bronze and Iron Ages, whereas at lower latitudes, such as in England, it appeared the glaciers had retreated much sooner and correspondingly the boulder clay was overlain by deposits containing vastly more primitive stone tools.[24] In France, pioneering prehistorian Boucher des Perthes largely focused on tools found in gravel deposits—which were also determined to be of glacial origin—while prehistoric studies in Switzerland in the mid-nineteenth century studied ancient dwellings submerged in the shallows of glacier-fed lakes.

Part of the challenge in determining the age of these artifacts lay in the relative narrowness of the Pleistocene as a geological epoch, which belied attempts at breaking down the period into more usefully fine-grained periodizations. The Danish antiquarian Christian Jürgensen Thomsen had first proposed the division of prehistoric artifacts into the successive Stone, Bronze, and Iron Ages at the beginning of the nineteenth century.[25] Even at the time, he recognized that tools belonging to previous eras had continued to be produced and used in later periods, as well as the fact that types of tools continued to change and evolve within these typological categories—both facts that complicated any effort to use the presence or absence of the tools to do any more than the most rudimentary time-telling.

Like many more remote geological eras, the Pleistocene was commonly differentiated based on its characteristic fossils; unfortunately for the purposes of relative dating, tool cultures appeared to evolve far more rapidly than species did. The French paleontologist and cave explorer Eduard Lartet attempted to subdivide the Stone Age, or Paleolithic, period with reference to the major fossilized fauna he had discovered in deposits in France, designating (from oldest to most recent) the ages of the great cave bear, the mammoth, and the giant reindeer. Lartet's scheme failed, since these faunas often appeared in strata together, and the first two were also rarely discovered in the caves or glacial gravels that produced the majority of paleolithic tools.[26] The reindeer, by way of contrast, most often appeared in association with prehistoric human settlements—suggesting that they were kept by prehistoric humans as a source of antlers, bone, and food.

The French prehistorian Gabriel de Mortillet instead turned to classifying the types of stone tools themselves, rather than relying on the imperfect and inexact temporality of faunal associations alone and using a combination of stratigraphy and the cultural evolution of the form and sophistication of the lithic technologies to characterize successive prehistoric periods. Following the method of geologists, who named eras and characteristic deposits after their type locations, de Mortillet named the various Paleolithic cultures after the sites in France where they were first found—Chelles, Moustier, Solutré, La Madeleine—and then transformed them into adjectives for the sake of uniformity, giving us the successive Chellean, Mousterian, Solutrean, and Magdalenian periods, characterized by the distinctive tool types.[27] The Chellean was the simplest and seemingly the earliest, assuming as de Mortillet did that human culture followed a linear and unified pattern of cultural development, evolving along a single track.[28] But his judgment of the age of the Chellean was also due to its overwhelming abundance in glacial drift, the layer of gravel, clay, and other looser deposits that overlaid the compacted boulder clay one layer down.[29] This indelibly associated the Chellean—and thereby the earliest widely recognized prehistoric stone-tool culture—with a period immediately after the retreat of a glacier—although, if there had been multiple European glaciations, the trick now was to figure out which one.

The glacial association pinned the Chellean to a particular remote time, even if the precise antiquity of that period remained uncertain. But de Mortillet's assumption of a linear teleology of cultural development negated using the presence or absence of Chellean-type tools as the equivalent of a time-approximating index fossil anywhere outside of Europe. If all human cultures advanced along the same technological pathway, and if Europe was assumed to be the leader in human development, it was impossible to tell if a given Chellean artifact was indeed ancient (contemporary with the European Chellean) or if it was the first step undertaken by a less advanced human culture at a time much nearer to the present day.

In their study of Swiss glaciers, Penck and Brückner argued that de Mortillet's stages of paleolithic tool development were not only cultural but also temporal, and that their work demonstrated that Paleolithic man had been present not only after the most recent ice age (Würm) but also as a witness to the penultimate ice age, the Riss, with further tools found in deposits indexed to the Riss-Würm interglacial.[30] This question of a singular ice age or multiple ice ages, and whether all paleolithic finds were necessarily postglacial (a view held by many British researchers) or could be found between glacial deposit levels, as Penck and Brückner believed, was the leading question in European prehistory on the cusp of World War I.[31] It also opened up the obvious possibility that if an association could be established and a glaciation dated, then a culture could be dated too, at least in Europe.

Around the same time, two additional key points were raised that would play into the later studies of glacial prehistoric chronology. The first was the possibility raised by researchers such as British meteorologist C. E. P. Brooks that the past ice ages that were becoming well known in Europe might also be visible in the stratigraphy of other parts of the world, including in near-equatorial regions where the glaciations might be replaced by periods of heavy rain.[32] The second was raised by the eminent French prehistorian Henri Breuil, who argued in 1913 that Paleolithic classification schemes created by the previous generation were simplistic.[33] They proceeded from an assumption of autochthonous cultural development in Western Europe that could not be sustained in the face of evidence from prehistoric sites to the east, in central Europe and Russia, and to the south, in the Iberian peninsula and across the Mediterranean in North Africa, all of which suggested cultural and technical exchange between prehistoric Western Europe and other continents.[34] Breuil did not rule out the possibility of spontaneous evolutionary progress in one locale, but he hypothesized that several of the persistent questions of European paleolithic archaeology might only be resolved with comparative data from other regions.[35] Additionally, the increasing likelihood of the nonuniform complexity of prehistoric cultural development limited the utility of treating stone tools like index fossils to demonstrate the contemporaneity of different regional sites. But if the glacial-interglacial geological system could be reasonably extended through other parts of the world, it might be possible to prove approximate contemporaneity by other means.

Human Antiquity and the Interwar Chronologists

Through the 1920s and into the 1930s, geologists working across western and central Europe identified evermore precise gradations in the glacial stratigraphy of the Pleistocene. From Penck and Brückner's four divisions, they had further delineated two separate cold phases in each of the Günz, Mindel, and Riss glaciations, and three cold phases had been observed in the Würm, although the magnitude of glaciation Würm 3 appeared to be significantly less than Würm 1 or Würm 2.[36] Milankovitch's work and his curve describing the fluctuations in solar radiation provided a reasonable explanation for these variations—in fact, British meteorologist G. C. Simpson suggested in a review in 1945 that, in this case, the tail may have wagged the dog and that the field evidence might have been rather more convincing if it had been discovered before Milankovitch's radiation curves were first published, rather than after.[37]

Milankovitch's work did not explain every observed geological complexity, as the geologist and paleontologist Frederick Zeuner cautioned students at the University of London's Institute of Archaeology in 1935.[38] Nor did it fully explain why the Pleistocene ice ages had occurred in the first place. But it did generate concrete dates, by correlating solar radiation minima and maxima with the glacial and interglacial periods that had been so thoroughly worked

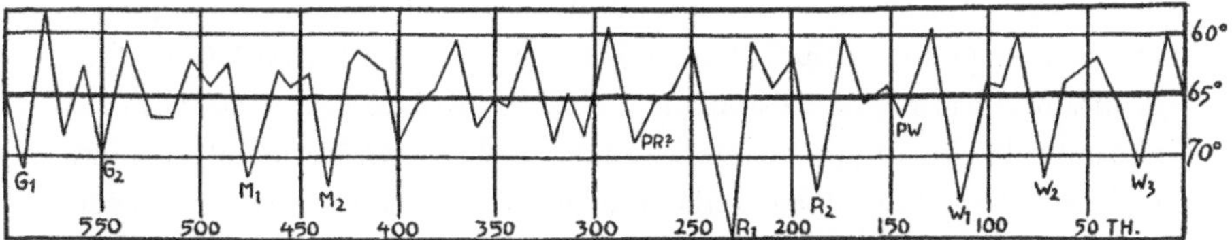

FIGURE 11.1. "Curve showing amplitudes of solar radiation at lat. 65°N during the summers of the past 600,000 years. After Milankovitch, 1930," in F. E. Zeuner, "The Pleistocene Chronology of Central Europe," *Geological Magazine* 72 (1935): 363. Courtesy of Cambridge University Press.

out in Europe. The three cold periods of the Würm glaciation, for example, could be neatly aligned with a trio of periodic solar radiation minima at 23,000 years (W3), 72,000 years (W2), and 115,000 years (W1) (fig. 11.1).[39] As Zeuner noted, "The linking up of our detailed stratigraphy with the curve of the solar radiation brought us an absolute chronology for the Pleistocene."[40] Correspondingly, this "absolute chronology permits more accurate dating of fossil man than has hitherto been possible, provided that the exact position of the remains within the detailed succession is known."[41]

Within the well-attested stratigraphy of European continental prehistory, the Milankovitch-informed chronology did not destabilize the existing sequences of stone-tool cultural development. However, solar radiation figures introduced a critical independent time scale into prehistoric dating calculations; no longer was it necessary, or indeed acceptable, to assume an unilinear evolutionary sequence of tool cultural developments and then use the presence or absence of specific kinds of tools to establish contemporaneity of deposits across European sites. "Types cannot at once serve to characterize cultures and to mark periods of time," wrote the Australian archaeologist V. Gordon Childe in his presidential address to the Prehistoric Society of Britain in 1935, where he repeatedly emphasized the criticality of a "time scale quite independent of the material" and the value of a timescale correlated with the patterns of the ice ages, even if they operated in unwieldy units that were difficult to index to the ordinary experience of human time.[42] An independent timescale allowed investigators to determine the direction of prehistoric migrations, or the transmission of cultural traditions between separate groups. It also might establish stronger historical relationships between prehistoric tool cultures and the remains of ancient European hominins like the Neanderthals, whose fossilized remains had been discovered with increasing frequency since the start of the twentieth century.[43]

World War I had disrupted the peacetime practices of European survey geology, as field sites became battle fields and as many geologists were drafted or enlisted either as ordinary soldiers or as consultants on the geological problems of prolonged and static trench warfare.[44] But the evolution of industrialized warfare had also made it clear to involved governments that it was critical to identify and control sources of oil and other mineral resources in the event of future conflicts. The result was an infusion of funding and personnel into existing national and colonial geological surveys, and the establishment of new surveys in overseas territories.[45] This also meant there were more geologists looking at new landscapes in close detail and more geologists who were deeply interested in correlation projects and in determining the ages of different geological formations to identify deposits of oil and minerals. But this economic work also provided opportunities and informal support for independent exploration that had little to no likelihood of contributing to the war economy of any nation or empire. Combined with the new insights in prehistory and ice-age geology emergent immediately before the war, this confluence of circumstances created an opening for the rise of a new and polycentric study of glacial geology and human prehistory. These studies extended far beyond the prehistoric bounds of the Alpine glaciers of central Europe, but they retained the language of the four Alpine glaciations while looking for evidence that the ice-age events had a global footprint.

In Uganda, the head of the newly created Geological Survey—and initially its only employee—was the British-trained geologist and casual prehistorian E. J. Wayland. Wayland was interested in the possible evidence of past glacial events in the Southern Hemisphere and the importance of gravel beds as a source of prehistoric stone implements, a field of research that first drew his attention while on assignment with a mining company in Ceylon before the start of the war.[46] Early in his tenure in East Africa, Wayland became interested in the stratigraphic evidence of alternating dry and rainy periods, or pluvials, which seemed to fall in the layers associated with Pleistocene fauna (fig. 11.2). These pluvial and interpluvial strata also frequently appeared to hold a succession of stone tools that increased in complexity and sophistication over time—and bore striking similarities to some of the earliest recognized prehistoric tool cultures in Europe.[47] Critical questions remained to be answered, however, about the precision and validity of these correlations—particularly given that Louis Leakey's East African archaeological expedition in neighboring Kenya claimed in the late 1920s that they had evidence of microliths in the deposits of the first pluvial, corresponding to the Günz-Mindel glacial periods—an impressively, or more likely improbably, early date.[48]

Wayland was cautious of overreading the evidence of "seductive" glacier-pluvial correlations, although he believed the evidence would probably hold up to scrutiny from geologists, meteorologists, and prehistorians in the long

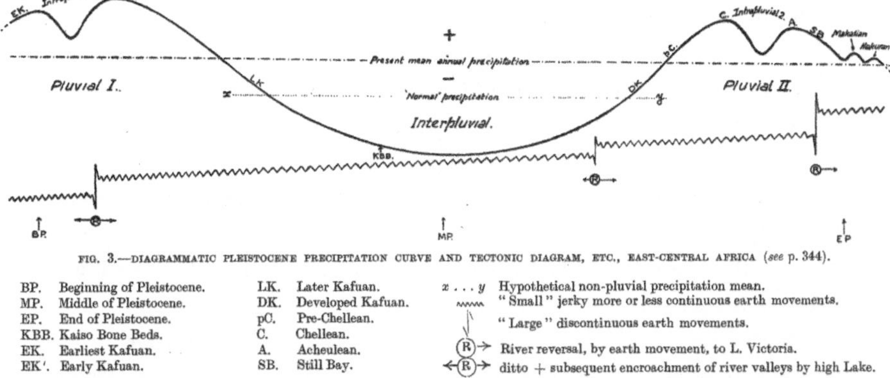

FIG. 3.—DIAGRAMMATIC PLEISTOCENE PRECIPITATION CURVE AND TECTONIC DIAGRAM, ETC., EAST-CENTRAL AFRICA (*see* p. 344).

BP. Beginning of Pleistocene.
MP. Middle of Pleistocene.
EP. End of Pleistocene.
KBB. Kaiso Bone Beds.
EK. Earliest Kafuan.
EK'. Early Kafuan.

LK. Later Kafuan.
DK. Developed Kafuan.
pC. Pre-Chellean.
C. Chellean.
A. Acheulean.
SB. Still Bay.

x . . . y Hypothetical non-pluvial precipitation mean.
"Small" jerky more or less continuous earth movements.
"Large" discontinuous earth movements.
(R)→ River reversal, by earth movement, to L. Victoria.
←(R)→ ditto + subsequent encroachment of river valleys by high Lake.

This curve has little numerical significance. We do not certainly know the mean annual precipitation at Pluvial maxima, nor whether it was practically nil at the Interpluvial maximum (although, if significant, it must have been extremely low). The points to which the curve should be taken in the intrapluvials is similarly uncertain. Only one date is palæontologically fixed: it is that of the topmost Kaiso bone horizon, which is early Middle Pleistocene. The beginning of Pluvial I is dated, with a high degree of probability on general geological grounds, as Late Pliocene. The better to accord with common usage, the Pleistocene is shown as ending with what has been regarded in the past as the end of Pluvial II, part 2; but the writer considers it probable that Pluvial II, part 2, is not yet complete, and that it is logical to regard our own time as Late Pleistocene. The line *x . . . y*, which is arbitrarily placed, indicates more or less the order of the mean annual precipitation to which (on the writer's supposition) the curve is, in an oscillatory fashion, descending.

FIGURE 11.2. Wayland's hypothesis of how rainfall changed in East Africa during the Pleistocene period. E. J. Wayland, "Pleistocene Precipitation Curve and Tectonic Diagram, etc., East-Central Africa," *Journal of the Royal Anthropological Institute* 64 (July–December 1934): 346. Courtesy of Royal Anthropological Institute / John Wiley & Sons.

term.[49] His understanding of the underlying climatology was built largely from the meteorological works of C. E. P. Brooks and G. C. Simpson's solar variation thesis, although Wayland may have later encountered Milankovitch's work for the first time in 1935.[50] However, by the end of the 1930s, Wayland was convinced that the evidence for the succession of pluvials in East Africa was sufficiently sound to allow him to advance another, even more controversial claim: that human antiquity in Africa was much greater than previously suspected and demonstrably greater than anywhere else in the world.[51] While wider scientific consensus on and subsequent public awareness of Africa as the human evolutionary "cradle" was a postwar phenomenon, Wayland's very early arguments make clear what was potentially at stake in these transcontinental glacial correlations: not merely a meteorological or geological solution but a total reevaluation of the evolutionary history of the human species.[52]

At the 1937 International Symposium on Early Man, held at the Academy of the Natural Sciences in Philadelphia, no fewer than three roundtables were organized to discuss the comparative chronologies of "early man" and Stone Age tool cultures in Europe and Africa (together), in Asia, and in North America.[53] A researcher from Minnesota, for example, reported that his state was experiencing a "regular deluge of skeletal discoveries" in Pleistocene glacial gravels owing to a massive spate of road building and resurfacing funded by the New Deal.[54] Determining the age and the appropriate glacial correlation for these remains was difficult, however, due to the flattened postglacial terrain of the region and to the inherent challenges of excavating at a roadside; at least one find was crushed by heavy machinery shortly after discovery.[55] Unfortunately, a question early in the session about the best method for correlating the historic ice sheets of Pleistocene North America with those of Europe was not discussed in detail because the audience (and the stenographers) appeared to have had difficulty understanding the geologist who tried to answer the question and because the session chair rejected the intrusion of European matters into his North American discussion.[56]

Nonetheless, throughout the symposium participants returned to the benchmarks of glacial chronology—gravel deposits, evidence of ice-sheet advance—while trying to sort out the geological, paleontological, botanical, zoological, and archaeological sequences of the Pleistocene period and understand what kinds of long-distance temporal comparisons were possible within this new frame. The symposium also represented a rare opportunity for direct conversation among researchers who were usually very widely distributed around the planet. Participants arrived from South Africa, Burma, Java, and China, as well as many parts of Europe and North America—although many of the Chinese researchers seem to have had their papers presented in absentia, probably because the Geological Survey of China was packing up their laboratories in Beijing ahead of the advancing Japanese forces.[57] Even as these researchers grappled with the history of the human species tens of thousands of years in the past, their capacity to conduct that research was strongly shaped by the material and political conditions of the present.

What did these correlation efforts mean for the study of human prehistory at the end of the 1930s? For researchers like Pei Wenzhong of the Geological Survey of China, they offered a means by which to arrive at a geological age for the critical hominin fossil specimens of *Sinanthropus pekinensis*, or Peking Man, unearthed in caves at Zhoukoudian, a quarry town not far outside Beijing.[58] In a detailed study bringing together the geology, paleontology, and prehistoric archaeology of the Quaternary Period (encompassing both the Pleistocene and the Holocene), he observed that the geological record of Europe showed four major cold periods that aligned with four distinct periods of alternating erosion and sedimentation in the physical geography of China.[59] To build out his correlation, Pei drew on paleontological evidence of European Pleistocene fauna species, comparing the presence or absence of index fossils with evidence from Chinese sites (fig. 11.3). He concluded *Sinanthropus* belonged to the Lower Pleistocene age, "in spite of the lithic industry which exhibits a few features which, in Europe, do not appear until

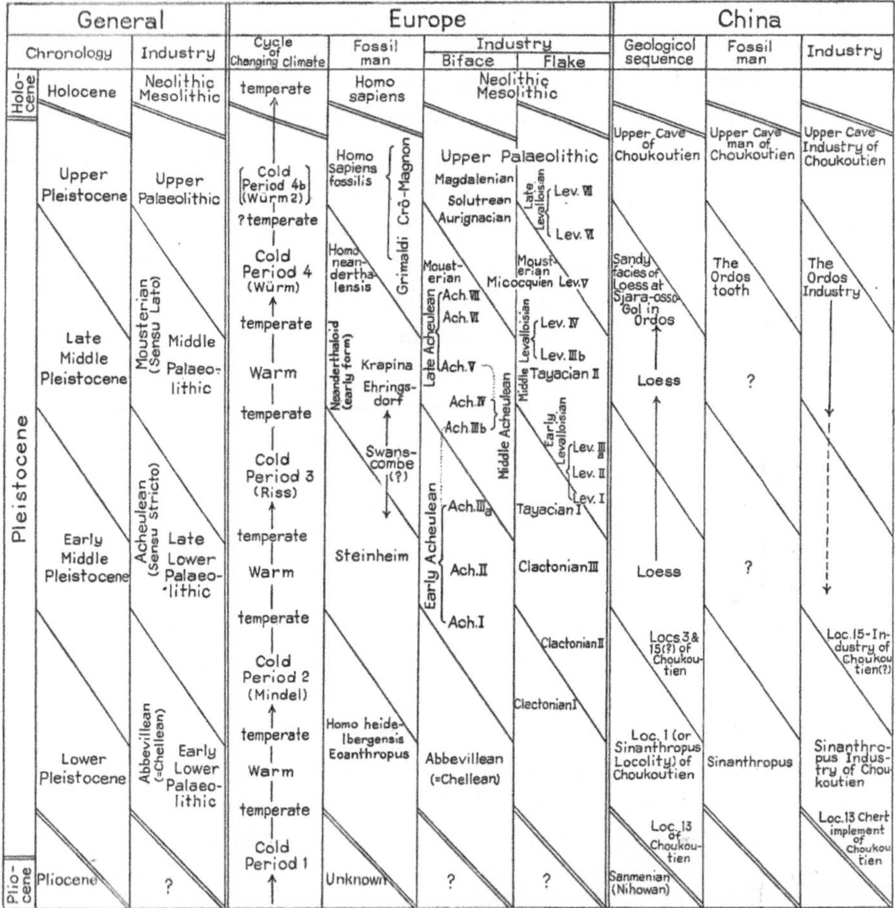

FIGURE 11.3. "An attempted correlation of geology, fossil man, and prehistory in the Quaternary periods of Europe and China," in Wen-Chung Pei (Pei Wenzhong), "An Attempted Correlation of Quaternary Geology, Palaeontology, and Prehistory in Europe and China," Institute of Archaeology Occasional Paper no. 2 (1939). Courtesy of the Institute of Archaeology, University College London.

later."[60] This made *Sinanthropus* contemporary with the very earliest horizons of European hominin fossils—*Homo heidelbergensis*, a robust hominin jawbone from Germany that was widely believed to represent the makers of the earliest Chellean-type stone tools—and the challenging *Eoanthropus*, also known as Piltdown Man, from England, which would later be revealed as a faked specimen.

However, the lithic industries at Zhoukoudian, later superseded by more advanced tool types made of stone and carved bone, suggested *Sinanthropus* might have been more technologically and culturally developed than its cousins on the other end of the Eurasian continent. Pei noted that "doubts have been raised by some European scientists as to whether *Sinanthropus* was the maker of this industry," or whether some other chronologically later and evolutionarily more advanced species (whose bones were not present in the deposit) might have made the tools and simply dropped them down a cave later in time. On this point, Pei noted that his colleague Franz Weidenreich had "carefully examined this question and come to the conclusion that these doubts are without foundation."[61] Similarly, the *Sinanthropus* sites showed the use of fire at consistently earlier periods than in their European equivalents, although Pei noted this was potentially due to the better preservation of the cave site.[62] Last, the Zhoukoudian materials also showed that there was simply a much wider range of prehistoric stone-tool types present at many stages in the Pleistocene in China that often did not map onto the known European sequences—although frequently those tools showed technological developments that generally appeared later in European deposits.

Pei's correlation approached glacial stratigraphy in two different ways. Predominantly, he relied on evidence from European studies of Pleistocene stratigraphy that had established a very detailed understanding of what fossilized fauna and what types of stone tools were characteristically found at specific levels and showed how the Chinese artifacts stacked up. But he also examined this material in relation to the four alternating periods of erosion and sedimentation that could be seen clearly in the stratigraphy of north central China. Part of the challenge that Pei faced, however, was that Pleistocene glacial action in China was a running controversy across the 1930s—one that, as Grace Shen has noted, broke down along largely national grounds.[63] In papers published in 1931, American geologist George Barbour and Canadian anatomist and paleontologist Davidson Black both took up the subject of paleoclimatology and the challenges of transcontinental correlation.[64] Barbour regarded Pleistocene China as having been relatively or almost entirely ice-free, making correlation hinge almost entirely on comparing index fossil species between East Asia and Western Europe; fortunately, increased surveying in the 1920s and 1930s had begun to fill some of those faunal gaps.[65] Both Barbour and Black also pointed to British meteorologist G. C. Simpson's 1928 work on solar fluctuation as a way to explain past climate shifts evident in the Chinese strata that aligned with the well-known glacial periods on other continents, but they did not suggest that glaciers might have covered China in the recent geological past.[66]

However, in 1934, Li Siguang (J. S. Lee) published a series of articles on Pleistocene glaciation in Lushan, in the Yangtze valley, pointing to scraped striations, boulder clay, and other geological signals that were difficult to interpret *except* as signs of prior glacial action.[67] Given the extensive gravel sheets found in surrounding regions, as well as the increasing evidence of widespread glacial action in many other parts of the world, Li concluded, "A general lowering of temperature in the Pleistocene time over the whole globe would thus seem to be more than probable."[68] In a reply, Barbour professed himself unconvinced by Li's data; if Lushan was a Pleistocene glaciation, it was a very strange one that did not fully conform to the evidence he would expect to see.[69] Li continued to conduct research both in Lushan and in neighboring regions, repeatedly returning to the point that there was very little reason to doubt that glaciations had occurred and that East Asia had experienced similar Pleistocene climatic swings like the rest of the Northern Hemisphere.[70] For Pei, writing in 1939, the most pragmatic and best-supported path forward seems to have been to emphasize the fossil faunal correlations first, the four-phase physiographic evidence second, and then to gesture at Li's glaciation work third—emphasizing that firmer comparisons would be possible pending future data.[71]

Getting that data, however, would be unavoidably delayed by the exigencies of world war in Europe, Asia, and even East Africa. In Uganda, Wayland remarked in a report, "Speaking selfishly, [the war] could hardly have come at a worse time for our archaeological work."[72] He continued, "Archaeology cannot expect to matter much in war time," but it was highly frustrating to have to pause field expeditions indefinitely in the middle of promising investigations into the Pleistocene climate and the prehistoric human world. Wayland, at least, ended on an optimistic note: "Here is the foundation, it is as sound as we, with our limitations, are able to render it; upon this, others may build."[73]

Conclusion

Glacial chronology brought together the climatological and geological history of the Pleistocene with the chronology of the human Paleolithic. But, even its most tenacious practitioners found themselves struggling to make their glacial geochronology practicable for addressing the problems they most wanted to solve. The resultant chronological divisions were shorter than geological epochs but still unwieldy, and many of the estimates of the age of glaciations turned out to be little better than guesses. Subsequent research established that the mechanisms of glacial advance and climatological shift are delicate and complicated, although Milankovitch's numbers have held up better than those of most of his contemporaries; today he is regarded as an early pioneer in the field of paleoclimatology, albeit with some modification to his original ideas.[74] In the case of Wayland in particular, many of the conclusions that he drew about the prehistoric climate of Uganda were later shown to be wrong—not the products of "pluviations" but instead caused by the tectonics of the nearby East African Rift.[75] And yet these efforts, imperfect as they were, helped create a modern and polycentric scientific consensus around a new way of seeing the earth in the past: as part of a climatological system that encompassed the planet and that made visible, in turn, a naturalistic globe-spanning human prehistoric chronology.

Geology is about time travel, a way to reach back into the planet's distant and mostly prehuman past. But for the most part, that time has been traveled only in relative increments rather than in absolute spans of years. To understand and speak about the geological past with greater precision, practitioners of the geo- and paleosciences at the beginning of the twentieth century turned to physical sciences, like astronomy and atomic physics, and in so doing, began to once again set boundaries on time. By folding human cultural time (in the form of paleolithic industries) into geological natural time, these researchers also opened the door to rejecting teleological racial hierarchies and singly determined cultural progression—writing a new story of humanity's deep history in the process.

12

Exchanging Fire

A Planetary History of the Explosion

Nigel Clark

"Earth-Shaking Invention"

"If Experience did not both Inform and Certify us, Who would believe, that a light black Powder should be able, being duly manag'd, to throw down Stone-Walls, and blow up whole Castles and Rocks themselves," pondered natural philosopher Robert Boyle late in the seventeenth century.[1] What shocked and intrigued Boyle, historian Haileigh Robertson notes, was not just the brute force of exploding gunpowder but the sense-defying rate of the reaction that took place.[2] Upon ignition, Boyle calculated, gunpowder expanded in an instant to some fifty thousand times the original size of its grains.[3] While Boyle was seeking to unravel the mysteries of the volatile compound, the explosive power of gunpowder continued to wreak destruction. Death or grievous injury delivered at imperceptible speeds by incomprehensible forces was becoming part of European life.

While speeds that defy the human senses have attracted much attention in the context of digital media, social thinkers engaging with environmental issues tend to reflect more on the challenges of slow and persistent change. Philosopher Isabelle Stengers urges her readers to embrace the painstaking working up of underrecognized problems rather than latching onto those blatantly disastrous events that already "have the power to force unanimous recognition."[4] Likewise, literary studies scholar Rob Nixon would prefer us to be moved more by the "slow violence" of long-term environmental degradation and toxicity than by the eye-catching spectacle of "falling bodies, burning towers, exploding heads, avalanches, volcanoes, and tsunamis."[5]

For social thinkers who are drawn to transformations at the planetary scale—the impacts on Earth Systems and lithic strata announced by the Anthropocene hypothesis—the inclination to attend to deeply protracted processes can also be strong. Such concerns are often framed by acknowledgment that our home disciplines have insufficiently prepared us for the extremely *longue durée*. I suspect I'm not alone in feeling the pull and poignancy of paleontologist Stephen Jay Gould's reflection that "deep time is so difficult to comprehend, so outside our ordinary experience, that it remains a major stumbling block to our understanding."[6]

Yet developments in the earth and life sciences over the last half century have also made it clear that there's more to earth history than immensely drawn-out timescales and durations: Gould's own theory of punctuated equilibrium contributed significantly to the idea that biological evolution combines gradual and rapid change.[7] With its focus on thresholds or tipping points in the operating state of planetary systems, Earth System science inherits and amplifies this concern with multiple tempos of transformation. In a related sense, the task of identifying synchronous and planetwide impacts of human activity in the earth's rocky crust—required for the formalization of the Anthropocene hypothesis—is drawing geologists into engagement with changes far more rapid than most have previously reckoned with. While it is still open to revision, the Anthropocene Working Group's preferred candidate for a signal that marks the end of the Holocene is radionuclide fallout from post–World War II testing of thermonuclear warheads.[8] Which is to say that a panel composed mostly of researchers who are "overwhelmingly concerned with ancient, pre-human rock and time" may well pivot its case for a new geological epoch around the consequences of events that are over in microseconds.[9]

There is a backstory to the interest of Anthropocene exponents in big explosions. Atmospheric chemist Paul Crutzen, who named and helped frame the Anthropocene concept, was one of the progenitors of the nuclear-winter hypothesis. A nuclear war, he predicted, would result in massive wildfires generating photochemical smog that could "change the heat and radiative balance and dynamics of the earth and atmosphere" with devastating impact on surviving humans.[10] This scenario was an important precursor of the idea that human action could not only impact on the overall Earth System but could do so abruptly. It's also worth recalling that much of the scientific evidence leading to the confirmation of the plate-tectonic hypothesis in the early 1960s came from seismographic stations set up to detect Cold War nuclear explosions. At the same time, tracking radioactive debris from nuclear weapons testing as it moved through the atmosphere, hydrosphere, and biosphere made a major contribution to understanding the interconnectivity of the Earth System.[11]

While a radionuclide marker may fulfill the criteria of a clear geosynchronous signal in nascent geological strata, Anthropocene scientists need to be careful about the way this evidence is framed and presented. Activists and their academic allies will likely ask what the mobilization of radioactive traces to authenticate an epochal scientific claim means for place-based communities who have suffered the lasting ecological, physiological, and psychic effects of superpower military-industrial competition—what its implications are for the Pacific Islanders, Native Americans, Aboriginal Australians, Kazaks, and others upon and above whose unceded customary lands atomic weapons were detonated. We can anticipate such questions because social thinkers and critical practitioners have already been taking the science of the Anthropocene to task for both inadequately considering its own situatedness and partiality, and for failing to give enough credence to a multiplicity of other ways of experiencing or knowing the earth.[12] In short, while social critics may affirm the need for disclosing the physical violation of planetary processes, they have called out Anthropocene science for its own implication in the epistemic violence that has long characterized encounters between the West and the wider world.[13]

In this chapter, I take the event of the explosion as an occasion for engaging at once with physical and epistemic violence. But I want to put a twist on this by focusing not on the brutal era of European colonial conquest, or the subsequent planet-threatening superpower rivalry, but on an earlier set of life-threatening encounters that are a condition of possibility of the heavily weaponized global histories that followed. There is no A-bomb without "conventional" explosive weapons, and there are no conventional missiles, bombs, and guns without gunpowder. And this is a line of development that takes us far from Europe.

Historians trace the discovery of a chemical mix that ignites rapidly enough to create an explosive release of energy to ninth-century China.[14] Not only did the Chinese concoct gunpowder they also—decisively and extensively—realized its potential as a weapon (fig. 12.1). When historian Joseph Needham opens his magisterial forty-three-year study of explosive technics with reference to the "earth-shaking invention of gunpowder,"[15] he is well aware of the work of Boyle and compatriot researchers—and of the tumultuous impact of explosive weaponry on the battlefields of Europe. His primary concern, however, is with the invention and deployment of gunpowder in China.

My aim here is to take the "earth-shaking" power of gunpowder quite literally—as Needham seems to have intended. In conceiving of the explosive force of gunpowder as a geological or planetary event, I draw on certain insights of the contemporary earth and life sciences. At the same time, by following weaponized gunpowder from East to West and tracking its experiential and subjective impacts on a pyrotechnically ingenuous Europe, I want

FIGURE 12.1. *The Great Victory of Qurman* (1760), fragment of painting by a collaboration of European Jesuit and Chinese artists that depicts a formation of soldiers with muskets and camels carrying artillery. Courtesy of the Picture Art Collection / Alamy.

partially to unsettle the epistemic confidence of the modern West that still underpins today's big stories about the how the earth works and what to do about it.

My case for the geological eventfulness of gunpowder emerges out of what I take as a central provocation of Anthropocene geoscience for social thought: the incitement to think with and through a dynamic, self-differentiating earth and to acknowledge that our species or genus is capable of bringing new kinds of functionality to the operation of this planet.[16] More specifically, I pick up on writer Jack Kelly's observation that the split-second combustion of gunpowder is an entirely new kind of fire: one that "does not exist anywhere in nature."[17] Attending to gunpowder not simply as a significant juncture in military or technological history but as an event in planetary history helps us to grasp its potential for disrupting earth and life processes. More than this, an appreciation of the utter novelty of explosive fire alerts us to the challenges it poses to the human sensorium and to the cultural or cosmological orderings through which subjects collectively seek to make sense of their worlds. And it is in this regard, I argue—with the help of military and pyrotechnic historians—that the unexpected arrival of gunpowder weaponry and its compacted development in Europe relative to China had profound repercussions on European subjectivity and epistemic formations.

I will not be making a case that the material traces of exploding gunpowder in either its military or civil applications would make a good marker for the onset of the Anthropocene. But I do want to suggest that thinking through the movement of a novel anthropogenic fire from East to West complicates the tenaciously Western-centered narrative of most Anthropocene science—while also potentially contributing to its laudable concern with planetary dynamism and change. In a related sense, I argue, facing up to the shock and trauma arising, especially but not uniquely, from the abbreviated experience of gunpowder in Europe offers insights into the ways Europeans later amassed and unleashed firepower. Finally, circling back on questions of unfolding climatic and Earth System change, I suggest that the fraught experience of learning to function in proximity to the force of the explosion might offer clues about the apparent willingness of so many people to tolerate the threat of runaway planetary heating.

Genealogy of the Explosion

Philosopher Yuk Hui's recent study of the relationship between cosmology and technology hinges around China's humiliating defeat by the British in the mid-nineteenth-century Opium Wars.[18] Hui depicts the Chinese experience of being outgunned by the European power: "the two Opium Wars in the mid-nineteenth century had destroyed the civilisation's self-confidence, and thrown it into a whirlpool of confusion and doubt."[19] The extended period that followed became known in China as the century of humiliation. More immediately, Hui recounts, some Chinese intellectuals responded to the rout of the Opium Wars with a slogan of "learning from the West to overcome the West."[20]

In an important sense, however, the West's moment of military triumph was itself a learning from the East that eventually enabled a partial "overcoming" of the East. As Hui notes in passing and other scholars have examined in detail, gunpower was a Chinese invention that calamitously returned home, intensified and augmented, some half a millennium after it had found its way to Europe. While hardly cause to belittle a civilization, the idea that the Chinese squandered their concoction of volatile black powder on fireworks—widely circulated in the nineteenth and twentieth centuries—has been roundly dismissed as a myth that served to bolster Europe's sense of its own technological bravado. "Early modern warfare," counters military historian Peter Lorge, "was invented in China during the twelfth and thirteenth centuries."[21]

The conventional Western narrative has it that ninth-century Taoist alchemists stumbled across the volatile mixture while seeking elixirs of eternal life. However, neither the serendipitous nor the supernatural aspects of this storyline should go unquestioned.[22] What we do know is that by combining charcoal, sulfur, and nitrates in the right proportion and exposing the mix to flame, researchers concocted a fire that burned with extreme speed. The pyrotechnic compound came to be known as *huo yao*, or fire drug, suggesting that medicinal uses were at least in contention.[23] But Chinese military engineers, already masters of flaming arrows and other incendiary weapons, were quick to apply the exceptional flammability of the black powder to the demands of warfare.[24]

The escalation and differential development of explosive weaponry is a staple of military history. Historians and social theorists, however, have devoted less effort to situating gunpowder within a more general history of fire or combustion—perhaps reflecting a relative paucity of an integrative concern with fire in Western thought. "Fire" is the vernacular term for the chain reaction, triggered by an ignition source, through which chemical energy is converted into thermal energy in an oxygen-saturated environment;[25] "deflagration" is the technical term for the heat-releasing or exothermic process whereby heat produced ignites still more fuel.[26]

In the case of gunpowder, sulfur and charcoal provide the volatile compound with fuel, producing heat that causes the nitrates found in saltpeter to let loose their oxygen atoms. In the right proportions, this release of pure oxygen accelerates the conversion of fuel into hot gas in a few thousandths of a second, resulting in deflagration of such rapidity that it has no natural equivalent.[27] When this extremely high-speed exothermic reaction takes place in a confined space, the result is sudden, vigorous release of energy, or an "explosion." Explosions routinely occur in the natural world, such as when volatile plant oils ignite during wildfires or when volcanoes build up uncontainable pressure, but like the split-second combustion that drives it, exploding gunpowder has no earthly predecessor. Indeed, with the help of environmental historian Stephen Pyne, we might think of near-instantaneous combustion and the rapid release of energy it entails as the first entirely new form of fire since terrestrial biomass began to burn during the Devonian Period some four hundred million years ago.[28]

Having first exploited its incendiary properties, Chinese military engineers began to explore both the propellant and explosive capabilities of gunpowder, or *huo yao*. But, as Lorge recounts, progress was initially slow on both practical and conceptual levels. "The intellectual component is important," he stresses, "because it required the acceptance of a completely new idea in weaponry: the explosion."[29] Once researchers began to embrace the multiple possibilities of split-second combustion, an extraordinary array of military uses for the pyrotechnic mixture was trialed—including gunpowder-delivering birds and kites, flaming rockets, exploding pots, and flame-spouting lances. The names given to these weapons—such as "flying incendiary club for subjugating demons," "ten-thousand fire flying sand magic bomb," and "burning heaven fierce fire unstoppable bomb"—convey at once the exuberance of this experimental wave and the shock effect the armaments were intended to produce.[30]

Over the turbulent centuries of the Song dynasty, this profusion of weaponry was narrowed down to what we would now recognize as guns, bombs, grenades, and rockets.[31] As historian Tonio Andrade sums up, "in the hundred years from 1127 to 1279, the second part of the Song dynasty, human beings went from primitive gunpowder weapons like gunpowder arrows to a whole array of more sophisticated weapons, including fire lances, proto guns, and, by the end of the period, true guns."[32]

As explosive weaponry was refined and standardized, it began to travel beyond its original site. How guns and gunpowder reached Europe and the Islamic empires remains uncertain, though it seems likely that the expansion of Mongol rule over Eurasia played a part.[33] Preceded by reports or hearsay of explosives, firearms seem to have arrived in Europe in the early 1300s. In the words of Needham: "all the long preparations and tentative experiments were made in China, and everything came to Islam and the West fully fledged, whether it was the fire-lance or the explosive bomb, the rocket or the metal-barrel hand-gun and bombard."[34]

Informed by Pyne's pyrocentric thinking, I am proposing that we view this event not simply as a case of accelerated technology transfer but as the arrival of a novel kind of combustion: a fire not simply strange to Europeans but relatively new to the planet. Such a reading suggests the value of reversing Hui's concern with the shockwave induced by European military technology in China in the nineteenth century—by inquiring into the impact of explosive firepower from China on a pyrotechnically naive Europe. Or as Needham sagely concludes his study, in the early fourteenth century "the Western world was set upon the fateful road to all the techniques of managing explosions."[35]

Explosive Exposure

It took time for the pyrotechnic powder arriving in Europe to fully ignite. Tracking the response of late medieval Europeans to the explosive mixture, historian Kelly DeVries suggests it was first apprehended as "a mysterious substance which imitated God's power."[36] As its use spread, scholars sought to integrate gunpowder more fully into the prevailing episteme—that encompassing field in which earthly processes mirrored heavenly ones and the microcosm reflected the macrocosm.[37] In the resultant accommodation, literary theorist Roy Wolper elaborates, "thunder and lightning are God's presence in the sky; gunpowder is God's presence on earth"[38]—a reception aided by the uptake of the fiery explosion into the popular spectacle of firework displays.[39] But, as Wolper adds, gunpowder's glaring destructiveness grated against easy assimilation. By the early seventeenth century, natural philosopher Francis Bacon was actively recontextualizing the explosive powder. Transplanted from battlefield to laboratory, gunpowder came to figure more generally for the hidden potentials of nature that waited to be unbound by Bacon's experimental method.[40]

When Bacon's major works were published, gunpowder had been working its way into European warfare for almost three hundred years, and of late had nearly seen off the British Parliament. Needham relays what he sees as a rather timeworn narrative that gunpowder weaponry's capacity to breech castle walls and unseat knightly cavalry played a significant role in the undermining of European military aristocratic feudalism. He is clearly more interested in reminding us "how unstable Western medieval society was in comparison with that of China":[41] a point later military historians have reinforced.

Resonating with Needham, Lorge argues that while the weaponization of *huo yao* in China took place in the context of an extensive, centralized bureaucratic system that was already over a thousand years old, the late medieval Europe into which gunpowder found its way centuries later was a fractious throng of principalities and kingdoms. His more general point is that weapons alone, wherever they are deployed, do not transform a society or polity: what matters most are the forms of social organization through which military technologies are adopted, developed, and deployed.[42]

> The wealthy, mature, and stable societies of Asia, though subject to political developments and upheaval, gradually incorporated the new weaponry without great social change. Matters were much different in the poor, undeveloped, unstable societies of Europe. There the introduction of new weaponry coincided with a dramatic period of demographic and economic growth, and political consolidation.[43]

But this comparison leads Lorge to at least partially revise his own prioritization of sociopolitical variables. For he goes on to suggest that in the European case the demands of gunpowder warfare pushed polities—unknowingly—in the direction of the bureaucratized institutions and logistic networks characteristic of China. "It took the invention of systems and practices similar to those of China," he concludes, "before European governments and armies began most fully to exploit the use of guns in war."[44]

What happens, then, if we take this idea of a certain convergence between emergent European governance systems and their much older Chinese predecessors and layer in the idea that the trigger event is a new form of terrestrial fire? Lorge is right to draw attention to the conceptual demands posed by the explosion as "a completely new idea in weaponry," but the challenge is still more profound if we conceive of the coming of ultra-fast deflagration as an event in *planetary* history: as a novel elemental force with which certain human subjects have had to learn to coexist. And it is in this regard, I argue, that the greatly accelerated European encounter with split-second deflagration relative to the Chinese experience matters.

Historians concerned with the transitions accompanying the full embrace of firearms by European armies have taken the Thirty Years' War (1618–1648) as a turning point, with particular emphasis placed on the introduction of volley fire by linear rows of infantryman.[45] Both the growth in size of standing armies and the intensification of gunfire added firearms' heft to a catastrophe that took the lives of some eight million combatants and civilians in central Europe, including a third of Germany's population. Philosopher Stephen Toulmin is far from alone in viewing the Thirty Years' War as one of the most brutal and unremitting conflicts in European history.[46] Toulmin proposes that the shockwave unleashed by this conflict was pivotal in a shift in European thought from the open-minded, inclusive, and frequently sensuous humanism typical of the sixteenth century to the more abstract, defensive, and constricted thinking of the seventeenth century. René Descartes's dogmatic quest for certainty, he maintains, is emblematic of this narrowing of reason.[47]

As Toulmin recounts, the Thirty Years' War broke out when Descartes was in his early twenties and ended two years before his death.[48] Curious about emergent military techniques and with an interest of his own in calculating the trajectory of moving bodies, Descartes joined Maurice of Nassau's army at the opening of the war.[49] He was subsequently present at the Battle of White Mountain, where four thousand Protestant troops were slaughtered in an hour or so. It was a month before this engagement, on the night of November 10, 1619, that Descartes experienced the sequence of dreams that he viewed as the inspiration for his celebrated scientific method.[50] As a contemporary biographer wrote, "He thought he heard a sudden, loud noise, which he took for thunder. Terrified, he immediately woke. Upon opening his eyes he noticed sparks of fire scattered about the room."[51]

Unsurprisingly, psychoanalytic thinkers have detected signs of post-traumatic stress in Descartes's visions and connected his renowned mind-body dissociation to its wartime context. Noting the young serviceman-savant's likely proximity to the discharge of explosive weapons, analyst Robert Withers adds that, being trained in medicine, Descartes was also likely to have encountered the physiological damage inflicted by cannon and musket.[52] What these considerations of Descartes's personal experience bring to Toulmin's more general account of repercussions of the Thirty Years' War is the reminder that by the seventeenth century physical and psychic exposure to explosive weaponry had become part of everyday life for many Europeans.

Compared with Descartes's subsequent drive for an unwavering, self-grounded cogito, Bacon's championing of hands-on, experientially led inquiry can come across as relatively modest[53]—though we shouldn't overlook his intimation that nature's truth is revealed through violence. Science studies scholar Donna Haraway's depiction of the Cartesian "god-trick" of scientific objectivity as a flight from the complications of embodiment, situatedness, and responsibility to "a realm above the fray" may be even more apposite than intended.[54] For fray it most certainly was—the relentless slaughter of the Thirty Years' War encapsulating all the horrors that a fleshy, impressionable observer might wish to flee from. More than just exposing the human body and senses to a new kind of threat, we might see the weaponized explosion at this juncture as beginning to shape an entire milieu of shocks and forces.

"The cannon," observes cultural historian Lewis Mumford, "was the first of the modern space-annihilating devices, by means of which man was enabled to express himself at a distance."[55] It's important to keep in mind that, whereas well-trained archers singled out targets, as late as the eighteenth century, historian Priya Satia insists, "firearms were for terrorizing at a distance with *unpredictable* fire."[56] While much has been said about the role of breaking down battlefield operations into discrete, rehearsable gestures in the shaping of the modern subject,[57] the paradox of this disciplining process was that it centered on training combatants to function in an environment configured by largely random death or injury—including that inflicted by the malfunction of one's own weapon.[58] This is an experience novelist Andrew Miller imparts through the voice of an early nineteenth-century British soldier, a brutal and brutalized infantryman who knows "what it was to stand in line while the enemy guns swept away the men on either side of you, made them non-men, butcher's trash."[59]

When they did not kill outright, as "blooded" soldiers knew all too well, gunpowder weapons produced horrific injuries: "wounds that could not be stitched up neatly like blade or arrow wounds."[60] The damage was not all visible. Well before formal psychiatric recognition of post-traumatic stress disorder or the "war neurosis" diagnosed by Sigmund Freud after World War I, psychic disturbance resulting from fire-armed conflict was well documented.[61] In *The Wealth of Nations* (1776), Adam Smith conveyed a sense of shock and derangement that overflows the field of conflict when he wrote of the "noise of fire-arms, the smoke, and the invisible death to which every man feels himself every moment exposed . . . a long time before the battle can be well said to be engaged."[62]

Nothing I have been reporting here is meant to imply that Europeans exposed to explosive weaponry were uniquely sensitive to its effects. Warfare in the East was frequent, brutal, and large scale, and as Lorge concludes of China and neighboring powers, "Asians were just as eager to kill each other with the most effective available weapons as were Europeans."[63] Europe's nineteenth-century superiority over China in firepower, as we will shortly see, had a lot to do with timing. But the different ways that Europe and China absorbed and processed the experience of gunpowder warfare, I suggest, also has much to do with broader and deeper "civilizational" framings of fire. And thinking through fire in this way neither starts nor ends with gunpowder.

The Field of Fire

I have been broadly following Lorge's argument that "without the Chinese revolution in warfare there could not have been a European revolution."[64] While Europeans and the Chinese themselves have made much of Western military supremacy in light of the disastrous encounters of the mid-nineteenth century, Lorge maintains that the technological gap at the time was still relatively slight and that the success of British aggression owed more to the political disorganization of the Qing government.[65] Where military disparities were most pronounced in the skirmishes of the first Opium War, some historians have argued, was in naval power. And much of this came down to the presence of the steam-powered gunboat *Nemesis*—whose shallow draft, maneuverability, and general efficacy as a "workhorse" looks to have been as significant as the armaments it carried.[66]

This connection of gunpowder-based weapons with steam power is of great significance, not just for military or even social history but also for the still-unfolding history of the earth. Just as Lorge and others have stressed the importance of situating weapons within their socioinstitutional contexts, so too do we need to consider the social and civilizational framing of this novel planetary fire. This means opening the complex issue of how Chinese researchers discovered a new mode of combustion in the first place. And, in a roundabout way, it brings us to the issue of how Europeans came up with new ways of deploying an older kind of combustion.

As Needham stresses, incendiary devices and mixtures played a major role in warfare in China centuries before the invention of gunpowder.[67] This in turn, he explains, was part of a more general concern with physicochemical experimentation in which fire was a key element. *Huo yao* was one of many discoveries arising from a spree of pyrotechnical exploration that spanned at least six centuries, a research tradition in which there was no pronounced separation between medical, alchemical, military, and ceremonial applications of incendiary discoveries—as the term "fire drug" indicates.[68] "Smokes, perfumes, hallucinogens, incendiaries, flames, and ultimately the use of the propellant force of gunpowder itself," concludes Needham, "form part of one consistent tendency discernible throughout Chinese culture from the earliest times."[69]

We should also consider China's early lead in high-heat technology. By 1500 BCE, observes historian Jack Goody, Chinese artisans were attaining kiln temperatures well over 1200°C. These unprecedented heat levels enabled manufacture of glazed stoneware and the casting of bronze and iron.[70] An entire industrial complex took shape around these pyrotechnologies, Goody observes, characterized by "a large scale, labour-intensive chain of production, with ore-miners, fuel gatherers, ceramacists and foundry workers."[71] Only much later, he adds, did the high-heat methods pioneered in China move westward. This millennium-and-a-half of developing high-heat practice helped equip the Chinese with both the technical capabilities for channeling the force of explosive deflagration and the institutional foundations for the extensive capitalization of these capacities.[72]

If only in a preliminary way, positioning gunpowder and the experiments from which it emerged in the *longue durée* of Chinese pyrotechnic exploration helps us to see why the shockwaves of explosive warfare, alarming even in China, were still more difficult for Europeans to absorb. It's worth recalling that despite recurrent hostilities in Europe, development of gunpowder weaponry was neither rapid nor inexorable—Kelly describes "the curious stasis that gripped military technology from the end of the Thirty Years' War in 1648 until well into the 1800s."[73] What may ultimately have been more consequential over this time were the more "existential" impacts I addressed in the previous section: the constricting of reason, the dissociation of mind and body, the hardening—we might even say brutalizing—of exposed subjects.[74] For this is a matter of global significance. Capacities, dispositions, and sensibilities that were honed and exercised in European domestic conflicts were also instrumental in Western colonial expansion.

As I touched upon earlier, describing the movement of projectiles played a significant role in the development of Descartes's scientific method: his interest in moving bodies falling somewhere between Galileo's insights on the parabolic curve traced by cannonballs and Newton's extrapolation from projectiles to planetary motion. In this regard, Kelly argues, concern with the accuracy of artillery played a significant role in establishing the Western scientific premise that the object world followed predictable pathways—although it took several centuries to markedly improve battlefield targeting.[75] In relation to fire more generally, what we need to recognize here is the distance thought and practice is traveling from the idea that fire is a versatile force that transmutes matter to the association of explosive weaponry with a single, determinable trajectory.

There is still a tendency, prevalent in Anthropocene scholarship, to put the stress on the massive expansion of power and capability that results from combusting fuels to drive heat engines. What the shift from a more metamorphic conception of fire to viewing explosive fire as the driving force of linear motion brings into relief, however, is the profound narrowing or constriction that is taking place in the imaginary of fire.[76] In the West, this momentous reduction reaches its fulfillment in the thermodynamic thinking that is closely associated with the ascent of new industrial heat engines. As physicist Ilya Prigogine and Isabelle Stengers observe:

> Fire transforms matter; fire leads to chemical reactions, to processes such as melting and evaporation. Fire makes fuel burn and release heat. Out of all this common knowledge, nineteenth century science concentrated on the single fact that combustion produces heat and that heat may lead to an increase in volume; as a result combustion produces work.[77]

Yet historians have also noted more direct links between engines that use fire to do "work" in this restricted sense with the European experience of exploding gunpowder. As Mumford observed in the 1930s, "the gun was the starting point of a new type of machine: it was, mechanically speaking, a one-cylinder internal combustion engine."[78] Needham fills out this storyline—tracking a history of projects aimed at applying gunpowder to useful weight-lifting or piston-driving tasks that goes back at least as far as the early sixteenth century.[79] Scientist-inventor Christiaan Huygens's project with the French Academy of Sciences in the 1670s is pivotal. Huygens wrote:

> The force of cannon powder has served hitherto only for very violent effects . . . and although people have long hoped that one could moderate this great speed and impetuosity to apply it to other uses, no one, so far as I know, has succeeded.[80]

It was Denis Papin and Gottfried Wilhelm Leibniz, initially working under Huygens on the *moteur à explosion*, who together recognized that steam power offered a safer route to creating the vacuum that could drive a piston.[81] Papin set research and development on a path toward external combustion—the use of fire-heated boilers to drive engines. Though not with gunpowder as its motive force, the internally combusting *moteur à explosion* would be momentously revived some two centuries later as the driving force of the automobile. But prior to the return of internal combustion, the external combustion of the steam-powered heat engine joined forces with gunpowder to devastating effect.

As military historian Martin van Creveld explains, "the invention of the steam engine . . . freed weapons from the limited power provided by horses, enabling their size, weight, and power to grow many times over."[82] Or, in the case of a gunboat like *Nemesis*, it greatly increased speed, work rate, and maneuverability. At the same time, steam power combined with explosive deflagration to dramatically accelerate mineral-energetic extraction. Steam engines pumped water from mines and transported extracted materials, and explosions opened rock faces and facilitated large-scale infrastructural development.[83] In turn, this compounding of two distinct forms of combustion amplified the ability of Western interests to forcibly gain access to the land and resources of other peoples across the globe. But we should also be mindful that the readiness to unleash this convergent and concentrated firepower on other life-worlds may itself manifest the tempering, brutalizing effect of the European subject's intensive exposure to the explosive inferno.

Anthropocene Echoes of the Explosion

As climate activist Greta Thunberg famously exhorted world leaders, "I want you to panic . . . I want you to act as if our house is on fire."[84] Timely words. But we might ask whether the apparent willingness of certain sectors of the global population to dwell in the shadow of a rapidly heating planet bears some relationship to several centuries of painful accommodation to an explosive milieu—to the risk of having their "house" not merely burned down but blown apart.

It's worth recalling that the historical juncture that Anthropocene scientists identify as the most likely threshold of the new geological epoch immediately follows the endpoint of what Andrade refers to as "the European warring states period" that he sees as stretching from 1450 to 1945.[85] By the time that a half-millennium of recurrent conflict came to a close, Europe's explosive firepower had reached such levels that entire populations found themselves inhabiting, in Walter Benjamin's words, "a field of force of destructive torrents and explosions"—a total environment of bomb blasts and firestorms capable of reducing entire cities to ash and rubble.[86]

Slow violence may be important, I've been arguing, but we still have work to do to make sense of an imperceptibly fast violence so deeply insinuated in the contemporary world that it is often barely registered. While trillions of tiny explosions still propel most of the global fleet of land-based vehicles, bigger but often unseen detonations continue to tear great volumes of rock apart. Anthropocene geologists point out that anthropogenic mixing, or "turbation," of rock fabric has so outstripped the impacts of any other organism that it has "no analogue in the Earth's 4.6 billion year history."[87] Again, Needham is prescient in appreciating the historical role of the fiery explosion in destratifying the earth, citing Boyle's observation that a "few barrels of gunpowder" suffices to blow up "many hundred, not to say thousand, tonnes of common rock."[88] By the mid-nineteenth century, commercial application of gunpowder for mining and civil engineering had overtaken military uses. Chemist Alfred Nobel's concoction of an explosive that was "more violent, brisant and shattering than propellant gunpowder" greatly accelerated this trend,[89] and the dynamite industry, Kelly adds, "grew faster than any other business in history."[90]

This of course is not simply "anthropic" rock turbation but a very particular social interaction with geological formations—as we can glean from the horror and outrage with which so many of the planet's peoples have apprehended the Western eagerness to turn the earth inside out. But the traumatic history of Europe's truncated encounter with gunpowder prompts us not simply to ask how its mighty earth-moving capabilities were achieved but to also consider the connections between lived exposure to explosive violence and the unprecedented intensity of the West's drive to blast open the earth. This question also relates to the social toleration of exceptional fatality rates of mine workers, both within Europe and its colonial extractive frontiers.

Explosive extraction has disaggregated the earth's lithic strata to such a degree, geoscientists have observed, that it is unsettling the logic of superposition—the sequential lithic layering through which geologists conventionally interpret the earth's deep history.[91] In this way, planetary changes triggered by a part of humanity undermine our very ability to make sense of the events scientists are attempting to measure and model. At the same time, the power of the weaponized explosion has underpinned the global imposition of social and material relations that have eroded the capacity of many peoples worldwide to live with and respond to the changing conditions of their own environments.[92]

While the planetary volatility that Anthropocene science seeks to substantiate may have begun to react back upon its own onto-epistemological surety, critical social thinkers allege that it has not yet impacted sufficiently to destabilize the geosciences' assumption of global authority. In short, the Anthropocene geostory has yet to properly connect the physical violation of the Earth Systems and structures it documents with the epistemic violence that accompanies speaking on behalf of a world of exposed and marginalized others—a charge that may well be intensified by the choice of a radionuclide marker and the way this is presented. But the "earth-shattering" explosions that the Anthropocene hypothesis is foregrounding, I have argued, open a history of fiery exothermic reactions, a genealogy that points to a deep, enduring entanglement of physical and epistemic violence in the becoming dominant of the West.

We cannot replay European history without gunpowder. And perhaps Western thought itself still lacks the conceptual tools for making sense of the sudden arrival of a new kind of planetary combustion—not least because this thought is itself forged and scarred by the confrontation with an alien fire. Jacques Derrida writes of knowledge beginning from incomprehension and exposure, of "the wound or inspiration which opens every speech and makes possible every logos or every rationalism."[93] But some rationalisms, some epistemes, may be more scarred than others. Something deep, disturbing, and persistent occurred when a strange new fire descended on European lands that had no sociocultural or cosmological niche for it. And if the coming of the anthropogenic explosion was indeed a geological event, then learning to manage explosions otherwise, which must include managing with a lot fewer runaway exothermic events, might itself constitute another event in earth history.

PART IV

New Geo-Temporalities

13

Holocene Time Perspective

Perrin Selcer

"Paul Josef Crutzen was born in Amsterdam in 1933, and embarked on a cosmopolitan journey in 1957, when he left The Netherlands for Sweden."[1] The pilgrimage from Amsterdam to Stockholm might seem more like a commute than the first step toward world citizenship, but Hans Joachim Schellnhuber's memorialization of his friend, fellow disciple of Earth System science and prophet of the Anthropocene, pays heartfelt tribute to cosmopolitan ideals that are foundational to the earth sciences. Indeed, the cosmopolitan motto "unity in diversity" has guided the practice of natural history so profoundly for so long that it has been naturalized—or, just as felicitously, socialized.[2] Despite the homogeneity of earth scientists, then, the ideology of earth science fits comfortably within this book's endeavor to collect diverse cosmologies between the covers of a single volume. In this chapter, I provide a critical interpretation of a self-consciously cosmopolitan origin story traditionally told by the scientific tribe for whom Stockholm remains the cultural center.

The Stockholm cosmogony begins twelve thousand years ago. Before that, the world was crushed beneath a kilometer of ice. This is the story of the Holocene, although that name was not common before the second half of the twentieth century; earlier decades preferred the Human Period, Recent, postglacial, current interglacial, or, for Northern Europeans, Flandrian.[3] Geologists did not officially ratify the Holocene as the second epoch, following the Pleistocene, of the Quaternary Period until the first decade of the twenty-first century—by which time the "wholly new" epoch was overshadowed and its future curtailed by excitement over Crutzen's claim that we had already left it and entered the Anthropocene. The Holocene narrative's twelve-thousand-year timescale predated its official ratification by a hundred years, however, and so this chapter surveys the century-long process through which a local periodization became global.

Critics complain of the Anthropocene's anthropocentrism and Eurocentrism but fail to recognize that these critiques equally apply to the "Human Period" it supersedes. If, as Alison Bashford writes, "the Anthropocene is modern history," the Holocene is "Western Civ.";[4] after all, the ten-thousand-year story of the rise of civilization is as old as the Holocene. The epoch's timescale thus reproduces all the problems of teleology, Eurocentrism, and stadial cultural evolution that confront world history instructors every semester. Here, I mostly bracket these critical dilemmas to focus on the challenges of working with a global timescale that lies at the intersection of human and earth history—of "earth time and lifetimes," as Emily Kern puts it in chapter 11.

The challenges are great. In the Holocene, the Scylla of pedantic particularity and the Charybdis of hand-waving generality sing with equal temptation. Its temporal scale lies between the geologist's unfathomable deep time and the historian's familiar generations. From inferences based on "proxy data"—ratios of atoms containing an extra neutron (i.e., isotopes), pollen grains, plankton shells—paleoscientists reconstruct the global climates and regional ecosystems of the Quaternary Period's lost worlds. Technical virtuosity and rigorous standardization are coupled to breathtaking speculation and sketchy correlations to tell grand stories of humanity's Pyrrhic triumph over the great forces of nature. This oscillation between the general and the particular makes the Late Quaternary timescale ideal ground to explore tensions between unity and diversity of a cosmopolitan cosmology.

The meaning of the Holocene epoch depends on the storyteller's perspective, of course. Scientists' identities provide standpoints from which they perceive the past, as does the disciplinary training through which they learn to mitigate the problem of subjectivity and recognize important research questions. It is not surprising that the most recent epoch of the geological time scale (GTS) should prove particularly meaningful to its creators. But it is still important to remember the provincial origins of global classifications and to explore their contemporary cultural and political implications. This chapter, however, is not (or not only) a humanist's undressing of the insidious universalism of hegemonic natural science. I am just as interested in reflecting on what historians can learn from paleoscientists about how to work with timescales and material traces of the past.[5] As befitting disciples of a cosmopolitan cosmology, Quaternary researchers have been attentive to local differences and developed a more pluralistic repertoire of periodizations and creative metaphors than suggested by critiques of Western science's universal linear time and deterministic cycles.[6]

Geological Time Perspective

Temporal perspectives are not just in the eye of the beholder. They are also material properties of objects. Archaeologist Geoff Bailey's "time perspectivism" clarifies distinctions between two aspects of scale: span and resolution, both of which have intuitive but tricky relationships with spatial scale.[7] Span measures length and size. Different things—a hearth, pine tree, bubble of ice-trapped air, marine rock—encode past conditions at time spans ranging from a season to millions of years and geographies ranging from a hillside to the globe. Resolution is the clarity and tempo of increments within the span. The hard dentin of a mastodon tusk formed throughout the animal's decades-long life, ranging over perhaps a few hundred kilometers (span); changes in the tusk's density and molecular composition reveal daily growth increments and preserve a seasonal diary of an individual life (resolution).[8] Three-kilometer-long glacial ice cores are so valuable because they combine enormous span (up to eight hundred thousand years) and high resolution (annual layers with atmospheric data smeared over decades).[9] Ocean currents mix slowly sinking foraminifera shells across hundreds of kilometers before they settle on the seafloor, where burrowing critters further mix them to create strata with data time-averaged over a few centuries to a few millennia. A comprehensive collection of eighty "high-resolution" proxy data sets for temperature spanning fifteen thousand years of the last deglaciation had a median resolution of two hundred years.[10] Although researchers crave ever-higher resolution, they find details "noisy" and so routinely smooth data to produce legible patterns.[11] Climate, for instance, could be defined as low-resolution weather.

Different disciplines work with different things and, therefore, develop different time perspectives. At the timescale of current events, what appears as a cause from an annual perspective (e.g., a flooding river destroyed irrigation infrastructure) becomes an effect over a longer span (e.g., irrigated agriculture caused flooding). Over larger timescales, the fundamental properties of things (say, rocks) appear different (rigid solid during epochs; viscous liquid over eras). And although everyone agrees that interesting questions demand multiscalar analysis, integrating evidence and arguments across scales is easier to prescribe than perform. Indeed, Bailey argues that time perspectivism means different disciplines' techniques and concepts are often incompatible; the best we can strive for is clear-eyed recognition of what particular perspectives reveal and conceal.

The distinction between span and resolution is key to understanding the GTS. The planet is so old and our geological history so detailed that the GTS cannot be depicted to scale—or, more accurately, geological timelines graphically represent relative resolution better than span. Geologists classify time according to a hierarchical scheme of eons (lasting from over 500 million to nearly 2 billion years), eras (66 million to nearly a billion years), periods (2.6 million to 300 million years), epochs (11,700 to 45 million years), and ages (several thousand to 20 million years). Prior to the explosion of richness in the fossil record that marks the Cambrian Period, the lower units of the classification drop away; the disappearance of ages and epochs and then periods and finally even eras signals a dramatic reduction in resolution. Geologists represent this loss of resolution through graphic compression of space; on paper, the official International Chronostratigraphic Chart sanctioned by the International Union of Geological Sciences gives earth's first eon, the six-hundred-million-year-long Hadean eon, an area equal to the 2.7 million years of the Pliocene epoch. The Pliocene lies at the threshold of another jump in resolution; its upper boundary marks the beginning of the Quaternary Period 2.59 million years ago, of which the Pleistocene epoch makes up 99.5 percent. The remaining one-half of 1 percent of the Quaternary Period is the Holocene epoch, the 11,700 years of which occupy a row of the International Chronostratigraphic Chart the same size as a five-million-year Triassic age.[12] The point is not that the timescale distorts span but that it accurately represents relative resolution.

The Holocene only becomes perceptible from a Quaternary time perspective. We may roll our eyes at atavistic young earthers and chuckle at naive predecessors who, just a couple of hundred years ago, thought history began *only* in 4004 BCE. Yet, as Martin Rudwick reminds us, early modern Europeans found these "fifty or sixty *centuries*" an "almost inconceivably lengthy history of the world."[13] The Holocene's 117 centuries—on the same order of magnitude as Bishop Ussher's biblical chronology—still subtends the periodization of Western civilization's origin story and remains mind-bogglingly long in terms of lived experience, policy-making, and historical practice. Today, however, the GTS's 4.56 billion years renders 11,700 years imperceptibly short. The greater challenge is seeing the Holocene from the perspective of earth history.

The distinction between size and resolution became tractable in geological time with the introduction of credible calculations of absolute time around the turn of the twentieth century. Before that, geological ages were relative, estimates varied wildly by discipline, and the fragmentary state of the stratigraphic record made distinguishing span from resolution guesswork—a well-preserved section of ancient rock, after all, appears infinitely bigger than a missing section, although geologists might speculate that such a sedimentary "hiatus" represented entire ages. No wonder the authoritative *The Geologic Time Scale 2012* named Arthur Holmes "the father of the Geologic Time Scale."[14] A British prodigy who matured into one of the century's leading theorists of plate tectonics, Holmes was an undergraduate when he began refining techniques for turning newly discovered radioactive minerals into chronometers. In 1913, he published *The Age of the Earth*, which provided a comprehensive critique of previous attempts to quantify geological time based on uniform rates of change in biology (evolution), physics (planetary cooling), and geology (erosion, sedimentation, and marine salinization) before demonstrating that ratios of uranium to lead and radium revealed a reliable planetary "hourglass." From the metronomic action of radioactive decay, Holmes dated Archean rocks to 1.6 billion years and assigned dates to Carboniferous rocks that match my wallchart today. Geologists, the upstart proclaimed, had gone from a dearth of time in which to move mountains to "an embarrassing superabundance." Rather than reinforce extreme Darwinian gradualism, Holmes reasoned that the earth's excessive age suggested present rates of geological change could not have been continuously sustained. Instead of witnessing "average conditions," he concluded, we must live in an unusually rapid phase of cyclic fluctuations in the rhythm of planetary time.[15] Both uniformitarianism and catastrophism characterized earth history; the difference depended on the geologist's time perspective.

Unfortunately, the resolution of Holmes's absolute chronometer was too coarse to perceive the rhythms of the Quaternary ice age that preoccupied the great geological theorists of his day.[16] Given the half-life of uranium, all Recent geological history fell before his chronometer's first tick. Relative timescales faced similar challenges. For older periods, the blurry temporal and spatial resolution of the marine fossil record assured geologists that they were observing global change. Biological evolution moved too slowly to inform interpretations at the high resolution of Recent time, however, and so naturalists turned to the progressive evolution of archaeological artifacts to establish postglacial chronologies.[17] The Human Period was thus an apt moniker; even geologists marked Late Quaternary time according to the rhythms of cultural history. But chronologies based on preliterate artifacts (like those based on fossils) only indicated relative time and remained open to dispute.

Locating the Origins of the Holocene

Fortunately, field scientists did not have to wait for their lab-bound colleagues' invention of radiocarbon dating after World War II to develop credible chronometers of absolute time that worked at a Late Quaternary timescale. At the

FIGURE 13.1. Sandy and clayey varves (2 and 3) exposed by road cut in Stockholm area, and a section of tumbled small rocks (1). Varve 1073 marks the transition "when the Baltic ice-lake was drained and salt water entered through the passes N of Mt Billingen, causing a sudden change in the type of sedimentation," which would become a marker of the Pleistocene-Holocene boundary. Gerard de Geer, *Geochronologia Suecica Principles* (Stockholm: Almqvist & Wiksells Boktryckeri-A.-B., 1940), plate 13.

Eleventh International Geological Congress in Stockholm in 1910, Gerard de Geer—president of the Congress, University of Stockholm professor, prime minister's son, and member of Parliament—described a natural hourglass that, "by actual counting of annual layers, [established] a real geochronology, for a period reaching from our time backwards some 12,000 years."[18] De Geer's layers were "varves," laminated clay sediments deposited by the retreating Scandinavian Ice Sheet's annual summer meltwaters and now exposed by construction ditches around rapidly expanding Stockholm (fig. 13.1).

De Geer had first realized the potential of these "varves" three decades earlier through an analogy to tree rings, another chronometer that unlocked Recent annual chronologies (fig. 13.2).[19] Just as dendrochronologists extended their chronologies by matching the patterns of good and poor growing seasons recorded in the tree rings of overlapping generations of logs, de Geer matched patterns of sediment layer thickness at roughly one-kilometer intervals along the path of glacial retreat to produce an annual record of deglaciation. He correlated the disappearance of particular horizontal varves with the vertical lines of miniature moraines left at the edge of most winter's small readvance to demonstrate the location of the glacial headwaters that had deposited the sediments. Thus, he produced time-maps of the ice's eight-

FIGURE 13.2. "Normal clay varves in regular deposition . . . Essex Junction, U.S.A. The first varve locality measured in N. America, viz. by G. de Geer in 1891, remeasured in 1920. Photo Ebba Hult de Geer, 1920." Gerard de Geer, *Geochronologia Suecica Principles* (Stockholm: Almqvist & Wiksells Boktryckeri-A.-B., 1940), plate 1.

hundred-kilometer flight north. From this chronology, he divided the late glacial Quaternary into three subepochs: a Daniglacial, when the ice receded from Denmark into Skåne; a Gotiglacial, for the slow retreat across Southern Sweden; and a Finiglacial, during which the ice swept past Stockholm and Uppsala at the startling rate of one hundred to four hundred meters per year to Jämtland. There the sheet bifurcated, which de Geer identified as the beginning of the postglacial. Altogether, this deglaciation had taken five thousand varve years.[20]

In the August before the Congress, de Geer and his wife, Ebba Hult de Geer, investigated a twenty-four-meter-thick section of ancient Lake Ragunda's bottom. Since Ragunda was situated just fifty kilometers east of Jämtland and the annual varve deposition had terminated only with its accidental draining in 1796, the section represented essentially the entire postglacial. Based on average varve thickness, the de Geers estimated that seven thousand layers had accumulated since deglaciation had created the lake. They correlated the "beautifully laminated" bottom layers of the lake with the uppermost varves of Gerard de Geer's earlier five-thousand-year deglacial chronology to produce the "real geochronology" of the last twelve thousand years that he presented at the Eleventh International Geological Congress.[21] The Northern European ice age, he concluded, had ended just seven thousand varve years before present, considerably less than many contemporary geologists estimated—and a truly biblical timescale!

De Geer was so impressed with his innovation that he considered himself the founder of geochronology. He had planetary aspirations for his science from the beginning. In 1910, he extrapolated to the regional scale and exulted that such a "climatic curve for Northern Europe implies the possibility of comparison and correlation with similar curves from other formerly glaciated regions as especially North America." Such correlations could begin to resolve the question of whether "the much discussed glaciations in different regions were synchronous and due to general climatic causes, or, on the contrary, were of local origin."[22]

To explore global-scale causation, the organizing committee of the Eleventh Congress asked national delegations to submit reports on postglacial climates, which it published in a companion to the conference proceedings. European and North American contributions dominated the volume, but it also included impressionistic reports from Swedish geologists who had spent a few weeks in Persia or Patagonia and from colonial officers from low latitudes where postglacial chronologies implied 250-million-year timespans dating to the end of the Permian glaciation. This Gondwanan postglacial corresponded to the rhythms of Holmes's tectonic time perspective.[23]

Over the next three decades, de Geer and his acolytes extended his geochronology a few thousand years and refined regional varve correlations into the standard "Swedish Time Scale." They became ever bolder with their correlations; they moved from tracking the retreat of a single ice face at one-kilometer intervals to discovering teleconnections in postglacial varve pat-

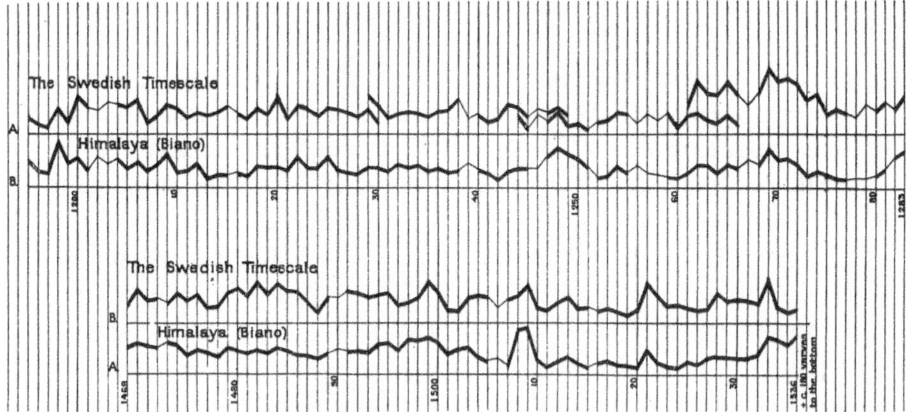

FIGURE 13.3. Correlations between the standard Swedish Timescale and a Himalayan varve series. The x-axis is time, with vertical lines representing annual intervals, and the y-axis is the relative (and corrected) thickness of sediments accumulated during each varve year. Matching patterns formed "constellations," which de Geer claimed proved climatic teleconnections and enabled the Swedish Time Scale to serve as a global standard. Gerard de Geer, *Geochronologia Suecica Principles* (Stockholm: Almqvist & Wiksells Boktryckeri-A.-B., 1940), 223.

terns that matched the Swedish Time Scale in other Northern European countries, across the Atlantic to New England and Canada, across the equator to Argentina and New Zealand, and even in tropical British East Africa, where layers corresponded to wet and dry seasons rather than summer meltwaters. De Geer termed patterns of thicker and thinner layers "constellations" and, in explicit reference to Linnaeus's earlier universalizing triumph, named them for regionally representative animals, culminating in *Homo Africanus*, an especially long sequence that enabled matching patterns over so many years that it brooked "no doubts concerning [the teleconnection's] validity."[24] Gone was the possibility of asynchronous, regional change; a multitude of apparent differences—a missing year, a too thick layer—were interpreted away as mere "local deviations" from a global story dictated by the whims of a single ultimate cause, variable solar radiation. Unfortunately, even de Geer's devoted protégés doubted the validity of these teleconnections.[25] Globalizing the Swedish Time Scale proved much more complex than identifying a universal climatic curve (fig. 13.3).

In the century following the Eleventh International Geological Congress, scientists standardized geological time. In 1969, the Commission on the Study of the Holocene of the International Union for Quaternary Research (INQUA), founded in 1928, resolved to place the Pleistocene-Holocene boundary at a round ten thousand years before the present, corresponding to the end of the Younger Dryas cold event, the last of the periodic abrupt temperature swings that characterized the Pleistocene. The Commission assigned nomenclatural precedence for the episode to the aptly named Nordmann. At the Eleventh International Geological Congress, Nordmann had presented a climatic periodization scheme that classified the millennium-long cold snap by the appearance of the Arctic *Dryas* flora in Denmark.[26] The flora's characteristic species, *Dryas octopetala*, a diminutive but hardy white rose, beautifully represents the Scandinavian origins of the Holocene epoch.

Contemporaneously with INQUA's resolution, however, stratigraphers proposed formally defining the GTS's units through Global Stratotype Sections and Points (GSSPs). Ideally, a GSSP is a marine sedimentary rock outcrop rich in fossils with a complete stratigraphy showing the before and after of a transition (the section) and a point that marks the boundary (the "golden spike"). The Holocene's brief history and high resolution rendered the search for a viable section of solid rock and biostratigraphy moot. For two decades, a borehole and varved lake bottom from de Geer's old stomping grounds were the leading GSSP candidates. In the mid-1970s, University of Stockholm geologist Nils-Axel Mörner proposed core B 873 in the Gothenburg Botanical Garden, which appeared to check all the boxes: marine sediment rich in pollen and foraminifera that was easily accessible because isostatic rebound following deglaciation had lifted the land above rising seas (and the hole was a few paces from the garden's entrance). Geologists located ten thousand years BP in a boundary 3.35 meters below the surface, correlated it with Swedish Time Scale varve 1073, which corresponded to the ingression of salt water into the Stockholm peninsula that signaled the start of the Younger Dryas, and demonstrated climatic change through biostratigraphy, isotope analysis, and sedimentology (note varve 1073 in fig. 13.1). Unfortunately, their peers worried that marine sediments did not afford the necessary precision for Late Quaternary temporal resolution and the proposal stalled.[27]

In 2008, the International Union of Geological Sciences finally ratified the Pleistocene-Holocene boundary point 1,492.45 meters—or 11,700 (+/− 99) annual layers—down a Greenland ice core. The golden spike recorded an "abrupt shift in deuterium excess values," which served as a proxy for sea surface temperature. The two to three per mille decrease in heavy hydrogen isotopes implied a *decline* of two to four degrees Celsius in ocean temperature, but paleoclimatologists attributed this to a northward shift in the source of Artic precipitation, which corresponded to "a sudden reorganisation of the Northern Hemisphere atmospheric circulation," which in turn signaled the onset of warming of ten degrees Celsius in Greenland in the decade following the Younger Dryas.[28]

The golden spike's location, coring technology, and isotopic geochemistry were all technoscientific testaments to the wizardry of the twentieth-century military-industrial-academic complex, but the 11,700 years were calculated by a more expensive version of de Geer's method: counting annual layers.[29] And although paleoscientists now dated the end of the Pleistocene to the final phase of what he had termed the late glacial, the ratification of the golden spike confirmed de Geer's "real geochronology." Fittingly, the GSSP ice core is archived at the University of Copenhagen, the first major city to emerge from under the ice during de Geer's Daniglacial Subepoch twelve thousand years earlier.

The Swedish Time Scale subtends Holocene geochronology. What makes this genealogy of the Holocene true and significant here, however, is not the significance of the Younger Dryas to climate history but the enduring importance of Stockholm in global environmental science and politics. As Sverker Sörlin and Eric Paglia argue—from the United Nations Conference on the Human Environment in 1972 (the Stockholm Conference), which elevated global environmental issues onto the international agenda, to hosting the secretariat of the International Biosphere-Geosphere Program, which inspired the Anthropocene proposal, to the moral authority of Greta Thunberg in climate justice debates today—Stockholm has retained its centrality to the cosmopolitan community that finds Holocene time meaningful.[30]

Diachronous Boundaries

In December 1909, de Geer proclaimed to a crowd gathered to inaugurate a University of Stockholm building that only "9,000 years had elapsed since the building-ground of the new house became free from the land-ice."[31] As Alexis Rider shows in chapter 9, the specter of potentially resurgent ice haunted popular scientific fantasies in de Geer's day. Teaching environmental history on a peninsula between two Great Lakes that were relatively recently under a kilometer of ice, I empathize with this excitement (even if I fear glacial retreat not resurgence). Still, Ann Arbor, Michigan, became a postglacial landscape thousands of years before the deuterium-depleted snows marking the start of the Holocene fell in Greenland. From de Geer to today, in fact, all reconstructions of the last deglaciation emphasize that "the timing of advance and retreat were both remarkably variable across the ice-sheet area."[32] Even a measure as definitively global as the Last Glacial Maximum (twenty-three to twenty-one thousand years ago) merely infers total planetary ice volume from marine oxygen isotopes; differences of thousands of years in the maximum of various lobes and glaciers characterized the great northern ice sheets.[33]

More fundamentally, at my previous job, two thousand kilometers to the south in Texas, the whole story of deglaciation, not just its timing, would not have resonated so strongly. Not only did ice never cover the region but its temperature only weakly correlates with Greenland's.[34] In the Southern Hemisphere, local Last Glacial Maximums appear antiphase or determined by the idiosyncrasies of local environments at millennial scale. Even the Younger Dryas was only a dramatic "cold event" in the high northern latitudes—scientists infer an unimpressive drop in the global average of 0.6 degrees Celsius, with no trend in temperature change (but significant precipitation anomalies) in the tropics and up to two degrees Celsius of *warming* in the Southern Hemisphere.[35] In Africa, the rhythm of Late Quaternary climate history manifested in changing monsoonal patterns. From about 14,800 until 5,500 years ago, Africa was much wetter, including a "green Sahara," and the Younger Dryas features as a severe drought in the context of this longer history of the African Humid Period. At a continental scale, desiccation was a "time transgressive" process occurring over millennia.[36] Finally, in Australia, historians and naturalists find the arrival of humans fifty thousand years ago a compelling beginning;[37] 11,700 years ago, nothing especially notable left a trace.

Given global climate variability, paleoscientists disagreed on the value of fixing a precise Pleistocene-Holocene boundary. In 1980, Herbert E. Wright Jr. and Richard A. Watson argued that, with rare exceptions that proved the rule, the boundaries between Quaternary time units—indeed, between all chronostratigraphic units—were inherently diachronous. This was not a dismissal of the Holocene's utility. From his base at the University of Minnesota, Wright had been collaborating for decades on interdisciplinary Quaternary research and co-led the Cooperative Holocene Mapping Project, the first major attempt to model climate after the Last Glacial Maximum.[38] True, they acknowledged, the Holocene might resemble just another Pleistocene interglacial, but "if nothing else, the pace of cultural change sets it apart." Their point was that fixing a golden spike in a single time and place rendered the classification artificial, whereas they favored a system representing how geologists used the GTS in practice: "natural time-transgressive geologic time units" materialized in strata that recorded a distinctive "group of geologic events, processes, and environments."[39] The Holocene began at different times in different places. An artificial system could be marvelously precise, but that precision came at the price of arbitrariness. And with Quaternary researchers routinely radiometrically dating sites, precise global geochronological classification became superfluous to correlation. A natural classification might be messy, but that reflected the reality of global change.

The Late Quaternary's high resolution clearly revealed the reality of diachronous boundaries. Watson and Wright framed their polemic as a response to the landmark *International Stratigraphic Guide*, which, following two decades of international consultation, the International Union of Geological Sciences published in 1976 to standardize practices. The *Guide*, however, had already flagged the problem: "Only units of the higher rank in [the GTS's] hierarchy lend themselves at present to worldwide application." Systems (the stratigraphic equivalent of geochronologic periods) covered "a time span sufficiently great so that they serve as worldwide chronostratigraphic reference units" and were the most widely used.[40] The geological record's resolution was too coarse for global correlation at the scale of epochs and ages. Temporal uncertainties exceeded epochs' time spans, but the *Guide* expressed hope that future techniques would improve chronometric precision to make these units global. The precision of Quaternary science, however, showed that enhanced resolution could also undermine a worldwide application of GSSPs.

As the ratification of the Holocene's golden spike proves, Watson and Wright lost the debate. The victory of global isochronous boundaries represented the triumph of absolute over relative time and artificial over natural classification in the GTS. When pressed, proponents of universal periodization acknowledge its conventionality. Rhodes W. Fairbridge, Columbia University professor, chair of the INQUA Subcommission on Shorelines of Northwestern Europe, and advocate for locating the Holocene's golden spike in Swedish sediments, for example, conceded that the effects of Late Quaternary climate change were time transgressive at a millennial scale, with "tropical climate indicators" showing that "present temperature levels [were] reached about 13,500 BP." He simply asserted, "Mid-latitude scientific requirements dictate the need for a more convenient date, and for this reason the warm-up following the last major stadial (Younger Dryas/Preboreal transition) was chosen."[41] Yet he rejected the diachronous nature of global chronostratigraphic boundaries.

Fairbridge rejected Watson and Wright's argument for diachronous boundaries because he was more concerned with causes than effects. "They ignored the philosophic concept of determinism," he explained, and so their scheme obscured the essential interdependence of all environmental change.[42] Diverse, time-transgressive effects were all results of the same ultimate cause, which for Fairbridge was subtle reverberations between the orbital cycles of Jupiter, Saturn, Earth, the moon, and the sun.

The search for the Holocene's proximate cause focused scientific attention on the North Atlantic, where, the story went, glacial meltwater had triggered a restructuring of oceanic and atmospheric currents that precipitated abrupt climate change. In 2020, however, paleoclimatologists reported a new high-resolution, "millennial-scale" dating of sediments from the Gulf of Alaska that demonstrated the trouble with precise global boundaries. Ice-sheet discharges into the Pacific had preceded those into the Atlantic. Moreover, they found that the release of iceberg "armadas" into the Pacific followed shifts in the oxygen isotope record derived from a Chinese cave speleothem, which represented changes in the Asian monsoon, and preceded a spike in standard Marine Isotope Stages derived from benthic boreholes, which tracked global ice volume. This new geochronology flipped the script, suggesting "low-latitude processes are an important driver of high-latitude climate."[43] An accompanying commentary began by stating that "science is theoretically objective, but biases and paradigms often originate from something as fundamental as field site accessibility, data density, or publication date" and concluded that "paleoclimate and paleoceanographic records far from the North Atlantic argue that it is time to revisit this paradigm central to paleoceanography and paleoclimate studies since the first high-resolution ice core was drilled in Greenland."[44]

The middle of Greenland is not an inherently accessible field site, of course. A century-scale history of Late Quaternary geochronology helps explain why scientists were so eager to collaborate with the national security state to locate the origins of the Holocene there, in a layer of ice formed midway through the last deglaciation. The episode neatly illustrates one way the present becomes the key to the past.

Correlating Proxies

The problem with accepting that the boundaries of geological time are just conventions is that it makes the GTS boring. In 1948, New Zealand geologist Robin Sutcliffe Allan admonished his American colleagues: "Modern stratigraphy, is uninspired, dull, and dead." The problem was that too many stratigraphers "serve Mammon—the oil companies, the quarry owners, and the others who exploit the raw materials of the Earth's crust." Economic geologists had the "temerity" to use fossils merely as "recognition symbols and time markers" and had abandoned the "nobler task" of historical geology to "decipher the story of earth history."[45] Allan extolled the paleoecological approach, which used knowledge of present-day organisms' relationship with their surroundings to reconstruct past environments.

Americans were positioned to lead this geological Renaissance, Allan assured his readers, because they were already at the forefront of paleoecology. He highlighted the work of former director of the Scripps Institute for Oceanography, T. Wayland Vaughan, which correlated current three-dimensional ocean temperature and seawater composition data with foraminifera species distribution so that the planktons' fossilized ancestors could be used to infer ancient marine environments.[46] Americans also led in terrestrial paleoecology. Frederic Clements coined the term in his famous study of succession, which established the ideal of plant assemblages developing into "climatic climax communities" that lived in dynamic equilibrium with their environments.[47] Clements's theory, however, built on Scandinavian studies of pollen and plant macrofossils preserved in bogs, which paleobotanists used as proxies to infer the region's postglacial climate history. Indeed, at the Eleventh International Geological Congress, where de Geer had presented his twelve-thousand-year geochronology, an embarrassingly vitriolic debate erupted between Gunnar Andersson, the secretary-general of the Congress, and Rutger Sernander, a member of the executive committee, over whether these bogs revealed regular climatic rhythms or the whims of local fluctuations.[48] The stakes were high: pollen data were a foundation of European postglacial geohistory and, as historian Melissa Charenko argues, predictable cycles boosted confidence that past patterns foretold future ones.[49]

More than deciphering the evolving rhythms of planetary change over time, however, ecologically reconstructing lost worlds depended on correlating data from the same time. Happily, Allan explained, because paleoecology demanded that correlations make ecological sense—that ancient polar bears weren't fishing tropical reefs—a renewed commitment to historical geology ensured more rigorous interpretation.

Correlation is perhaps the most interesting and indispensable word in the paleoscientific lexicon. Holmes ended *The Age of the Earth* not with a paean to the vastness of geological time but with a rousing call to correlate: "With the acceptance of a reliable time-scale, geology will have gained an invaluable key to further discovery. In every branch of science its mission will be to unify and correlate, and with its help a fresh light will be thrown on the more fascinating problems of the Earth and its Past."[50] Absolute geochronologies facilitated communication and collaboration between paleoscientific specialists, who could now confidently correlate their diverse data to reconstruct ancient worlds. Thus, while the Anthropocene's main value may be as an effective boundary object (i.e., something that brings disciplines into a common conversation), this always has been the primary function of the GTS's units. And like the Anthropocene, the inclusion of the human sciences distinguishes Quaternary interdisciplinarity from earlier units.[51]

Ironically, Holmes's "mission . . . to unify and correlate" demonstrated that different data sets produced different chronologies. The introduction to "Principles of Stratigraphic Classification" in the second edition of the *In-*

ternational Stratigraphic Guide (1994) put the problem bluntly: "The stratigraphic position of change for any one property or attribute does not necessarily coincide with that for any other. Consequently, units based on one property do not generally coincide with units based on another, and their boundaries commonly cut across each other. It is not possible, therefore, to express all the different properties with a single set of stratigraphic units; different sets of units are needed."[52] Climatostratigraphic, biostratigraphic, lithostratigraphic, chemostratigraphic, and magnetostratigraphic units each record distinct rhythms. Such syncopation meant that, even in the same locale, chronostratigraphic boundaries never matched every property's beat.

But how long did the record have to run to reveal that all the themes were elements of a single tune? At the Quaternary timescale, geologists were divided. Since Quaternary chronostratigraphy tracked climate cycles rather than biostratigraphy as earlier units did, *A Geologic Time Scale 2004* demoted the period to an "informal climatostratagraphic unit."[53] Unlike the recently relegated Tertiary and long forgotten Secondary, however, the demotion didn't stick; the Quaternary had already produced a vibrant interdisciplinary research community, which wasn't about to let its organizing concept get canceled.[54] Reincorporating it as a period in 2009, now including the just ratified Holocene epoch, required pushing the beginning of the Pleistocene epoch back eight hundred thousand years—within the uncertainty margin of some earlier periods but nearly a third of the Quaternary.

Periodizations frame narrative possibilities. The tradition of privileging climate in Quaternary geochronologies primed the international scientific community to perceive climate as the primary indicator of global environmental change, and thus the central concern of international environmental politics. The implicit synchronicity of the Holocene, the Neolithic Revolution, and "New World" cultural history, for instance, renders evidence that falls outside the received chronology suspect and scientists who argue that an object is a surprisingly ancient American artifact rather than an ecofact potentially kooks. The units of the GTS, in other words, are functional equivalents to Kuhnian paradigms, enabling and constraining the practice of normal paleoscience.

The blinkering effects of the GTS should not be exaggerated, however. The science of time is hardly naive about the effects of timescales. Geologists have created an international bureaucracy to evaluate periodization schemes, and Quaternary researchers enjoy a plethora of chronologies. Flipping over the International Stratigraphic Chart reveals a Global Chronostratigraphic Correlation Table for the Last 2.7 Million Years, which juxtaposes fourteen standard chronologies with the Quaternary: five regional glacial periodizations, two plankton biozonation schemes, Marine Isotope Stages, Antarctic oxygen and deuterium isotope records, Chinese Loess and Lake Baikal proxy-data sequences, Italian Marine Stages, and paleomagnetic chronologies. Each region and topic has its own specialized chronology. Paleobiologists have standard land mammal ages for each continent; paleobotanists have their pollen zones. Archaeologists employ fraught stadial chronologies of the Paleolithic, Mesolithic, and Neolithic, as well as metallic ages and ceramic periods. The logic of paleoecology, the science of correlation, dictates that the database of proxies, each a potential chronology, continuously expands as paleoscientists painstakingly produce novel sequential data sets of sediment layers, oxygen, hydrogen and carbon isotopes, pollen grains, lake levels, dust, and so on.

Each data set represents a unique palimpsest. Like a clay tablet on which a message has been smoothed over and a new text written, the effects of time have partially erased and overwritten the original signal. Each object requires specialized expertise to track back from a final quantitative value through the transformations, contaminations, and erasures inherent in analytical practices: slicing, grinding, dissolving, mixing, vaporizing, freezing, passing through magnetic fields, magnifying, scanning, cleaning, storing, labeling, shipping, coring. Each lab, each apparatus, each standard operating procedure produces idiosyncratic effects that must be accounted for, but with practice, the laboratory's uncertainties and errors can be calculated and controlled. Indeed, technical virtuosity may produce misleading precision, for far greater uncertainties were generated by millennia of exposure to heat and moisture, curious rodents and industrious worms, intense pressures and porous membranes, shifting currents and eroding banks, cosmic rays and chemical reactions. And then there are the still greater complexities of the processes that led to an object's deposition at a particular site. Accounting for these effects requires constructing a compelling biography of a molecule, fossil, or artifact—perhaps developing a formal "proxy system model"—to back calculate an original value.[55] Finally, comparison to a modern analog serves as a baseline from which to infer the meaning of this value. Closing the circle linking the present to the past, what archaeologists call "middle-range theory," is fraught with anxiety—perhaps the analogy is false because a molecule behaved differently under past conditions, an organism evolved new tolerances, or cultural adaptations vanished.

Whole regions are palimpsests too. Drawing again from Bailey, landscapes formerly covered by ice sheets approximate "true palimpsests" from which glaciers have "removed all or most of the evidence of the preceding activity." Landscapes that were never glaciated resemble "cumulative palimpsests": material traces of the past "remain superimposed one upon the other . . . but are so re-worked and mixed together that it is difficult or impossible to separate them out into their original constituents."[56] The crisp resolution of the true palimpsest of high northern latitudes make the last deglaciation appear a natural temporal unit, while no such break can be found in the cumulative palimpsest landscapes to the south.

From the planetary to the atomic scale, it is palimpsests all the way down. For Fairbridge, the "philosophical principle of determinism" dictated that, ultimately, each of these data sets was part of a single, unified story. But whatever their personal commitments to determinism, in practice, paleoscientists distinguish between local, regional, and global processes; they specify whether patterns unfold over diurnal, decadal, or millennial timescales. If climate history motivates research, local effects obscure a global signal—so much noise to be filtered out; the middle of an ice sheet and the bottom of the sea provide excellent global proxy data because their local histories were boring. If biological or cultural history inspires research, spatiotemporal averaging effects likely limit interpretation. Every specialty produces its own bespoke chronology, and at the Quaternary timescale, they do not all meaningfully correlate.

"Planetary multiplicity," to borrow the language of Nigel Clark and Bronislaw Szerszynski, is a feature not just of exoplanetary comparison or contrasting eras of earth history; it is driven by multiple stories unfolding simultaneously in the same place.[57] As David Turnbull writes about the peopling of South America, this proliferation of subplots undermines "the orthodox narrative" that different discipline's stories "can be mapped onto each other in a single, unified, linear, and chronological narrative."[58]

Such planetary multiplicity complicates determinism but is no cause for despair. It suggests that earth history is an endless creative resource, continuously renewed by the interests of the present. Rather than a single, linear narrative, paleoclimatologist Julia Kelson describes her practice, motivated by anxieties over future droughts, of reconstructing past hydrological cycles from the isotopes of curious soil carbonates as tracing ∞, an infinite time loop connecting an uncertain past to an uncertain future with an ever-turning twist at the present.[59]

The Holocene Baseline

From the perspective of the Pleistocene, the Holocene appears doubly anomalous. It is an unusually stable interglacial and also a period of unprecedentedly rapid anthropogenic change. From the perspective of the Anthropocene, the former appears the necessary condition for the latter. The Stockholm Resilience Center's "Planetary Boundaries" framework, for example, famously describes the preindustrial Earth System as a steady-state "safe operating space" for humanity.[60] The implications are doubly ironic. First, Late Quaternary climate history, including especially the start of the Holocene, reveals an extraordinarily sensitive Earth System subject to catastrophic nonlinear change; stability may be a laudable social ideal but is an illusory natural baseline. Second, prophets of the Anthropocene celebrate the concept's power to dissolve the divide between natural and human history, but their nostalgic vision of the stable Holocene renders the passivity of the environment the precondition for human agency. No wonder the Anthropocene framing frustrates scholars who have spent their careers exploring ancient human-environment interactions; using the inert Holocene as a baseline to measure Anthropocene acceleration turns places humans have reconstructed through millennia of habitation into landscapes without history.[61]

Turning an entire epoch, the whole ten-thousand-year history of civilization, into a single baseline is an impressive "trick of perspective."[62] It is performed by privileging a few proxies of the global-scale environment and tuning out data that tell alternative stories. The editors of a science and technology studies special issue on environmental baselines explain that selective bias based on a "balance-of-nature" model is inherent in the practice of baselining. Instead of imagining the past as a single fixed line, they draw on the work of environmental anthropologist Tim Ingold to propose evaluating the present and envisioning the future as if they emerged through a "meshwork"—an entanglement of "trails along which life is lived." The meshwork metaphor, they argue, can turn baselining into a humble act of "tentatively making paths that 'cut across the terrain of lived experience'" connecting the past, present, and future. Because meshworks emphasize the web of alternative lines that could be traced, they are "politically contested and ethically charged from the start."[63] Meshworking offers the potential to honor and engage a more cosmopolitan collection of diverse earth histories as we negotiate trajectories into the future.

And yet even these champions of multiplicity insist on the urgency of their critique because, in the Anthropocene, "we no longer have the *holocenic* option of remedying the problems of baselining by tinkering"—changes to "the planet's whole ecological status" mean that conditions that used to be "taken as a given . . . no longer apply."[64] The Holocene's global baseline is the indispensable heart of contemporary environmental politics. Retracing the local origins of this geochronology makes it possible to situate the road from Stockholm as one well-traveled trail through a cosmopolitan meshwork.

The ease with which prophets of the Anthropocene render the twelve-thousand-year history of civilization a moment in time is worth reflecting on too. Bailey explains that moments, including the present, must have duration, since we can only know them through enduring material traces. The duration of a moment is partly determined by the scale these traces happen to record and partly by the interests of the observer. For historians like me, comfortable in the weeds and trained to reject simplification, zooming in feels like an epistemic virtue, even a necessary move in the pursuit of justice. Yet large-scale time perspectives obviously reveal meaningful patterns invisible from up close. And so, in this chapter I've worked against my scholarly instincts to smooth a century of earth science—a timespan equal to the margin of uncertainty proposed for the Holocene's golden spike—into a moment in time. Admittedly, this time-averaging distorts the history of a century that experienced revolutionary changes in concepts (plate tectonics, Earth System science), techniques (mass spectrometry, ice and deep-sea drilling, computer modeling), and institutions (security state funding, increasing disciplinary specialism)—and the Great Acceleration of planetary time itself. But amid celebrations of paradigm shifts that promise (yet another) second Copernican Revolution, I am struck by the endurance of the questions, concerns, and practices that drive earth science.[65]

14

"American Blitzkrieg" or "Ecological Indian"?

Inequalities in Narrating Environmental Degradation through Deep Time

Melissa Charenko

The Anthropocene poses a challenge to formulations of human agency.[1] Proponents of the concept suggest that humans have become "a major environmental force" such that a new geological epoch has begun.[2] Most of the Anthropocene discourse in the global north implies collective human responsibility for a planetary-scale problem.[3] On this interpretation, all humans are implicated in modifications to earth's systems regardless of immense inequalities in planet-altering powers.[4]

Many scholars have rejected the universalizing and homogenizing tendency of the Anthropocene's grand narrative by specifying the who, what, when, and where of this new geological epoch. Such responses typically single out rich and industrialized nations for causing much of the environmental degradation, although others have moved away from the nation-state to focus on the prevailing economic system of extraction: capitalism.[5] Still others have called attention to gender and race as key factors in the unevenness of the Anthropocene.[6] In these formulations, humans do not share equal responsibility for modifying the planet. Instead, particular logics and systems imposed by the powerful give rise to significant environmental changes.

Specifying culpability helpfully situates human agency in a set of historical conjunctures that the universalizing narrative of the Anthropocene fails to capture. But, as these specific historical narratives of responsibility form, it is useful to remember that not everyone can participate in developing them. Just as there are immense inequalities in humans' planet-altering abilities, there are immense inequalities in who can give voice to the causes of such transformations. This chapter demonstrates how dominant power structures prevented some voices and ways of knowing from interpreting the past. By rejecting certain knowledge systems, a group of scientists controlled the narrative about those particularly responsible for environmental degradation, which recreated colonial myths.

This chapter reflects on narrative inequalities and their consequences by focusing on a late twentieth-century debate about the causes of extinctions that had occurred thousands of years prior. Akin to the dominant Anthropocene narrative, one of the most popular theories about the causes of these extinctions, the "overkill hypothesis," pointed to collective human responsibility for extinction as humans arrived in new parts of the globe. But there were also more specific narratives of responsibility that developed in the late 1950s. These accounts blamed Indigenous peoples for the extinctions of large fauna in the Americas as the last ice age ended about 11,700 years ago.

To single out Indigenous peoples as particularly responsible, American commentators drew on settler-colonial science despite known shortcomings. Scientific narratives, with their resonances with the environmental movement and discussions of human violence, ultimately had explanatory power that scientists denied to oral traditions. Scientists devalued Indigenous ways of knowing, casting aside Indigenous peoples' narratives of the past. Commentators then used the scientific narrative, which placed responsibility for past environmental degradation on Indigenous peoples, to deny the indigeneity and sovereignty of Indigenous peoples. The thinking went that, thousands of years ago, Indigenous peoples had come from elsewhere and mismanaged the environment such that they could not be trusted to govern the land today.

Discussions of culpability for former extinctions remind us that there is a politics to the deep past, especially when the past is extended to the present.[7] Denying Indigenous peoples' participation in twentieth-century debates about the causes of ice-age extinctions perpetuated power imbalances between scientific ways of knowing and oral traditions and between settler-colonizers and Indigenous peoples' narratives of the past and present. In a current moment where scholars are negotiating the causes of environmental destruction by weighing universalizing versus specific narratives of harm, it may be easier to continue to erase the power geometries that elevate certain narratives about the deep past instead of disrupting the systems and stories that are serving, and have served, some quite well.[8] Scholars would do well to avoid these silencing tendencies in twenty-first-century narratives of the Anthropocene lest they wish to refashion colonial myths.

The Overkill Theory of the Pleistocene Extinctions

As the glaciers retreated 11,700 years ago, signaling the end of the Pleistocene (at least in the Northern Hemisphere, as Perrin Selcer reminds us in chapter 13), many large mammals went extinct in North America. Gone were the woolly mammoths, mastodons, giant beavers, and thirty-two other genera of large-bodied mammals. Many scientific theories prior to the middle of the twentieth century suggested that climatic changes at the end of the last ice age likely caused this mass extinction event.[9]

In contrast, the overkill theory of the Pleistocene extinctions emerged from Paul S. Martin's work in the late 1950s. Martin, who had trained in zoology, biogeography, and pollen analysis before being appointed to a position at the University of Arizona in the geosciences, questioned climatic theories. He believed the fossil record did not support climatic theories because that record indicated a differential loss between large and small animals: large mammals experienced a near-total decline while smaller genera went virtually unscathed. Climatic change should have affected species regardless of size. Martin thus thought that climatic explanations did "injustice to the temporal and ecological record."[10]

While Martin found little support for climatic theories, he began to posit that the arrival of people along an ice-free corridor at the end of the Pleistocene had played some role in the extinction event.[11] The glaciers had waxed and waned several times without large-scale extinction, but scientists had recently agreed that humans had arrived in North America for the first time as the ice age ended. As Martin said, "the late Pleistocene environment has some unique features. Man is the only one clearly identified."[12]

Martin advanced a stronger version of his hunting theory in the mid-1960s when he claimed that human predation was likely *the* cause of the Pleistocene extinctions, including those outside the Americas. In 1966, he published the "overkill" hypothesis in *Nature*, which explored how the peopling of different landmasses at different times explained global extinction patterns.[13] On his view, human hunting accounted for exterminations of large fauna in the Western Hemisphere, Australia, Madagascar, and on other islands. It also explained why large species survived in Africa where they had evolved alongside hominids and adapted to human predation.

For overkill in the Americas, Martin's thinking went that the " 'consumer-oriented' rush through the virgin continent" coupled with "powerful behavioral reinforcers associated with the excitement of the chase" would easily have turned the new arrivals into a "superpredator . . . a species which kills for more than food alone."[14] Given humans' destructive nature, in the blink of geological time, the megafauna went extinct. By the 1970s, Martin likened these extinctions to a human blitzkrieg.[15]

Scholars have demonstrated that Martin's theory resonated with the postwar rise in ecological consciousness. As they show, "the overkill argument captured the popular imagination during a time of intense concern over humans' destructive behavior towards life on earth."[16] Scholars cite published examples of scientists accepting overkill by linking past destruction to modern destruction amid the homily of ecological ruin. This line of reasoning is visible in numerous letters to Martin, where scientists wrote expressing support for overkill given contemporary extinctions. For example, biologist Steven Christman wrote in the 1980s, "I refuse to believe that prehistoric people practiced conservation of their resources any better than modern man. . . . When man first arrived anywhere, he was immediately a predator to be reckoned with. . . . But as weapons and hunting skills improved, even more species of prey became vulnerable. The first people in Florida clubbed the giant tortoises to extinction, later they wiped out the Carolina parakeet, and they're still working on the sea turtles."[17] In this formulation, humans, both prehistoric and modern, left environmental destruction in their wake.

Overkill resonated with more than just ideas of anthropogenic environmental destruction; the theory reflected Cold War discussions about human capacities for destruction writ large. In the 1960s, the American public—already gripped by Cold War anxieties about nuclear annihilation, the war in Vietnam, urban unrest, and race riots across the United States—came to accept humans' capacity for large-scale destruction. They learned that humans' evolutionary lineage diverged from other apes because of humans' capacity for violence. Popular works by zoologists and ethologists Konrad Lorenz and Desmond Morris, as well as playwright and science writer Robert Ardrey, espoused these ideas.[18] Symposia, classroom activities, movies, and other exhibits followed. For instance, in 1966, the University of Chicago hosted a symposium entitled "Man the Hunter," which resulted in a book of the same name published in 1968.[19] The book brought together recent ethnographic research on hunter-gatherers and argued that hunting separated humans from apes in their "biology, physiology, and customs."[20] It further claimed that hunting was once a universal way of life, necessary for human survival. Given the importance of hunting, it must have been "both easily learned and pleasurable."[21] By seeing hunting as an enjoyable genetic and social characteristic that had been shaped by selective pressures, the speakers argued it was difficult for humans to exercise restraint.[22]

Martin's thinking reflected these ideas about human aggressiveness. In 1973, Martin published results from a computer simulation of overkill. It posited that one hundred "Paleoindians," as he now commonly referred to the first people in North America, arrived near present-day Edmonton, Alberta, approximately 11,500 years ago (fig. 14.1).[23] Martin then posited that the population moved south at a rate of about thirty-two kilometers per year and killed one animal per person as they traveled. As they migrated and hunted, they also reproduced, doubling their population every twenty years. With those

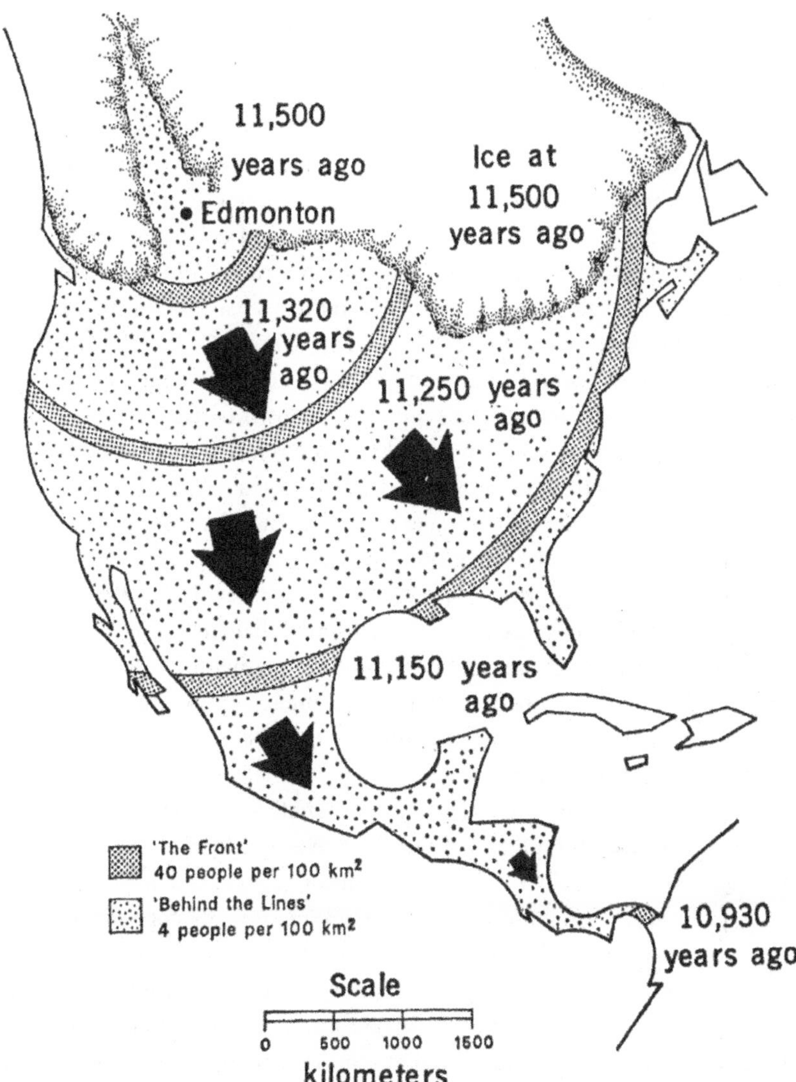

FIGURE 14.1. Martin's model showing the spread of the first people in the Americas, who, in this model, would have swept through the Western Hemisphere and decimated its fauna within about a thousand years of their arrival. Paul S. Martin, "The Discovery of America," *Science* 179, no. 4077 (1973): 972.

estimates, in three hundred years, Paleoindians, whose population would have grown to one hundred thousand people spread over sixteen hundred kilometers, would have killed over ninety million megafauna. Martin wrote, "We need only assume that a relatively innocent prey was suddenly exposed to a new and thoroughly superior predator, a hunter who preferred killing and persisted in killing animals as long as they were available."[24] On this model, Paleoindians would have extirpated the megafauna from North and Central America about six hundred years after the first hunters went south from Edmonton. The animals would be extinct from South America about five hundred years later.

Critics seized on Martin's assumptions about birth and migration rates, arguing that both were much too high.[25] Given these critiques, Martin and computer modeler colleague Jim Mosimann teamed up in 1975 to produce models that assumed slower birth and migration rates. All but one of these more conservative estimates resulted in Paleoindians reaching the tip of South America a thousand years after they arrived in the Americas, with the megafauna hunted to extinction as they migrated.[26] The model that failed "is one in which the hunters manage to regulate their own density and develop a conservation program."[27] Martin and Mosimann rejected this model because sustained yield practices did not align with the archaeological record. They also rejected it by positing that humans, due to their killer instinct, lacked the capacity to regulate their consumption, an assumption which they did not feel the need to support, and which went unquestioned by critics. Humans clearly "preferred killing" and hunted beyond their immediate needs.

With few critiques of assumptions about human nature, Martin was able to use the Pleistocene extinctions as a prominent example of human destructiveness. In 1969, Martin discussed human aggression at a symposium on "violence on the American scene and war as a form of violence."[28] In his opening address, Martin cited the Pleistocene extinctions as well as evidence of crushed baboon skulls found alongside the bones of early man (which Martin called an "assassination") to argue that "you can make a case for man having been a killer ape." He went on to claim that it was time to stop "cling[ing] to the notion that prehistoric man was an innocent animal, capable of no wrong, with no destructive tendencies."[29] At humanity's root, Martin asserted, was a capacity for violence or, as a newspaper covering the symposium wrote, Martin's overkill model "indicated that man may be by his nature a violent creature."[30]

Given humans' inherently destructive natures fueled by the thrill of a hunt in a landscape full of possibility, Martin proclaimed that humans were killer apes. In the mid-1970s, newspapers from the *Los Angeles Times* to the *Chicago Tribune* ran headlines on overkill, declaring ice-age man "guilty of overkill" and "Man Always a Natural Killer."[31] The Associated Press reported that "primitive man slaughtered far more of these great beasts than he needed for food because he had an innate desire to kill, an urge which has persisted through the ages and plagues society today."[32]

The early understandings of overkill thus capture the universalizing tendency present in Anthropocene narratives. Overkill relied on widely held beliefs about the destructive tendencies inherent in human nature to make claims about both the past and present. Scholars have pointed out that this homogenizing formulation may be unhelpful to current environmental problems. By seeing human dominion over nature as a panhuman trait, humans are always outside the environment, and human presence will always be a significant threat to biodiversity.[33] With overkill supporting the belief that human actions always negatively impact the environment, there is little room for actions that might lead to sustainability or resilience. Despite this critique, observers ranging from Elizabeth Kolbert to Yuval Harari to John McNeil suggest that humans were "overkillers" from the start by referencing the Pleistocene extinctions.[34]

Scientific Critiques of Overkill

While overkill captured imaginations during its formulation and continues to resonate as a prime exemplar of human destruction in Anthropocene narratives, many scientists have disagreed with Martin's theory. Archaeologists and anthropologists, in particular, were and remain unconvinced by the evidence.[35] They have long noted that there were few kill sites, places where human artifacts were found in association with the remains of extinct fauna. As anthropologist Grover Krantz, who developed a multicausal account of the extinctions where hunting only played a minor role, put it in the 1970s, "the absence, in most cases, of the remains of horse, camel, mastodon, or pronghorn from kill sites and their occurrence in natural deposits hardly support the contention that man killed them off."[36]

This absence was particularly striking when archaeologists compared the American and European records. In eastern Europe, scientists had found megafauna remains alongside art objects and other cultural artifacts, but there were neither cave paintings nor household objects in the North American archaeological record. There were few sites that indicated any associations between humans and many of the extinct species of megafauna. At North American sites where scientists found the extinct genera in archaeological contexts, scientists found only thirteen of the thirty-five genera that Martin had argued went extinct due to human hunting. The small number of sites, along with the small number of species represented in them, led to a concern that the evidence only supported anthropogenic explanations for the extinction of mammoths and mastodons, rather than the whole host of species implicated in Martin's hypothesis.[37]

Martin, however, did not view the paucity of sites as a reason to call overkill into question. Instead, by the 1970s, he was using his simulation models to explain the absence of human-megafauna associations in the fossil record. As he reminded a graduate student several years after he first adopted this line of reasoning, "a very narrow extinction 'window' makes it easy to explain the absence of extinct fauna in archaeological contexts."[38] As he explained, Paleoindians had rapidly killed and butchered the megafauna as they populated new territory. They would not have had time for cave paintings depicting the new beasts they encountered. They were too busy hunting, migrating, and reproducing. Further, Martin believed that Paleoindians killed animals as they found them, meaning that kill sites would be scattered and ephemeral. Paleontologists would not see the distinctive remains such as those found at buffalo jumps, the cliffs where later the Indigenous peoples of the Great Plains drove bison to their deaths en masse. Since Martin posited that the human population was relatively small, not very dense, and on the move, they would have been unlikely to leave much archaeological evidence. Given these assumptions, Martin claimed, "the only remarkable aspect of New World archaeology is that *any* kill sites have been found."[39] This claim that overkill would have left little archaeological evidence effectively made Martin's hypothesis untestable. Archaeologists were unlikely to find evidence to support Martin's views. It was now the lack of evidence that buttressed his claims.

Many archaeologists and anthropologists involved in the critique against overkill were concerned about how Martin used negative evidence to strengthen his theory. In their view, Martin was not practicing good science. Anthropologist Shepard Krech III, for example, lamented, "If only there were numerous archaeological sites with associated extinct megafauna to test Martin's thesis of overkill. But there are only fifty or so sites—a mere handful."[40] Archaeologists Donald Grayson and David Meltzer argued that Martin's move to turn the lack of evidence for overkill into empirical support "removes the hypothesis from the realm of science and places it squarely in the realm of faith."[41] These critics found it "amazing" that Martin would continue to support overkill.

Given these archaeological and anthropological critiques, we should be wary of importing overkill narratives of human destructiveness into our big histories. Using overkill as a key example of human destructiveness ignores ambiguities in the fossil record. The story is likely more complex than any soundbite story of overkill by Paleohunters.

More importantly, adopting overkill wholesale discounts Indigenous ways of knowing the deep past, even though there is empirical support for oral traditions. Despite the strength of Indigenous arguments, Martin and others were not willing to engage with Indigenous critics. Martin had engaged with anthropologists and archaeological critiques of the hypothesis, offering a new explanation that accounted for the lack of evidence. When it came to Indigenous concerns, Martin, and others, dismissed them out of hand. Martin believed that Indigenous views were no more than politically motivated narratives that relied on unsubstantiated evidence. Denying Indigenous peoples' oral traditions deprived Indigenous people the agency to participate in discussions of the deep past. Because narratives about the past shaped understandings of Indigenous peoples and their ability to manage the environment in the present, scientists denied contemporary Indigenous peoples a voice.

Questioning Native Americans' Destructive Tendencies

Up until this point, I've suggested that the overkill hypothesis—despite its shaky evidence—resonated with concerns about humans' destructive tendencies. This reading of the deep past made sense when Martin first proposed the theory, and it resonates today amid the universalizing narrative of the Anthropocene. But overkill was not always used to universally blame all humans; some formulations blamed specific peoples. Beginning in the 1980s, scientists singled out Native Americans as agents of overkill in the Americas.

When Martin originally proposed overkill, it would have made little sense to identify Native Americans as destructive agents. Ongoing ideas of the primitiveness and the "vanishing Indian" supported claims that Indigenous peoples would not have been technologically capable of such widespread destruction.[42] With the rise of environmentalism in the 1960s, Native Americans gained traction as the most ecologically minded human beings. Indigenous peoples were depicted as "ecologically noble savages" who lived, as they always had, in perfect harmony with the environment.[43] The famous antilitter campaign featuring a "crying Indian" shedding a tear for the wanton environmental destruction brings to life this assumption.[44]

Scholars have shown that the "ecological Indian" set an impossible standard for Indigenous peoples: it allowed Euro-Americans to judge how "authentic" Indigenous peoples were in living in a harmonious relationship with the environment. Indigenous peoples could not reach this ideal, especially since white scholars judged their actions against modern, settler-colonial standards of conservation and environmentalism. These standards had roots in the Judeo-Christian tradition that privileged sacrifice and self-denial rather than excess and self-indulgence.[45] Environmentalists had clashed with Indigenous peoples over sealing, whaling, and fishing. Indigenous peoples were supposedly extracting from the land in ways inappropriate to the vision of white environmentalism. The possibility that Indigenous peoples were responsible for widespread extinctions thousands of years prior suggested a level of environmental mismanagement that would place Indigenous peoples' actions on par with modern, settler-colonial society. In some cases, writers viewed Indigenous peoples' actions as worse than those of contemporary Western societies.

Jared Diamond was among the first to begin transforming the image of the "ecological Indian" into the "unecological Indian" using overkill as his prime example. Diamond had originally studied physiology but had expanded his scientific interests into evolutionary biology and biogeography by the 1980s.[46] He was concerned about the depletion of resources and questions of colonization, including why some populations were able to dominate others, which he explored in his first popular book, *The Third Chimpanzee* (1991). Related to these interests were two articles in *Nature* in the early 1980s. They discussed global extinction patterns, including overkill. He warned that "man the exterminator risks eliminating more species in the next few decades than in all his previous history."[47]

In *Discover* magazine in 1987, Diamond wrote enthusiastically about overkill for popular audiences and singled out Indigenous peoples. The article began by claiming that "even before archaeology uncovered evidence that the first Indians had emigrated from Asia, many scientists suspected that that's what happened, because the Indians of today look so much like contemporary Asians."[48] The article then explained how these people could wipe out North America's megafauna using the mechanism proposed in Martin's models, which had simulated "overkill by Paleoindians" (fig. 14.2).

Indictments like Diamond's encouraged Native American activists, led by Vine Deloria Jr., to critique overkill. Deloria was a member of the Standing Rock Sioux who had attempted to demythologize how white Americans thought of Native Americans since the publication of his first book in 1969. He objected to how people like Diamond used the overkill hypothesis to suggest that Native Americans had mismanaged the environment.[49] In Deloria's estimation, this conclusion supported "conservative newspaper columnists, right-wing fanatics, sportsmen's groups, and scholars" who believed that Native Americans lacked "moral fiber and ethical concern for the Earth."[50] Deloria objected to this line of thinking, but rather than critique Native Americans' so-called destructive nature, he most strongly condemned the hegemony of scientific theories, which he claimed had little empirical support compared to Native American knowledge systems. Deloria used the overkill case as a prime example of this problem in his 1995 book *Red Earth, White Lies: Native Americans and the Myth of Scientific Fact*.

In *Red Earth, White Lies*, Deloria began by showing how archaeologists

FIGURE 14.2. Excerpt from Diamond's article depicting mammoth hunting and moralizing about early hunters' lack of conservation compared to modern humans' ability to learn from the past. Jared M. Diamond, "The American Blitzkrieg: A Mammoth Undertaking," *Discover*, 8 (June 1987): 88. Illustration by David A. Johnson; reproduced by permission of the illustrator.

and anthropologists had struggled to find evidence of large-scale human-caused extinction at the end of the Pleistocene. Deloria wondered why this ill-supported scientific theory had gained popularity given these problems, especially when Native American oral traditions provided strong evidence that climatic change caused these extinctions. He especially objected to Diamond's statements in *Discover*, such as, "it's highly suspicious that the sloths and the goats disappeared just after Clovis hunters reached Arizona. Juries have convicted murderers on the grounds of less compelling circumstantial evidence."[51] Deloria thought that Diamond's comments were "symptomatic of the manner in which scientists have tried to indict Paleo-Indians for the massive extinctions." Deloria likened the condemnation of Indigenous peoples to a "southern lynching," where a single accusation "was enough to ensure a conviction," even though there was "little evidence that Paleo-Indians hunters did anything more than occasionally catch a mammoth at a watering hole."[52] By invoking this example, Deloria showed how race influenced narratives of culpability, with a dominant group able to blame a marginalized group with thin evidence.

Deloria continued his critique by exploring Western science's hegemony over Indigenous cosmologies. Deloria's critique of science related to broader issues that overwhelmed North American archaeology and anthropology beginning in the 1980s. The issues became more vexed in the 1990s when Indigenous scholarship took a decolonial focus and advocated self-determination in land claims. The central question was whether Native Americans would be able to present their history or whether others would act as their spokespeople, a question Indigenous scholars had begun asking in the 1960s.[53] Native Americans complained that anthropologists had long believed that they "could know more about Indian tribes than the Indian" and maintained ownership over Indigenous artifacts and history by creating a homogenizing Euro-master narrative.[54] Inspired by Deloria and others' work, Indigenous researchers and communities came to insist on a research ethic responsible to Indigenous communities and research useful and valid for Indigenous peoples. But Deloria thought that more needed to be done to take Native American voices seriously and ensure that scholarship on Indigenous peoples would stop playing a political role in legitimizing, perpetuating, and sustaining imperialist attitudes toward Indigenous peoples.

Deloria's critique of science as a purveyor of truth in *Red Earth, White Lies* fits within this larger discussion over who could tell which stories, what evidence counted, and what researchers' responsibilities were to those they studied. In an interview, Deloria stated: "I'm not anti-scientific, although I write in that style. What I'm trying to push is that to be truly scientific means you have to be empirical; you have to have data and facts to support your beliefs."[55] He thought that the indictment of Paleoindians lacked empirical support. For example, Deloria claimed that Martin's model of a front of migrating hunters was not sensitive to how animals move and how hunters hunt. He advanced this idea in an interview in 2004: "They say a hundred Indians came down from the north and killed a mammoth a week each and as they marched south they left no animals behind them. That's stupid. Animals have rhythms and rotations. You could kill a whole bunch this summer and they'd just sneak around and go back up north or back west or away from the direction you were. If you could clear an area of animals, American white hunters would have done this with the deer 200 years ago. Those animals aren't bound to be in an area. Now, hundreds of Indian scholars out of their own personal knowledge could easily refute all of that. But there are no Indians there saying, 'Wait a minute . . . .'"[56]

In Deloria's estimation, the unfalsifiability of overkill and its difficult-to-support claims should have encouraged scientists to rethink their theories. Martin had done this with anthropological and archaeological critiques of his theory, but he did not do so given Deloria's concerns. Deloria thought it would have been easy to look at other evidence, especially from Native American knowledge systems, rather than discounting this evidence. By "casting aside" the "Indian explanation . . . as a superstition," Deloria wrote, "scientists have maintained a stranglehold on the definitions of what respectable and reliable human experiences are." Native Americans had explanations of "their origins, their migrations, their experience with birds, animals, lands, waters, mountains, and other peoples," but, in a settler-colonial society, these were never considered valid.[57]

Deloria claimed that the archaeological and geological record could support Native American oral traditions that suggested climatic explanations for the extinctions. He pointed to many examples of this kind of corroboration. In a Native American account discussing the origin of the Chief's Face, a rock formation on Mount Hood, Oregon, an elder commented: "In those days [early times] the Indians were also taller than they are now. They were as tall as the pine and fir trees that cover the hills, and their chief was such a giant that his warriors could walk under his outstretched arms." But the mountain exploded, and the people could not live near it for a long time. When they returned to the area, "the children, starved and weak for so long, never became tall and strong as their parents and grandparents had been."[58] Deloria demonstrated how this tradition described a condition of malnutrition that might arise after several generations of food scarcity. The same could be true of the megafauna, which, forced out of favorable living conditions as the ice age ended, became smaller and more like the animals that currently roam North America.

Deloria further showed how other Native American traditions invoked climatic explanations. In one story about bones at Salt Lick on the Ohio River, a Delaware chief recounts, "In ancient times a herd of these tremendous animals came to the big-bone licks, and began a universal destruction of the bear, deer, elks, buffaloes, and other animals: that the great Man above, looking down and seeing this, was so enraged that he seized his lightning, descended on the earth, seated himself on a neighboring mountain, on a rock of which his seat and the print of his feet are still to be seen, and hurled his bolts among them until the whole was slaughtered, except the big bull."[59] Native American commentators took this to mean that climate, not hunting, caused these extinctions. Frustrated that Native Americans' explanations were rarely accepted, many began to corroborate these myths with the geological record and scientific findings, locating evidence of these more catastrophic climatic changes in the geological record.[60] But science, they argued, was not the only way to know; Native Americans should be taken seriously without resorting to Western standards of truth.

As Native Americans critiqued overkill, they were expressing sentiments about hegemony and limits of science, but Martin and others did not see their critique in this way. Soon after the publication of *Red Earth, White Lies*, Martin responded to an interviewer's question asking if Deloria's concern was about the lack of archaeological data. Martin disagreed, arguing that Deloria condemned overkill for political reasons.[61] As Martin explained, "Vine Deloria's trashing of overkill reflects (in my view) his lifelong war against anything that he believes might reflect badly on Native Americans and/or upset those who do not believe that Native Americans were always here (in the New World). I doubt he was ever seriously interested in the extinctions during near time or in the problem of what caused them."[62] This response resulted from Martin's view that the evidence from models and the geological record pointed to hunting as the cause. He thought that unqualified people like Deloria were participating in identity politics rather than raising legitimate questions about the indirect scientific evidence that supported his theory.[63] Martin had engaged archaeologists and anthropologists by developing a new narrative about the absence of archaeological evidence when they had described the limited evidence. He would not do the same for Native Americans who pointed to climatic causes in their oral traditions. Martin dismissed these out of hand, allowing the myth of "Indian" savagism to reappear in this new formulation of overkill.

The image of the "savage Indian" featured prominently in Shepard Krech III's book, *The Ecological Indian: Myth and History*, published in 1999. The widely acclaimed book detailed the ecological sins of Native Americans, opening with a chapter on the Pleistocene extinctions. The book uses colonial sources to demonstrate that Native Americans exhibited destructive behavior against the nonhuman world. As Gesa Mackenthun points out, "Krech was obviously reacting to what he perceived as an unacceptable romanticization of Native Americans' environmentalist habitus disseminated through activities of the environmental movement."[64] For the environmentally inclined, it was not uncommon to "play Indian," which included adopting idealized versions of Native Americans' ecological practices.[65] But Krech suggested that Native Americans contributed to species extinction, undermining the image of the "ecological Indian."

The modified narratives of culpability that Krech, Diamond, and Martin presented resonated with readers trained in the colonial mythology of "Indian" savagism. Such attitudes devalued evidence from oral traditions. Overkill by Paleoindians discounted Indigenous cosmologies and denied Indigenous peoples' rights to the land. As one journalist wrote after reading of the overkill theory in Krech's book, "Canadian Indians should not be accorded the superior sanction of high-minded environmentalism in negotiations of land settlements. . . . It should also mean much more balance in responding to native demands and needs simply because discussion of them no longer should be burdened with the guilt piled on the whites for devastating a noble people whose societies once lived—and might do so again—in perfect harmony with nature. Indians are neither more noble nor ignoble than other people—in their blood, or in their history."[66] Some of Deloria's concerns about the way an unsupported theory about the deep past could be used to deny present-day land claims had materialized: Martin, Diamond, and Krech's privileging of supposedly apolitical scientific evidence over supposedly political oral traditions meant that Deloria and others had little ability to respond.

Conclusion

As this volume makes clear, earth knowledge has always existed in multiple traditions. In the case of the Pleistocene extinctions, conservationists largely followed Martin and made humans the agents of extermination. Archaeologists and anthropologists preferred climatic and multicausal explanations where humans played only a minor role. Native American oral traditions spoke of climatic factors as a key cause. And yet, only one of these narratives—overkill—resonated widely. The multicausal explanations of archaeologists demanded refutation, but the earthly cosmologies and oral traditions of Indigenous peoples did not warrant engagement. Scientists denied Indigenous narratives about earth's archives, along with Indigenous narratives about human agency and culpability. Instead, the power geometries and colonial politics were left alive and well in formulations of overkill that blamed Native Americans. Accepting overkill masks these ongoing colonial legacies.

The overkill theory is similar to readings of earthly archives in this volume where the geologically ancient and political geology of modernity collapse together. But the merging of geological evidence with modern politics has long had critics, albeit ones who have been silenced by the prevalent way of knowing and the ready resonances of dominant ideas with popular political views. As scholars navigate questions of causality in the Anthropocene, they must consider the presence and dominance of certain cultural perspectives along with the absences the dominant perspective can perpetuate. Erasures exist in universal narratives of human culpability *and* silences can abound in more specific narratives of responsibility. Scholars have tried to move away from the universalizing narrative of the Anthropocene because of concerns about power, but, as I've shown here, specific narratives of human actions can maintain power imbalances as well.

In readings of the past, myths and stereotypes can be buttressed to read the geological record in multiple ways. Narratives that evoke "ecological Indians" and "savage Paleoindians" reinforce settler-colonial ideas about the environment and Indigenous peoples. These narratives then support power imbalances by denying that there are multiple ways of knowing the deep past. Only the dominant tradition is given a voice.

Most scholars considering agency and responsibility in the Anthropocene are attuned to power imbalances, but the overkill example demonstrates that it is all too easy for particular readings of the past to reinforce the inequalities many are trying to overcome. We do not just need specific narratives to overcome the universalizing and homogenizing narratives of the Anthropocene. We also need diverse voices and sources to narrate the past.

15

Imperial Melancholy and the Subversion of Ruins in the Amazon

Raphael Uchôa

> Not even the simple and modest moss that as a symbol of melancholy covers the ruins of ancient Roman and Germanic grandeur stretched over the remains of that South American antiquity: there dark and ancient virgin forests hid the monuments under the humus and dead debris of the peoples of long gone and everything that the hand of the man of old created is covered by layers of incalculable decomposition.
>
> CARL F. P. VON MARTIUS, 1832

> The Indians have been resisting for five hundred years, I am concerned with whites, how are they going to do to escape this.
>
> AILTON KRENAK, 2020

Ann Stoler speaks about thinking of the *ruins* "as sites that condense alternative senses of history" and *ruination* as "a process that weighs on the future and shapes the present."[1] In this chapter, I explore the possibility of alternative meanings of history, particularly related to the notion of ruins and its consequences. The first notion of ruins analyzed here was apprehended in the nineteenth century by the German naturalist Carl F. P. von Martius and transformed into a hermeneutical tool to impose an interpretation of the supposed biological and civilizational state of the Indigenous peoples of South America in the nineteenth century.[2] The above quotation by Martius is the sign of the overlapping of a melancholic feeling of ruins—evoked from the European romantic aesthetic of the late eighteenth century—with the supposed ruined and decomposed state of the Indigenous people of South America. This sense of Martius's metaphor was adopted as authoritative by several scholars in the global north in the nineteenth century, including the British geographer Clements Markham, the Scottish physician John Macpherson, and the American physician Samuel Morton.

In contrast, the second sense of ruins was conceptualized by two Indigenous thinkers from Brazil: Ailton Krenak and the Yanomami shaman Davi Kopenawa. Since the end of the twentieth century, both thinkers have (re)conceptualized the notion of ruins as a derivation of capitalist modernity. Between the two historical moments under analysis—Martius and other Europeans in the nineteenth century and Krenak with Kopenawa in the twentieth century—several authors have explored the analytical power of the notion of ruins and others have legitimized for the West the epistemic participation of populations historically relegated to being objects of natural history and anthropology. In the first case, Stoler's works inspire the present analysis insofar as she works with the notion of the "ruinous process in imperial formations," instead of empire per se.[3] I discuss in another work how the relationship between the German political and scientific project with Brazil in the early nineteenth century did not follow a colonial model in the traditional sense of land occupation and control, as was the case of the Portuguese empire.[4] Instead, it was an *imperial formation*, or a *relation of force*, marked by alliances between European monarchical houses for the exploitation of South America and the conceptualization of Brazil as a nation.[5] This conjuncture resulted in exploratory and racialized policies that effectively resulted in the denial of the dignity and the rights of native people, which left profound marks on the Indigenous populations of Brazil.

In addition, perhaps more important for our analysis, is the notion presented by Stoler that "ruin is both the claim about the state of a thing and a process affecting it."[6] Martius's conceptualization, romantic and imperial, printed on pages widely read in Brazil and around the world, about the supposed ruined status of the "American race" (referring to all Indigenous peoples from the Americas), served for decades as a conceptual device in the hands of the Brazilian state—a state invented in the nineteenth century that had as one of its main functions the paternalistic and colonial intervention in the lives of the Indigenous populations to exploit their resources and territories.[7]

The denial of and resistance to this concept of ruination as a pillar of the Brazilian national project has been historically constant among the Indigenous populations in Brazil. This chapter also analyzes this resistance as a form of conceptual subversion of the rationality underlying the notion of ruins. To analyze this process and vault the European characterization of "the Indigenous condition," I turn to what Viveiros de Castro calls an anti- or counteranthropology emerging from ethnographic data and written narratives of a series of Indigenous people from the lowlands of South America.[8] In this chapter, I refer to *Ideas to Postpone the End of the World* and *Life Is Not Useful* by Krenak and *The Falling Sky: Words of a Yanomami Shaman* by Kopenawa.[9]

According to Viveiros de Castro, counteranthropology is defined as Indigenous anthropology, that is, anthropology from the point of view of the Indigenous people. This notion, derived from his concept of "Amerindian" perspectivism, has been deeply influential in studies that seek epistemological alternatives for the (re)conceptualization of Western constructs such as *nature*, *culture*, *society*, *human*, and *nonhuman*.[10] My analysis of the conceptualization of ruins from the Indigenous point of view is the foundation of one of the arguments of this chapter: that there was in the twentieth century not only the subversion of the notion of ruins but also of its *object*, which was reoriented from the Amerindians to the Europeans. From the Amerindian perspective, the feeling of melancholy of the ruins continued to exist, resignified with the notion of the Anthropocene and the resulting extinction of humanity, particularly Western humanity. It is in this sense that Krenak declares in the quotation at the beginning of this chapter concern for how the whites will escape extinction, in his view, an inevitable consequence of capitalism.[11]

From the European perspective, notions such as *extinction*, *degradation*, and *degeneration* were fundamental terms in the field of natural history, archaeology, and geology in the eighteenth and nineteenth centuries.[12] In fact, to testify the antiquity of the "American race," Martius used a geological language derived from debates on the natural history of the earth to characterize the Amerindian state as a geological strata and as "orographic formations, stratified, one on top of the other."[13] His reasons for such a supposed state were derived from a catastrophic view: the "American race," once a great civilization, was suddenly and violently destroyed by earthquakes at some point in time. Martius expressed this in the language of theodicy, arguing that such a state was somehow part of a bigger divine plan, and, therefore, the extinction of the "American race" would fulfill some God-designated purpose.[14]

A German Naturalist in South America and the *Ruination* of the Amerindians

Several European traveling naturalists described and depicted South American nature in the nineteenth century. These descriptions also included the Indigenous peoples, who were considered part of the natural history of the regions visited, alongside the flora and fauna. One of these naturalists, Martius, traveled to Brazil as a member of an Austrian expedition to South America in 1817. In Brazil, Martius and zoologist Johann von Spix traveled from Rio de Janeiro, the then imperial capital, through the hinterlands up to the Amazon basin. The expedition, one of the first to Brazil, ended in June 1820. On his return to Europe, Martius carried thousands of botanical, mineral, zoological,

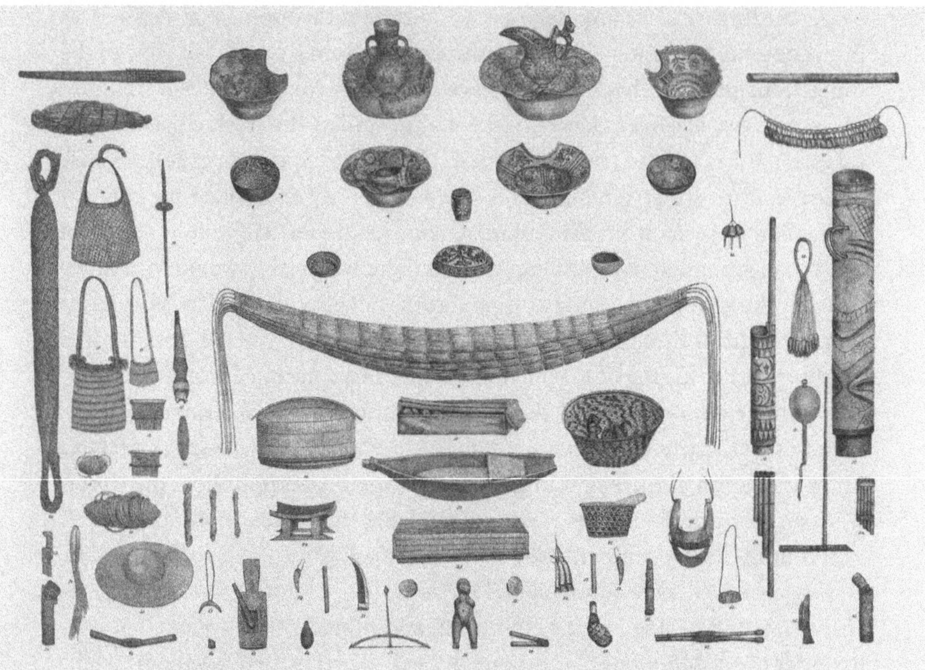

FIGURE 15.1. Illustrations of ethnographic specimens brought by Martius and Spix to Munich from Brazil, including finely painted pottery, woven bags and hammocks, a wooden boat, instruments for spinning and weaving thread and rope, carved bone and wood blades, a bow and arrow, a clay human figure, jewelry, and musical instruments. The caption translates as "Indian Instruments." Carl Martius and Johann von Spix, *Atlas zur Reise in Brasilien*, 1817–1820. Courtesy of Cambridge University Library.

and ethnographic specimens, some of which ended up at the Cabineten der Kunst Akademie der Wissenschaften zu München (fig. 15.1).

Martius's travels, publications, and the institutional spaces through which he circulated were sponsored by a complex transatlantic consortium of monarchs, including the Kingdom of Bavaria, the court of Habsburg Lorraine in Austria, and the Empire of Brazil led by the Brigantine court.[15] Martius's expedition to Brazil was in the context of the arranged marriage of Archduchess Maria Leopoldina, daughter of Francis I of Austria, to Pedro, the son of Dom João VI, king of Portugal, Brazil, and the Algarves—the future Brazilian emperor Dom Pedro I. Thanks to these political connections, Martius succeeded in transporting books, plants, and ethnographical material back to Europe, had access to Iberian libraries, and became the ultimate authority on Brazil and the "American race."[16]

The epistemological steps through which Martius built the notion of the "American race" are interwoven with the notion of ruins. In his work, Martius used the symbol of ruins to conceptualize the Indigenous population of the

Americas. By "ruins," he meant that the Indigenous peoples in his time were not in their original state of civilization but represented vestiges of an ancient people with an elevated understanding of nature. A key moment in the process of the emergence of Martius's vision on the status of the "American race" occurred during his visit to the Museo Nacional de Ciencias Naturales in Spain, in the context of his return from the Amazon to Bavaria in 1820. The return route, from Brazil to Bavaria, departed from the northern Amazon, near Belém, in the state of Pará. Martius entered Europe via Lisbon. He brought specimens of natural history and ethnography—destined for the Royal Museum of the Academy of Sciences in Munich—and a botanical collection with about sixty-five hundred species.

Upon returning from Lisbon, Martius was received by Bavarian and Austrian statesmen and by Hanseatic consul Lindenberg, who hosted the German scholars and took them, with their collections, to the Casa da Índia, the regulator of commercial activity, established in 1503 to manage Portugal's overseas territories.[17] Martius also planned to travel around Lisbon and meet local scholars, but the conditions of his stay were shaken the day after his arrival by the Liberal Revolution in Porto, an anti-absolute movement launched on August 24, 1820, in the city of Porto.[18] Martius, a loyal monarchist, felt the shock of the political tremor and went on to Munich, traversing the European continent through Spanish lands, rather than the Mediterranean Sea.

Contact with the former colonizers of the Americas, especially in Madrid, was a decisive factor in the elaboration of Martius's ethnographic and comparative thinking. It was in the capital of the Spanish Empire that Martius gained access to the natural remains cataloged in the Museo Nacional de Ciencias Naturales, especially the "documents of the civilization of the ancient peoples of Mexico and Peru."[19] Martius reports the presence of "urns, lamps, metal weapons, gold leaf idols, diadems, etc.," objects that, for him, constituted indexes of an advanced civilization. The contact with such objects in Madrid prompted Martius to theorize comparisons between the different Indigenous peoples of the Americas. He established a comparison between the Indigenous peoples he met in Brazil and those in Mexico and Peru, with whom he only had access through the objects stored in Madrid. This comparative device was fundamental in his framing of the Indigenous peoples of Brazil as part of a much older race.

In addition to the artifacts from Hispanic America, two conceptual elements form part of the theoretical framework that shaped Martius' reconceptualization of the "American race": the romantic roots of his natural history and the degenerative theories that informed his thinking. Martius's romantic notion of "ruin" is enunciated in *Frey Apollonio: Ein Roman aus Brasilien*, a fictionalized version of his travels in the Amazon, finalized in 1831 but published posthumously. In a passage, Hartoman, the protagonist, asks Friar Apollonio, a Jesuit tending to the Amazonian people, about his impression of "these poor savages." Apollonio replies:

> I have an analogy, which expresses my opinion: the American humanity seems to me to be an immense ruin. They are decadent remnants of a powerful construction erected long ago—probably thousands of years ago—colossal, fantastic, of a style only apprehensible in dreams.[20]

Martius regarded the *remains*—ancient architectural monuments in Hispanic America—as archaeological evidence that could be deciphered, together with mythology. He drew attention to the "colossal buildings, comparable to the ones of the Egyptians, such as those in Tiwanaku, near the southeastern shore of Lake Titicaca, which the Peruvians, at the time of the Spanish invasion, admired as remains of an extremely ancient and numerous population." Martius used architectural language to support the concept of the ruin and decay of the "American race": "These and similar ruins scattered through the two Americas bear witness to the fact their inhabitants in remote centuries had a moral force and a civilization that today are entirely lost."[21]

Martius also used the geological language derived from seventeenth- and nineteenth-century debates on the natural history of the earth to characterize the Amerindian state.[22] He compared the ancient peoples of the Americas to the "orographic formations which constitute the crust of our planet" that were "stratified, one on top of the other." While "many have disappeared entirely as if they had been buried by their successors, others appear to us as a mixture of primitively unequal elements, combined in different ways, then evolving into new combinations." Finally, he says, "The earliest legends and stories allude to a *few great human masses*," but "the closer we come to modern times, the more they appear individualized in language, morality and locality." The correlation between the antiquity of the "American race" and geological strata is clearer in the following passage:

> In deciphering such historical evolutions, the historiographer is bound by the same method as the naturalist, because as he investigates the age and succession of the geological formations by the remains of the disappeared organisms, he receives precious indications of the essence and state of the previous humankind by the language and various customs and habits which, from a remote, pure or altered past, have been conveyed to the lives of later people.[23]

On these grounds, Martius concluded, "Noting the extreme fragmentation in small peoples, tribes, and hordes, in complete isolation . . . they appear to us as a formation of men disaggregated by volcanic forces in ceaseless activity." Such a state "cannot be the result of modern catastrophes; [but] indicates the action of millennia."[24] He added religious aspects to this enigma. Acquainted with Leibniz's ideas, he hypothesized that the contemporary state of *degradation*, *dehumanization*, and eventual possible *disappearance* of the "American race" evoked a theodicy. He asks whether their condition could be "a consequence of inveterate and brutal vices with which does the genius of our race punish both the innocent and the guilty, and whose severity to all nature, to the superficial observer seems an incoherent cruelty?"[25] From a general and teleological plan, which he believed to be imprinted into the very fabric of the universe, Martius developed the idea that the "red" race had a "defect in its organization . . . and already bore the germ of disappearance, as if it were destined to play an automatic role in the great gear of the world, a mere step on the human evolutionary scale." Moreover, there was no doubt according to Martius that "the American race is about to disappear" from the world stage, in which it had played a significant role but was now in a state of ruin.[26]

The supposed state of ruin of the Indigenous peoples of the Americas defended by Martius had historical and conceptual roots in the debates on degeneration in the second modernity. In the eighteenth and nineteenth centuries, the concept of degeneration was elastic from a semantic point of view. It was a fundamental concept in works on the natural history of the period, more precisely the natural history of the human being. Johann Blumenbach, for example, understood that human varieties had been derived from a firstborn through a process of degeneration, which would have nutrition and hybridization among its causes, among other factors.[27] One of Blumenbach's main sources was the French naturalist Georges Buffon's works. According to Phillip Sloan, in the mid-1770s, Buffon was concerned with the puzzle of geographical distribution between the Old World and the Americas, comparing information about humans and animals reported by travelers and hypothesizing an ancient path from Asia to America, which would support the thesis of continuity between peoples and animals between the two continents.[28] By 1760, Buffon theorized a necessary degeneration from one species to another, based on geographical, climatic, and nutritional aspects. He posited that these factors could affect "organic molecules" and would have been the conditioning factors for physical and moral differences between the inhabitants of the Americas and those of the Old World.[29]

In the nineteenth century, the term *degeneration*, in the context of anatomical and physiological discussions about races, acquired momentum among European authors.[30] The journal *Nature* provides evidence of the relevance of the concept of degeneration in the period and how Martius's ideas on the subject were debated. In 1874, the journal published a debate entitled *The Degeneracy of Man*, which focused on Martius's degenerationist argument. Edward Tylor started the debate with a "Letter to the Editor" on June 15, 1874. Tylor began the letter by recognizing the great anthropological interest in Martius's work and highlighting his "well-known" thesis that "the savage tribes of Brazil were the fallen descendants of more acculturated nations."[31] For the analysis, Tylor mainly used Martius's work *On the State of Civil and Natural Rights among the Aboriginal Inhabitants of Brazil*, published in 1832.[32] Tylor argued that such a statement by Martius was embarrassing "for students of civilization (me for example)." As a provocation, Tylor stated that this could be expected from someone like Bishop Richard Whately, who "did not examine the evidence," but should not be expected from Martius, who was recognized as an "eminent ethnologist, intimately familiar with thought and wildlife." Tylor questioned how a scholar of such grandeur could support degenerationist theories or declare that the "current state does not represent a natural savagery, but traces of decay from an ancient high culture."[33]

Melancholy and the State of the Amerindian Existence: Martius's Ideas between England and Calcutta

The text quoted by Tylor from Martius's *On the State of Civil and Natural Rights among the Aboriginal Inhabitants of Brazil*, describes a profoundly tragic and melancholic picture of the existential condition of Indigenous peoples within the Amazon rainforest. In it, Martius laments: "there, dark and ancient virgin forests hid under the humus and dead debris the monuments of the long-lost peoples and everything that the hand of the man of old had created is covered by layers of incalculable decomposition."[34] In this text, Martius draws a parallel with the "ruins of ancient Roman and Germanic grandeur" covered by "modest moss as a symbol of melancholy." In this text, the concepts of *ruin* and *debris* emerge as a characteristic element of the "American race," as, on the one hand, it reflects its antiquity and, on the other, it characterizes its state of decay and decomposition. Tropes are extracted directly from the European aesthetic matrix—an aristocratic obsession with the artistic representation of classical antiquity in ruins—to characterize the Indigenous peoples of the Americas.[35]

Not surprisingly, Martius's contemplations of the monuments represented a significant locus for the establishment of conceptual bridges between the general view of the Indigenous peoples of the Americas and his interpretation, which linked ancient ideas about civilizations and their permanence in time. In this sense, saying that the Indigenous peoples were in ruins reveals connections with eighteenth-century neoclassicism, in which the concept of ruins constituted a link between the present and a romanticized past. This was particularly evident in the Germanic lands, where the idea of the *return to Greece and Rome* constituted not merely a search for intellectual inspiration in the works of Aristotle, Plato, Cicero, and Ovid but also a nostalgic ideal of a lost civilization, still present in the ancient ruins. In this sense, Martius, for example, reading Homer or Sophocles, drew clear parallels between the destruction of the "American race" and the destruction of "a race of heroes" in Troy or the destruction of Plato's Atlantis.[36]

In his work, *Nature, Diseases, Medicine and Remedies of the Brazilian Natives* (1844), Martius interprets the ruins as resulting from an "enigmatic situation," which filled him with sorrow and melancholy.[37] To the best of our knowledge, Martius never found an answer to the enigma he had created. However, the melancholic sentiment about the supposed ruined state of the Indigenous peoples of the Americas had an enormous influence on scholars in Europe and North America. It appeared in a text translated and published by Scottish physician John Macpherson in 1846, under the title "The Natural History, the Diseases, the Medical Practice, and the Materia Medica of the Aborigines of Brazil" in the *Calcutta Journal of Natural History*.[38] One of the published texts was Martius's presentation of the alleged enigma of Indigenous peoples in a state of ruin.

In 1845, the same journal published a text by American doctor Samuel Morton under the title "An Inquiry into the Distinctive Characteristics of the Aboriginal Race of America."[39] In it, Morton introduced Martius and his theory of the "present state of the American aborigines" as the "fragments of a vast ruin."[40] Such a state, in Morton's view, was representative of the moral state of the "American race" and could explain "their habitual indolence and improvidence, their indifference to private property, and the vague simplicity of their religious observances."[41] Thus, using the symbol of the ruins, Morton summed up medical, theological, and political convictions by resuming the Lockean notion that the Indigenous peoples of the Americas had no notion of private property as a way of justifying European appropriation of the lands and resources of the Indigenous peoples of North American.[42]

In the United Kingdom, geographer Clements Markham was one of the main propagators of Martius's views on the ruined state of the Indigenous peoples of South America. Markham's interest in South America was notorious, including his role in the propagation of the cinchona tree and its export from South America to India and his contribution toward the production of quinine.[43] Markham translated and edited numerous colonial Hispanic texts on South America. In one of these, *Expeditions into the Valley of the Amazons,*

1539, 1540, 1639, he began a series of surveys on the tribes of South America, published in the years 1859, 1865, and 1910, called "A List of the Tribes in the Valley of the Amazon."[44] Although the focus of the initial edition of 1859 was ostensibly to present reports of some of the main expeditions through the Amazon basin to the English public, Markham presents a list of more than five hundred South American tribes characterized according to the conceptions of Martius.

For example, in the 1859 edition, Markham presents Martius's interpretation of South American "Indians" as that "which gains most credit amongst the philosophers of Europe," and he quotes Martius's melancholic view of the Indigenous peoples' existential condition verbatim in the text:

> The present and future condition of this race of men is a monstrous and tragical drama, such as no fiction of the poet ever yet presented to our contemplation. A whole race of men is wasting away before the eyes of its commiserating contemporaries; no power of philosophy or Christianity can arrest its proudly gloomy progress towards a certain and utter destruction. From its ruins there arises, in the most motley combination, a new and reckless generation, anxious to estrange their newly acquired country from its former masters. The east brings blood and blessings; social union and order; industry, science, and religion; but with selfish views, only for itself; for itself it erects a new world; while the race of men, which was once here the master, is fleeting away like a phantom, from the circle of existence.[45]

Markham copied this passage from a context in which Martius is comparing the Iberian and Germanic colonization of South America. Martius projected two European-centered narratives on South America: one starring Latin or southern Europeans, who "caused the wounds of America," and another starring the Germans, as representatives of Northern Europe, who could either be spectators of the ravaging of the "New World" caused by the "gold thirst of empires" or who could act in a spiritual capacity as the "sowing," exploring the "new territories," since "we Germans, even without colonies, have only one property, the field of the spirit."[46] Since these alternatives were purely rhetorical devices, the developments covered by the terms *sowing* and "field of the spirit" also became a justification for economic gain, and the "American race" the model to confirm a priori ideas on the nature of nation, language, and race.[47]

The Ruin of the Whites, or the Subversion of the *Ruins*

The European narrative of the "ruined native," in the sense of the degeneration of an earlier form, was prominent in modern Europe. It served several political projects in South America and informed a colonial ideology justifying the plundering of the material and immaterial culture of Indigenous peoples in the Americas.[48] Museums were privileged sites for the accumulation of natural history materials in Europe, and Martius's ethnographic collection was no exception. In 1820, after Martius returned from Brazil, the collection resided in the Staatliches Museum für Völkerkunde (now called Museum Fünf Kontinente), and in 1862, it became one of the museum's founding collections.

The first Indigenous people to have access to such artifacts in Germany were two Indigenous children, a girl and a boy, taken by Martius in 1820 from the Amazon to Bavaria as part of the botanical and ethnographic collection—as living specimens. The girl, who belonged to the Miranha tribe, located in the middle Rio Solimões, Martius called "Isabella," and the boy, from the Tapuia tribe, he called "Johannes." Both died shortly after arriving in Munich in the middle of winter (fig. 15.2).[49] Almost nothing is known about what they thought about the artifacts in the collection, of Martius, or about Europe.

More than a century after Martius's travels in the Amazon, it is possible to learn from written texts what some Indigenous people in the lowlands of South America think about the notion of ruins invented in Europe. It is necessary to remember that this notion carried an eschatology about the inescapable end of the Indigenous people, since for Europeans such as Martius and Markham, the Indigenous people already represented reminiscences of a deep past. One of the most critical narratives of this perspective is found in the manifesto narrated by the Yanomami shaman Kopenawa, *The Falling Sky: Words of a Yanomami Shaman*, an Indigenous account of Amazonian cosmology and a masterful critique of Western materialism. In the manifesto, Kopenawa talks about his visit to the Musée de l'Homme in Paris in 1990:

> Another time, I was taken to visit a vast house to which the white people gave the name of museum. This is a place where they lock up traces of ancestors of the people of the forest who disappeared long ago. Here, I saw vast quantities of pottery, calabashes, and baskets; many bows, arrows, blowpipes, clubs, and spears; but also stone axes, bone needles, seed necklaces, bamboo flutes, and a profusion of feather ornaments and beads. . . . Also in that city's museum, I saw the stone axes with which the long-ago inhabitants of the forest cleared their gardens, the hooks made of game bone with which they fished, the bows with which they hunted, the pottery in which they cooked their game, and the cotton armbands they wove. It really made me sad to see all these objects left behind by elders who disappeared in the distant past. . . . Finally, having seen all these things in the museum, I began to wonder if the white people had also started acquiring Yanomami baskets, bows and arrows, as well as feather orna-

FIGURE 15.2. Detail of epitaph (about 1824) for two Indigenous children, "Johannes" and "Isabella," who were taken by Martius in 1820 from the Amazon to Bavaria. Image © Münchner Stadtmuseum (Munich City Museum), Collection of Applied Arts.

> ments because we are already in the process of disappearing ourselves. Why do they so often ask us for these objects, when the gold prospectors and cattle ranchers are invading our land? Do they want to get them in anticipation of our death? Will they also want to take our bones to their cities? Once dead, will we also be exhibited in the glass cases of a museum? This is what all this made me think.[50]

Kopenawa's questioning of the structural reasons for acquiring objects for museums, whether by naturalists or by modern anthropologists, could not be more precise: "Do they want to get them in anticipation of our death?" From Kopenawa's perspective, the Indigenous people might disappear, as predicted by Martius, Markham, and other Westerners, but the cause will not be a biologically ruined and degenerated (Amerindian) nature, but the violent action of white people on their lands and their bodies. It should be remembered that the Yanomami are among the Amazonian tribes that have historically suffered the most from the search for minerals on their lands.[51]

It should also be remembered that the Yanomami associate mining activities (of gold and cassiterite) and the extraction of oil and gas with the weakening and rotting of the earth's layers, as well as the release of pathogenic effluents that disseminate epidemics and biological extinctions.[52] They understand that white people's ignorance of the agency of the spirits and shamans that sustain the cosmological status quo has already begun to unleash supernatural revenge, which has been causing droughts and floods in different parts of the planet. Soon, with the death of the last Yanomami shamans, evil spirits will take over the cosmos, the sky will collapse, and every being will be annihilated.

In this narrative, the foundation of Martius's notion of ruin is subverted. Many Indigenous people in South America transform the discourse of the ruins into criticism against white people (nicknamed giant armadillos or monstrous peccaries, for their incessant activity of digging and shaking the earth), against the West, against modernity, against capitalism, and against the State. As an emblem of this, Kopenawa ironizes Western pride in the ideal of technical and material progress:

> [The white people] probably thought: "There are many of us, we are valiant in war, and we have many machines. Let us build giant houses to fill them with goods that all the other peoples will covet!" Yet while the houses in the centre of this city are tall and beautiful, those on its edges are in ruins. The people who live in those places have no food, and their clothes are dirty and torn.[53]

In Kopenawa's sense, ruins are a sign of misery and poverty, resulting from the construction of cities and the filling of houses with material goods. Such signification, of ruin as misery, and capitalism, modernity, and consumption as its cause and catalyst, also constitutes the central criticism of the Indigenous writer Krenak, whose tribe was listed in Markham's work under the name Botocudo, a generic denomination given by the Portuguese colonizers to different Indigenous groups belonging to the macro-jê trunk (non-Tupi group) of diverse linguistic affiliations and geographical regions, whose individuals, in their majority, used lip and ear buds. Krenak writes:

> We are addicted to modernity. Most inventions are an attempt by us humans to project ourselves into matter beyond our bodies. This gives us a feeling of power, of permanence, the illusion that we will continue to exist. . . . This is an incredible drug, much more dangerous than the ones that the system forbids out there. We are so doped up by this nefarious reality of consumption and entertainment that we disconnect from the living organism on Earth.[54]

Thus, Kopenawa and Krenak represent the subversion of the notion of ruins, previously applied to their ancestors as an ontological category, toward a notion of *ruins* as *the result of* modernity and Western colonialism. The writings of these Amerindians subvert the object of the ruins. This implies that white people become objects of concern since they are the ones who will disappear. In his manifesto *Ideas to Postpone the End of the World*, Krenak explains, in reference to the end of the world: "In 2018 I was asked: 'What are the Indians going to do about all this?' I said: 'The Indians have been resisting for five hundred years. I am concerned with whites . . . How are they going to escape this?' "[55]

Kopenawa and Krenak use the trope of ruins to criticize white people and to enunciate melancholy as a reflexive platform for thinking about the end of a world, the Western world. This is recognized as a profound reflective exercise on the conditions of production of a modernity that is now in ruins. This conceptualizes the Amerindian populations no longer as the *melancholic moss* overlaying ruined remains of yore but as beacons of hope for a world in misery and decay. Anthropologist Viveiros de Castro apprehends this inversion of expectations:

> Those peoples we have been taught to see as survivors of our human past—peoples forced to "survive" in the present amid the ruins of their original worlds—are unexpectedly shown as images of our own future. Then, the notion of "survival" suddenly gains a whole other anthropological meaning, in the antipodes of that proposed by Edward Tylor.[56]

Such reflexivity derived from the teachings of Amerindians is enunciated in the work of Déborah Danowski and Eduardo Viveiros de Castro. In *Is There a World to Come? Essay on Fears and Ends*, the authors reflect on the present environmental and civilizational crisis from the perspective of a set of dystopian cultural productions (mainly fictional films) about the end of the world, as indexes for thinking about the "ruin of our global civilization" and the feeling of melancholy in the face of such a state. In another grammar, with similar semantics, the authors echo the sense of ruin expressed in Kopenawa's understanding. They point to the bankruptcy of the humanist optimism prevalent in the last three or four centuries of Western history, understanding the *ruin of the global civilization* as a consequence of its undisputed hegemony, and they predict that such a fall "could drag considerable portions of the human population with it. Starting, of course, with the miserable masses that live in the geopolitical ghettos and dumps of the *world system*."[57]

Conclusion

It was the beginning of the 1930s, about a hundred years after the first publications on the "ruined state" of the South American "Indians" by Martius, and

Claude Lévi-Strauss was leaving for Brazil to occupy the position of professor of sociology in São Paulo. Lévi-Strauss met the Brazilian ambassador, Luiz de Souza Dantas, at a ceremonial dinner in France and asked him how he should proceed to visit an Indigenous community in Brazil. The ambassador replied: "Indians? Unfortunately, dear gentleman, years have passed since they disappeared. . . . As a sociologist, you will discover exciting things in Brazil, but in the Indians, do not think more, you will not find even one."[58]

The vision of the Brazilian ambassador was mainstream in the thought and practice of the Brazilian elite between the nineteenth and twentieth centuries.[59] It is one of the clearest derivations of the supposed ruined state of Indigenous populations. I have analyzed the conceptualization of this notion based on the ideas of the German naturalist Martius. These ideas were well disseminated in the nineteenth century, in Brazil and the global north, and represented an ideological pillar of European colonialism and the process of territorial and racial domination of spaces in South America. In Brazil, this notion legitimized policies for the deterritorialization of populations from the nineteenth to the twentieth century in the context of mineral and vegetable extraction practices. The underlying ideology was precisely *the ruined* or *almost extinct* state of such peoples, who were to follow the natural process of disappearance or be incorporated into the fabric of Brazilian civilization.

Such an ideology of an alleged historical nature of the Indigenous condition was constructed based on certain comparative devices, among them the material culture of different peoples of the Americas. In particular, the Mexican and the Peruvian collections, which Martius encountered in 1820 in Madrid on his return from Brazil, served as an index of the *antiquity* of the American civilization. However, going deeper in time, Martius also relied on racial biogeographic notions as comparative devices. He believed in physiognomical and humoral continuities between the American and the Asian races, one of the five Blumenbachian racial types that Martius followed as interpretative gradient. In other words, for Martius, the key to a historical understanding of the Americas resided, to a large extent, in its comparison with the ancient Asian civilizations, which, at the beginning of the nineteenth century, were already understood as possibly linked to the ancient history of the Americas.[60]

Having established the antiquity of the "American race," Martius also explained the alleged state of being on the verge of extinction through the degenerative and catastrophic history of the Americas. He expressed such antiquity in a geological language: the "American race" was the last strata, placed on top of the other peoples and as the ruins of a humanity lost in time. However, Martius's vision went beyond the metaphor of the "American race" *as* ruins. He created a historical version in which the "American race" went through convulsions provoked by earthquakes, where the "soil opened and released sulphurous vapours and carbonic acid in such quantity and with such speed that the entire population had perished in the catastrophe."[61] For him, the Amerindians were the debris of such phenomena.

Currently, one of the peoples most affected by such thinking is the Yanomami. I have highlighted the figure of Kopenawa as one of the main representatives of a conceptual subversion in which the notion of ruins, generated in Europe, was ironically shown to be a mirror for Westerners, providing a critique of capitalism, modernity, and the very science that justified the plunder of biocultural objects that are today scattered among several museums in Europe. In the same conceptual movement, Krenak subverts white people into the objects of melancholy, no longer romantic and antiquarian but victims of a socioenvironmental disaster of their own making. This is not a superficial joke, assigning guilt and fatality to the West, as Indigenous peoples are fully aware of their own fatalities, past and present. It was Krenak himself who said more than once that "the end of the world was in the sixteenth century," indicating the end of the Indigenous world when the colonial period in South America began.[62] In the present time, this situation is perpetuated to the extent that the valley of the Doce River, in central Brazil, the region where Krenak was born, is continually affected by the activity of mineral extraction by companies financed by the capital of the global north.

Nevertheless, the conceptual subversion analyzed here—being one of many possibilities—brings some hope of transforming the colonial reality of an America historically conceptualized as a laboratory in which the global north was able to study its own history (e.g., Martius, Markham, and Morton), testing hypotheses about the "state of nature" and "ruins" of the Amerindians. Kopenawa and Krenak are representative of the resistance and subversion of such a project when conceiving a conceptual mechanism with the epistemological and political potential to generate rigorous and generous ideas—ideas that are now mimicked from Paris to London, from Amsterdam to Berlin, from São Paulo to New York, as operators of other worlds, of other social realities.

16

Gondwanaland Fictions

Modern Histories of an Ancient Continent

Alison Bashford

The historiography of modern geosciences includes a key line of inquiry that traces theories of ancient and global continental separation on a dynamic not a static earth. It is a twentieth-century history, to a large extent, analyzing Wegener's theory of continental drift, originally from 1912 and expanded thereafter as *The Origin of Continents and Oceans* (1915/1929). Wegener hypothesized that all the continents were once linked and that they drifted slowly apart, like icebergs.[1] South African Alex du Toit subsequently took a geological global south perspective and focused on the already-named and known Gondwanaland in *Our Wandering Continents* (1937).[2] The idea of "drifting" or "wandering" continents in distant time was ill-received initially, but by the mid-1960s, its refinement in theories that posited the earth's crust expanding in some places and shrinking in others, leading to plate tectonics, was generally accepted. Over fifty years, the geological and geodynamic history of the earth had been entirely reconsidered.[3]

Enabled by a century of earth scientists' research on wandering and drifting continents and poles, we can now enjoy the breakup of Pangea or Gondwanaland as a moment's internet entertainment on computer-generated imagery: two hundred million years or so in sixty seconds.[4] Watching one such reconstruction, YouTube viewers respond with sweetly humanized continents "breaking up":[5]

> Africa: i quit i wanna be my own continent.
> South america: yeah me to im going with north America.
> India: hello asia.
> Australia: its up to you antartica.
> Antartica: why:(

"Earth" finds the breakup too painful: "do you even know how much it hurts." But "Penguin" is happy: "Yay my home formed." Another viewer, however, gets more serious and doesn't want to play; all the flippant nonsense is brought back to earth. "Cool CGI . . . but I'm pretty sure this never happened."

Presenting whimsical twenty-first-century responses to Gondwanaland is more than a conceit. This self-consciously reimagined earth relationship follows a long modern tradition of fictionalizing, re-enchanting, and even sacralizing supercontinents and earth breakups. In the past, as now, there has been a marked cultural investment in this truly ancient geohistory. Gondwanaland has been proactively imagined and reimagined, even by those who claim it never existed. Human stories regarding long-past geohistories have wandered and drifted too, just like those many continents and their twentieth-century theorists. In modern cultural history, Gondwanaland has sometimes sat fictionally with Atlantis, and certainly with the lost land of Lemuria. Sumathi Ramaswamy has brilliantly analyzed the "fabulous geographies, catastrophic histories" of the latter, showing how, like Atlantis, the sunken Indian Ocean continent acquired its meaning from great discourses on vanished homelands and vanished people. Lemuria was culturally productive, enabling a poetics of loss. And loss, Ramaswamy concludes, is a necessary condition of modernity. "Lemuria" is part of an "off-modern" re-enchantment, one formed in the scientific metropole and reformed in the so-called peripheries of modernity, in this instance Tamil-speaking India.[6]

Should we read the modern imagining of Gondwanaland in the same way? Yes and no. Like Lemuria, Gondwanaland is the result of active place-making on the part of geoscientists as well as culture-producers across many genres; in travel writing, in science fiction, in children's fantasy, in ecopoetry, as well as in passing exchanges about a Gondwanaland computer-generated image. As this chapter shows, it is part of the realm of the fantastic and the mythic, the legendary and the imagined. Unlike Lemuria or Atlantis, however, Gondwanaland was real; its breakup did happen. And in another way, Gondwanaland is a quite different case. If Lemuria, as Ramaswamy argues, signals loss (even if a productive loss) "in the fascination with vanished homelands, hidden civilizations, and forgotten peoples and their ignored pasts," Gondwanaland is a world-making of a different order of absence and presence. Lemuria might enable the conjuring of "tangible presence out of sheer absence," but Gondwanaland is not the same; it is not sheer absence.[7] Rather it is a deeper and real geohistory that folds into the present in ways material and living.

Far from being lost, vanished, hidden, ignored, or even "past," Gondwanaland is present and manifest in at least three ways. First, it is present in Gond *Adivasi* politics in Central India. Gondwanaland the ancient supercontinent and Gondwana the region in Central India are intimately linked, but this connection is not always remembered or even in some cases known. Travel accounts, science fiction writing, and Gondwanaland-themed poetics reveal the stakes in the usage of Gondwana vis-à-vis Gondwana*land*. The modern history of Gondwanaland, the ancient megacontinent, always has Gondwana the Indian region at its modern geohistory heart, even if this is misrecognized or not recognized at all.

Second, Gondwanaland is present in a special biogeography of living ferns and trees. Remnant species still grow across these now-separated continents, creating unique natural heritage environments, some of which are world heritage sites, their Gondwanaland link acknowledged and protected. It is present, third, in the oil and coal that those ancient ferns turned into, fossil fuels now mined across the Southern Hemisphere: the "gifts of Gondwana," as some geologists have poetically called this geo-legacy.[8] These "gifts" fuel major extractive economies deriving from Gondwana coal basins.

In these ways, Gondwanaland holds more presence than absence or loss. And yet it is an imagined landscape that still signals far away in time and space and that conjures a prehuman earth time. Like a wilderness from long ago, the poetics of Gondwanaland is also part of the re-enchantment of the earth's past. As symbolic capital, however, "wilderness" no longer operates or is even findable in the Anthropocene. Not even Antarctica is available as purely extra-human or prehuman. Such place-times are now lost to us, as humans are everywhere, impacting everything. Gondwanaland, I suggest, does some of the cultural work that "wilderness" and wildness did in mid-twentieth-century high modernity.

These are the many dimensions and uses of a very present Gondwanaland. We might say that the more than two-hundred-million-year-old megacontinent has a modern history that began circa 1870, one that is economic, environmental, geopolitical, intellectual, colonial, postcolonial, and more.[9] This chapter is a contribution to the cultural history of modern Gondwanaland, part of the new histories that are being written of and for an ancient continent.[10] It examines Gondwanaland storytelling and world-making, its place in and as travel literature, and within a science fiction genre that also conventionally travels to other place-times. Gondwanaland offers geological fantasias of deep time and a categorically prehuman deep time.

Travel Stories: Gondwana and Gondwanaland

When European geologists worked in Central India in the late nineteenth century, they found fossils mysteriously recognizable from far-flung places: in the first instances, Madagascar and southern Africa. The idea emerged that great continents and what are now islands were once linked by land (not yet a massive continent). The Austrian geologist Eduard Suess gave it the name "Gondwanaland," after the ground on which, and in which, he found those fossils, picking up the idea from H. B. Medlicott, who led the earlier Geological Survey of India (1872), and who had suggested "Gondwana."[11] The Central Indian region and land of the Gond people thus became the original intellectual center of Gondwana*land* the ancient megacontinent. Yet it was already the center of the world, and homeland, for the Gond who had lived there before the Austrian geologists, before the British colonists, before the Mughals and the Maratha. Indigenous to India, the Gond were governed by their own successive kingdoms of Chandra, then Garha, and by the sixteenth century, Mandla—histories and dynasties that came to enchant later touristic and intellectual outsider-travelers.[12]

By the time Medlicott, Suess, and other geologists and paleontologists arrived, Gondwana was ruled by the British as the Central Provinces, an administrative region created in 1861. Yet, despite the uptake of "Gondwanaland" in geological discussion across the world, especially after Wegener's thesis of a dynamic planet, Gondwana the region was possibly better known among readers outside India in a parallel publishing genre: the travel account. Many Anglophone travelogues and ethnographies of Gondwana did not mention the phenomenon of Gondwana*land* at all. One, however, circled back strangely to inform, even to underwrite, Gond revivalist connections to the idea of Gondwanaland from the 1970s, the ancient megacontinent was folded back into the land politics of Gondwana the region and of the Gond people.[13]

This was Eyre Chatterton's *Story of Gondwana* (1916).[14] Chatterton—the bishop of Nagpur—was entirely enchanted by Central India; his book was an orientalist love story devoted to Gond histories and religion, an account of dynasties and civilization, of language and poetry. His book was a restatement of a colonial search for Indian aboriginality. As such, it neither needed nor even mentioned Gondwana as formative of Gondwanaland. Thus far, the geological story is as nothing compared to the cultural and anthropological interest in the travelogues of Gondwana, the place and people. And yet anthropologist Mayuri Patankar has recently shown how Chatterton's book directly informed and shaped a revivalist Gond politics from the 1970s. "The revivalists," she explains, "claim to have found pieces of their lost history in Chatterton's 'Gondwana.' Its textual and visual content finds its way into popular posters and folksongs." A set of caves known as a place of origin of the Gond became a site for revivalist pilgrimage. In an interview with Gondi writer Sunher Singh Taram, Patankar explains that he "delightfully narrates how Bishop Eyre Chatterton's text, *Story of Gondwana* (1916), was instrumental in the revival of the Kachargarh fair. Kachi Kupar Lohgad (The Hill Rich in Iron Ore) in Chatterton's text is identified as Kachargarh by the revivalists." Elsewhere, the scientists Suess and Wegener are "eulogized with utmost devotion."[15] Something of the geological megacontinent, then, found its way into late twentieth-century Gond poetry, art, folklore and mythology.[16] And there Gondwana*land* became part of Gond geopolitics. This has underwritten claims for homeland, even statehood, especially over decades in which displacement for forest clearing and for mining has become both common and highly politicized. At the same time, Gondwana the region in India is on one measure more lost—vanished and hidden—than Gondwanaland the formal name of an ancient megacontinent. *Adivasi* activists seek a reclamation of Gondwana as a formal administrative unit, since after the disintegration of Central Province and Berar, Gonds are now scattered across Madhya Pradesh, Chhattisgarh, Maharashtra, Odisha, and Telangana.[17] And yet, as one activist writes, "the historical territory of 'Gondwana' is an ancestral territory of Gonds and was earlier ruled by Gond kingdoms of Kherla, Garha, Chanda, Devgarh."[18] It should be reinstated, argues Akash Poyam, as a homeland and as a linguistic, cultural, and historical state (fig. 16.1).

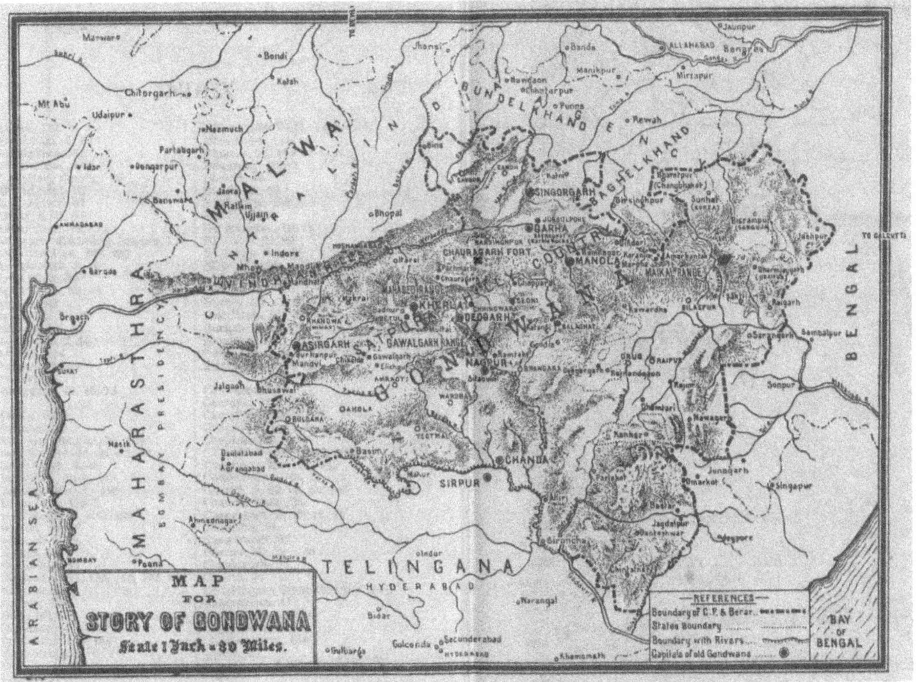

FIGURE 16.1. Foldout map, Eyre Chatterton, *The Story of Gondwana* (London: Sir Isaac Pitman & Sons, Ltd., 1916). Copy in author's possession.

A minor line of Gondwana travel writing unfolded through the twentieth century. Much of it was disinterested in Gondwanaland—but highly interested in the culture of the Gonds. Indrajit Singh's *Gondwana and the Gonds* (1944) followed Chatterton's example to some extent, exploring and explaining the Gond homeland in Central India and its history, landscape, and culture. It is anthropological in the context of contemporary Aboriginal politics in India, organized not least under what Singh set out in 1944 as the new movement "'Adi Basi,' or the children of the soil."[19] The term *Adivasi* was an invention of the late 1930s,[20] and while Gond people had not yet identified with or as *Adivasi*, they soon would. In the meantime, an anticolonial Gond nationalism of sorts was emerging, perhaps more sharply anti-Hindu than anti-British. Current Gond activist Akash Poyam, of Adivasi Resurgence, describes a Gond Mahasabha assembly of 1945: it was asked, "What is going to be the benefit of this independence? Even if India gets independence, Christians will leave, but the colonizers of our culture, traditions, and property—'Hindus' will stay right here and will keep ripping us off."[21]

Singh's *Gondwana and the Gonds* was not an account of contemporary political agitation. And it was less a travelogue and more a social anthropology of primitive economics, life, and culture, as the city of Lucknow's Radhakamal Mukerjee described in his foreword. It contributed, Mukerjee thought, to the "great problem of aboriginal India . . . to smooth and regulate its social and economic transition." His position was primitivist and anticolonial at the same time.[22] British game laws deprived the Gonds even of their bows and arrows; timber merchants penetrated the wilderness and exploited Gond labor in the very forests after which they were named; tenancy and land-alienation laws work against them; great forests were reserved for hunting by colonizing others.[23] The land, the geology (to some extent), the geography, and the climate were at issue, including the destruction of "what was once an interminable forest . . . virgin forest, intersected by high mountain ranges . . . broken only for small cultivation."[24] That the Gonds lent their name to Gondwanaland, as for Chatterton, was neither important nor interesting for Singh or Mukerjee. It was not continental drift geology that informed this Lucknow-produced work but social and cultural anthropology. Not Wegener or Alex du Toit, but Ruth Benedict, Raymond Firth, Sigmund Freud, and Bronislaw Malinowski.[25]

Lahar Singh's, *Gondwana: A Journey to the Centre of India* (2009) is a later twentieth-century twist on the touristic and the anthropological—another journey book, this time presented as a novel.[26] "Singh" was John Beaumont Ash, a linguist, photographer, and writer who first came to the Muria Gond village of Saratpur in 1973, learned various Gondi languages, and was eventually given the name Lahar Singh. His personal story crosses into travel, indeed into tourism, since in 2002 he established GreenGondwana, leading visitors across the region—an ethnotourism as much as a green tourism venture, trading on aboriginality.[27]

In other renditions in the travel genre, Gondwana and Gondwanaland are connected but often oddly. Tahir Shah's *Beyond the Devil's Teeth: Journeys in Gondwanaland* (1995) is a first-person travel account that begins with the (Gond) Tale of Lingo, with the traveler-writer's "fascination with the Gonds," the place, the people, the legends, and the histories of now-fallen kingdoms. He is told by an academic he meets on a train (and thus tells the reader):

> "The Gond kingdoms have fallen, their people live dispersed in poverty: the teak trees and the jungles have been cleared . . . but the importance of the Gonds must not be forgotten!"
>
> "What importance is that?"
>
> "Gondwanaland!" shrieked the weary academic.

And thus Shah's scholar explains, with woeful inaccuracy: "Long ago, India, Africa and South America formed one supercontinent. . . . This colossal landmass, known as Gondwanaland, was named after the Gonds and Gondwana."[28] Colossal and important but attenuated, Gondwanaland in this book turns out to be a travel route, with an antiquities-searching subplot that sends Tahir Shah from Central India to Africa to South America.

In such travel accounts, Gondwanaland is a deeply human present-day region that is sometimes connected to Central India, and at other time stands for the Southern Hemisphere. Its deep history is about primitivism and indigeneity, sometimes colonial, sometimes nationalist, but always human. Likewise for the Gond; their homeland is a politics of deep place to which Gondwanaland the ancient continent now adds immeasurable depth, perhaps even expediently underwriting a belonging and a claim in iterations from the Gondwana Homeland Party to the Humans of Gondwana Facebook group. It is pre-Vedic, but it is not prehuman. It is Gondwana, not Gondwanaland.

Far more geologically expert than these travel writers on Gondwanaland the great and ancient continent is a 1931 geographer's account of what lay at the ocean's depths as he sailed across the Indian Ocean. Both Gondwanaland and Lemuria were on his mind. The geographer Charles Albert Fenner, going by the pseudonym "Tellurian," wrote short travel pieces for the *Australasian*, including one from the ship's deck in November 1931. All the passengers were "aweary" from the turbulent storms. Like them, he longed for land, and he started to think about ancient land, now on the seafloor: "the mind of the geologist turns to the fact that what is now turbulent ocean was once, or more than once, dry land." In 1931, the *Glossopteris* fossil was known to be in Gondwana in Central India, in South Africa, and in Australia. But what this all added up to was still unclear and confused, especially regarding the prior continental connections: "There is no certain evidence that this was the case, but a great many things point in this direction. It is thought, for instance, that in Carboniferous times, which began 300,000,000 years ago, there was a great continent in the Indian Ocean, linking up portions of India, South Africa, and Australia."[29] Fenner nominated the megacontinent as "imaginary": "This great imaginary continent has been named Gondwana Land. It is also thought that at a much later date, just a few million years ago, there was a continent in the north-west Indian Ocean that has been called Lemuria." Indeed that was how he located himself at his time of writing: "We are somewhere over the lost continent of Lemuria at present."[30] In 1931, even for a geoscientist, Gondwanaland was still a fantastic place, as mythic as Lemuria, though not quite so mythic as Atlantis. "Whether or not such a continent really existed, and what were its boundaries, we cannot say for certain."[31]

Gondwanaland Science Fiction

Science fiction is also a genre of journeys. Sometimes travels are outward from the earth to other planets, sometimes they are oriented inward, where other place-times are to be found deep inside the earth. We find the latter at the conventional origin of science fiction, in Louis Figuier's *La Terre avant le Déluge* (1863) and Jules Verne's *Voyage au centre de la Terre* (1864). They have usefully been named "geological fantasias."[32] And unexpectedly we find Gondwana in Verne's *La maison à vapeur* (1880), at one point the plot relying on challenges posed by the "fierce tribes of the Gondwana." Predating Suess's "Gondwanaland," Verne may have known of the Gondwana series identified in geological discussion of the 1870s. Equally, it may be that the reference to the tribes of Gondwana was a primitivist not a geological or paleontological reference, notwithstanding Verne's penchant for the geologically fantastic. Either way, over the twentieth century, Gondwanaland was reimagined through science fiction on the one hand and children's fantasias on the other. In the process, it became almost entirely separated from *the* Gond or even India.

Time travel, time machines, and time devices are standard in science fiction, and unsurprisingly Gondwanaland versions redeploy the mechanism of falling into the earth, and thus through time. Yet this device holds a stratigraphic and a geohistorical truth, as Jules Verne well knew. Deep earth and deep time became materially and epistemologically connected in his nineteenth century. As the earth ruptures and cracks, other dimensions of past earths are literally revealed. By the early twentieth century, earth time was still expanding at a dizzying rate.[33] In 1929, Juliette Huxley expressed an overwhelming intellectual vertigo when she was escorted down a real earth rift by paleoanthropologist Louis Leakey into the Olduvai Gorge, part of the Great Rift Valley in Tanzania: "I am losing ground; I am falling down those cliffs at a thousand years a second." Just as her husband, Julian, and her brother-in-law, Aldous, were both writing various kinds of futurist science fictions, Juliette Huxley put this experience of geological space-time vertigo in words: "I humbly reflect on the relentless evolution which has linked me, across this million-years gulf, with the white bones under the cliff."[34] And just so, in Gondwanaland science fiction and fantasy, people fall into the earth, and into another place-time sometimes called "Gondwana" and sometimes "Gondwanaland."

In J. B. Rowley's contemporary children's series, *Trapped in Gondwana*, protagonists fall through cracks in the earth in an unimaginative, if effective, repetition. "Eleven-year-old Nellie Russell is just an ordinary girl until the day she becomes trapped on the ancient supercontinent of Gondwana. Now she must find the courage to fight fearsome Gondwanan animals including a Demon Duck, striped wolves and a tree climbing crocodile." Gondwana, here, is ancient forest. "Nellie journeys through the misty forests where unseen creatures, spirits of the underworld and troll-like monsters lurk. Along the way, she triumphs over tests, collects adventures, stones and deciphers riddles. But she must reach the centre of Gondwana before nightfall or she will never see her home and family again." The book's jacket entices young readers into the quest genre: "Will Nellie overcome these final obstacles to make it out of Gondwana and back to her home?"[35] "Gondwana" in this children's fantasia is forested and spirited and folkloric. But it is resolutely not home, even if the center of Gondwana is the quest place of magic release, a kind of land bridge to home. It is a difficult read, given the wider geopolitics of Gondwana. We know that even as this fantastic forest Gondwana is conjured as antihome, as alien other-place, a Gondwana homeland movement is hard at work trying to establish a state-recognized home, sometimes with reference to a "Gondwanaland" that is also imagined and mythic, if in a different register and with much higher stakes.

In Yves Sente's comic *Le sanctuaire du Gondwana* (2008), we are taken, once again, far inside the earth, through a secret entrance inside the lake-crater of an old volcano in Tanzania. This is a Mortimer and Blake paleo series that includes in this instance, and strangely, the characters of Mr. and Mrs. Leaky, paleontologists. In this rendition, the center of Gondwana is also a quest place. The Gondwana Shrine marks an origin of anthropocentric life. Raiding mythic and real geologies, and academic disciplines that deal with space and time—archaeology, paleoanthropology, geology—*Le sanctuaire du Gondwana* follows *Atlantis Mystery* (1957), *The Time Trap* (1962), and *The Sarcophogi of the Sixth Continent* (2003). Perhaps it was only a matter of time before "Gondwana" was discovered and deployed.

Time travel to Gondwanaland is unmistakably to a deep past not a science fiction future. Nicholas Ruddick has offered a historical account of the emergence and significance of "prehistoric fiction," French-born, like science fiction. He has analyzed 150 years of Francophone and then Anglophone stories set in that nebulous period called "prehistory" between (he problematically instructs) the first upright hominid and the invention of writing. Aligned with, and allied to science fiction, this is a speculative genre that treats of hominization in other worlds. In Arthur Conan Doyle's *The Lost World* (1912), for example, protagonists travel back in time to a South American mountain, communing both with dinosaurs and with "ape-men."[36] Yet the deployment of Gondwanaland in science fiction characteristically does something more and different to prehistoric fiction. *Le sanctuaire du Gondwana* aside, "Gondwanaland" is usually (and rightly) a prehuman space and time. Fictional travelers might meet "prehistoric" monsters and animals, but they do not travel to be confronted by prehistoric humans or hominid others. If prehistoric fiction is driven by a fascination with versions of us, there were no versions of us on earth 180 million years ago, and there is something in that fact that fictionalizers of Gondwanaland sustain, which they use. This is not, in other words, part of the culture of the ape's reflection,[37] caught definitively in *Planet of the Apes* film series of the 1980s. What, then, drives the deployment of "Gondwanaland" as a place and time for a deep past and prehuman science fiction? Increasingly, and broadly, it is environmentalism. Gondwanaland becomes detached from the Gond, from human-land politics, from homelands, from forests that humans cultivate, and reattached to an environmentalism that seeks and needs a more pristine nature. Over the 1980s, Gondwanaland came to mean something like "wilderness forest."

Craig Robertson's "science fantasy" was published as *Song of Gondwana* (1989). This is "set in and dedicated to Victoria's Otway Ranges and wild southern coast, and to its bird life." Here, Gondwana forests are renationalized as part of an apparently pristine Australian environment, but it also draws weight from its place in the ancient southern supercontinent. "As it disintegrated millions of years ago, the final great schism came when Australia broke away, along its southern rim, from East Antarctica. As Antarctica died beneath the ice, the Australian ark drifted north into the heat." Fully dissociated from India, "Gondwana" serves as an ancient place of danger for protagonist Lane, as Robertson the author leaps from fact to fiction:

> The forces unleashed by this cataclysm remain to disturb the life of Lane, an amateur scientist. He begins an ambitious study of birdlife based around an old farmhouse in the Otway Ranges. Deep in the rain-soaked forest he finds a dangerous mystery. There is a lethal presence somewhere in the very air around him, and a peculiar hidden valley. He meets Katrina, an obsessed recluse who is performing strange experiments with birdsong. His fascination with her leads him and his friends into a mythic realm that threatens to destroy them all.[38]

Like Rowley's children's fantasia, Gondwanaland here is ancient forest, pure and primeval, far more biological than geological.

In the 1980s, a different species, so to say, of Gondwanan "children of the soil" were born. As rainforest action groups undertook direct actions to save parts of the Australian wet tropics, the significance of various species' ancient connections to Gondwanaland crossed from evolutionary ecology to environmentalist activism and thereafter to an everyday, street-level public recognition. This was the decade in which what are now called the "Gondwana Rainforests of Australia" were successfully lobbied onto the World Heritage List (1986). "The evolution of the tropical, sub-tropical and temperate forests of Australia and their past links with Gondwanaland are . . . of great scientific interest," the International Union for Conservation of Nature advised UNESCO (the United Nations Educational, Scientific and Cultural Organization).[39] The Gondwana Rainforests are now greatly valued and protected: "Few places on earth contain so many plants and animals which remain relatively unchanged from their ancestors in the fossil record. Some of the oldest elements of the world's ferns and conifers are found here and there is a concentration of primitive plant families that are direct links with the birth and spread of flowering plants over 100 million years ago."[40] "Gondwanaland" is thus broadly understood and recognized in Australian culture, although its connection to the Gond of Central India is not. Indeed, in the strangest twist, "Gondwana" in Australian culture is linked vaguely to Australian Aboriginal culture, often presumed to be an Aboriginal term, one that is traded on regularly in enterprises that range from children's choirs to (rainforest) botanic gardens to energy companies driving coal and coalbed methane gas extraction. The slippage here between Aboriginal association, attachment to land, environmentalism, and (strangest of all) a loose but entirely perceptible Australian nationalist *ownership* of both "Gondwana" and "Gondwanaland" is one of the most curious twists in Gondwanaland's modern global history.

It is marked, but typical, that Craig Robertson's science fantasy has nothing to do with India. Gondwana is forgotten; at least it was in 1989. Nearly two decades later, however, the author set out on his own travels to find the real Gondwana for himself. In fact, his early geological training *had* made this link: "I was told as a geology student in the 1960s, over 50 years ago, that Gondwana meant land of the Gonds, and that they were a tribal people living somewhere in India. The idea of going there one day stayed in my mind over all the years and many travels. It was this that I set out to find in early 2007."[41] Working his way through old Gondwana as well as through old books, Robertson is still piecing together the history of the term.

Elsewhere in its modern cultural diaspora, "Gondwana" comes to mean not "Australian rainforest" but "Antarctica." In yet another children's story, *Flight to Antarctica* (1985), not only is Gondwana actually the name of a fantasy Antarctica but the protagonist children meet "Gonds," who are cast now as the Indigenous people of the icy continent. "Gondwana as mapped by the Gonds" is the frontispiece, a familiar quest-map, but bizarrely also a map of present-day Antarctica (fig. 16.2). When the children find themselves in this strange and icy place, a "Gond" explains: "This is Victoria Land, on what we call Gondwana, part of the most ancient continent of Earth." In the chapter, "In Which the Children Meet a Gond," they are told definitively that it's not Antarctica. "Gondwana is the name of this ancient place, the country of my people. It wasn't always as you see it now. It was once free of snow and ice—quite a hot place in fact, where lovely flowers and trees grew and bountiful rivers flowed to the sea."[42] The peopling of Antarctica with its own Indigenous people is perhaps the greatest fantasy of all. On the other hand, the author Margaret Andrew worked at the Scott Polar Research Institute in Cambridge in the late 1970s and early 1980s, and she accurately described the rainforest that did once grow in Antarctica. Real remnants of just that rainforest are precisely what is still protected in the Gondwana Rainforest World Heritage Areas of eastern Australia.

Gondwanaland reappears in German storytelling of the 1980s as well, again linked broadly to an environmentalist ethics, overlaid with a pacifist and antinuclear politics of that decade. A German fantasia "Gondwanaland" (1986) is subtitled, "A Story for Children and Those Who Want to Become Children Again" (fig. 16.3). It is the opposite of a dangerous place-time in this rendition. Far from being an antihome, "Gondwanaland" is the ideal and aspirational home for the family of all humanity, united in its difference, an expression of the ecological and cosmopolitan "oneness of mankind" that Radhakamal Mukerjee conjured in the 1960s.[43]

Far from Gondwanaland being found deep inside the earth, here it is found in the full moon—and in the protagonist's boyhood dreaming. Konstantin sits at his windowsill in the German woods, staring at the full moon. There he sees the earth's continents reflected. "Yes, he saw it very clearly, for he had often

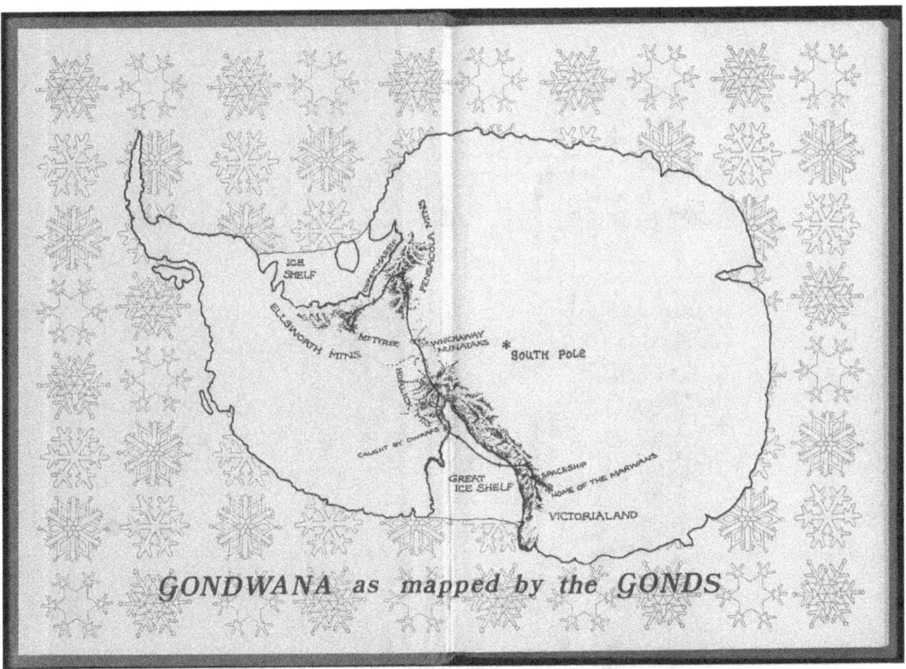

FIGURE 16.2. In a 1980s children's story, Gondwana is Antarctica, peopled by Gonds, here cast as indigenous to that continent. Frontispiece to Margaret Andrew, *Flight to Antarctica* (Cambridge: Burlington Press, 1985). Copy in author's possession.

studied the big world map that hung over his father's desk. Even when he was still little, he already knew that the Earth was made up of four large pieces, and he could now clearly make out these continents in the moon." Africa was the shape of his kite, America looked so thin in the middle that it could break. In his lunar dreaming, he perceived boys and girls just like himself, but living tantalizingly different lives across those great continents. If only they could get to know one another, but the continents are so far apart. Look at those vast oceans that separate children of the world. But wait! "His father had once told him that clever adults believe that, in the past, many, many thousand years ago, the continents once all fit together. The seas washed around this great clump of earth, which the adults called GONDWANALAND."[44] The idea of dynamic earth—its separations and possible reconnections—is easily reused. In Konstantin's dreaming, a powerful "crash and crunch" heralded the continents moving toward each other again "and they suddenly hooked into each other with a mighty roar." Konstantin looked at the moon again: "there it was again like earlier—the old GONDWANALAND, the one great protocontinent, the continents fused into a single clump, washed by a single great sea." The

FIGURE 16.3. The cover of *Gondwanaland: A Story for Children and Those Who Want to Become Children Again* shows Gondwanaland as the "one-world" of German environmentalism, circa 1985. Cover, Rolf and Michael Schmitt, *Gondwanaland: Eine Geschichte für Kinder und solche, die wieder Kinder werden wollen* (Bonn: Simon Verlag, 1985). Copy in author's possession.

people of the earth were reunited, the children riding their llamas, koalas, and giraffes across the world to become a multicultural, ecological one—a diverse unity. This is what the human earth *should* be like. But alas, he knew it was not: "Konstantin became sad. All of the good and beautiful, what had happened on and with the Earth, the ancient GONDWANALAND—should that all really just have come about from his dreamy imagination?"[45] Here, Gondwanaland is utterly human. It stands for One World, environmental cosmopolitanism 1980s-style. Gondwanaland's "environmentalism" in this German version is linked far more to an antinuclear and antiwar politics of the 1980s than a wilderness/nature-quest that governed Australian environmentalism in the same decade. It is the latter desire to conjure a pristine, prehuman landscape that has endured in a twenty-first century Gondwanaland poetics.

The Poetics of Gondwana/land

Why does an imagined Gondwanaland have such a presence across these genres? What cultural work does it do in an Anthropocene-aware era? Gondwanaland poetics carries forward this tradition of ancient continental place-making, and it is invested specifically in pure, nonhuman and prehuman landscapes. Gondwanaland serves as a site for conventional nature writing, for ecocriticism, and for experimental ecopoetics. Here Gondwanaland *is* a lost land, a lost way of being, like Lemuria. Gondwanaland serves as a profound before. It has been repurposed as a landscape born of Anthropocene-human desire for a pure time and space, an earth without humans that is no longer caught by or accessible in the "wilderness" on which we used to rely. The closest we get, or got, is Antarctica, but that too now needs to be remembered in the past, only its remnants saved in the present as "heritage." Gondwanaland, indeed, sometimes stands in for that lost Antarctica.

Stuart Cooke's "Echoes of Gondwana" (2016) is a nonfiction essay: part Gondwanaland travel piece, part Gondwanaland ecopoetics.[46] For him, those *Glossopteris* fossils are found living in *Araucaria* species. The "only surviving ancestors" of fossilized *Glossopteris* ferns are "like the petrified residue of a long-dispersed echo." It is a musing on Australian-Chilean connection and separation after Gondwanaland broke up. Long ago connected (as the *Araucaria* species shows), how did they go on to be so in colonial history and in biogeography? Gondwanaland becomes a kind of eco-diaspora. Cooke, we are told, "adopts a macro-perspective, writing from an understanding of place that spans millennia."[47] With the place-world "Gondwana," the poet can work up a transcultural ecocriticism, into what he calls "ethological poetics," or methods for understanding the creative compositions of non-human species.[48]

Cooke wants to put geological, biological, and Indigenous histories "in place."[49] As he explains in a recent article in *Landscapes*, his poems embody the recognition that any transcultural poetic vision taking an anticolonialist stance must attempt to eschew Western visions of categorization and subjugation and strive for multivocal expressions of a complex location.[50] Yet here again, "Gondwana" becomes magically available to and claimed by all, even those who purposefully pursue "de-colonial" writing. Cooke, as ecopoet, is glowingly reviewed as "creating a new vision of how to speak about the temporal, geographic, ecological and cultural histories of land and, in the process, expands our lexicon with which to express them."[51] Yet the cultural history of land is an attenuated one, a history of Gondwanaland without Central Indian Gondwana.

Gondwanaland offers a great deal for this ecopoet: it is a dreaming place more than a real place. A word expert, Cooke returns time and again to "Gondwana" but misreads it, or at best partially reads it, as "motherland of Australia," not homeland of the Gonds. It is ironic, then, that Cooke's poetry is often about forests. His poem "Fallen Myrtle Trunk," he explains, "passes through different levels of observation/perception, starting with a broad consideration of time and space, including Gondwanan or evolutionary space-time."[52] Gondwanaland proves to be a useful if strange "planetary locality," reinvented in a tangle of postcolonial and environmental purpose, and colonizing misunderstanding. Unwittingly perhaps, inside a professed transculturalism lies an almost nationalist environmentalism.[53] The apparently outdated geologist-student and novelist Craig Robertson did far better in his etymological travel-searches for the Gond in Gondwana.

For distinguished French American poet Nathaniel Tarn, *Gondwana* (2017) is equally devoid of India and becomes, once again, Antarctica. His long poem was written there in 2008. The magic of the present-day icy continent seemed to be "*Finis Terre*, final of earth, or / 'end of the world' they call it here." Yet understanding Antarctica thus seemed insufficient, and Tarn finds an even more profound space-time in the Gondwanaland beyond. Alternately, this is the beginning of the world, the end of the world, or the center of the world.

> Toward center,
> Center of center,
> Where earth-mind turns solid

On this poetic Antarctica/Gondwanaland, there is a timelessness, a purity, but also a nothing-ness. Unrecognizable from the wet rainforests of Gondwanaland World Heritage sites, this poetic "Gondwana" is like a lifeless beginning.

"Naked you go into this continent / in endless search of cleanliness." And "You can imagine white / drawing in your colors, / all body differentiae, / until you walk as a ghost, / as someone who has crossed a limit on no map."[54] This is not a rich wilderness of life but a blank-white start, or end, of life, of earth.

Conclusion

Once the heart of Gondwanaland, even Antarctica no longer serves as enchanted beyond, the out-of-time, beyond-human continent it once was. "Gondwanaland" has become available in its stead, stepping into that politico-cultural space. This appetite for whole-earth, geologically defined, deep-time places is motivated by Anthropocene cultures. Put another way, the very uptake of "Anthropocene" itself, its sudden and great purchase, has a by-product: a nostalgia for a land before human time. In Gondwanaland poetics, ancient land does offer a modern re-enchantment.[55]

And yet we still need to grapple with Gondwanaland in the present. If Ramaswamy writes of Lemuria's invention in the geological knowledge-making metropole, and its travel to the "putative subaltern (and subalternized) margins under the pressure of the spread of colonial rule, the attendant consolidation of global capitalism, and the enabling conquest of our earth by the natural and human sciences,"[56] Gondwanaland's story is equally entangled, if not more so. It is part of colonial capitalism in every literal and material sense—Gondwana coal driving a modern carbon economy in its place of origin, so to say: in India 98 percent of the total coal reserve *is* Gondwanaland fossil, two hundred million years and more in the making.[57] And India is second only to China in leading coal production across the world. It is the strangest world/environmental history irony that Gondwanaland thus produced the fossil-fueled Anthropocene. This makes an Anthropocene-era nostalgia for enchanted and pure Gondwanaland both impossibly poignant and highly apt. At least Gondwanaland coal is safe in the ground in Antarctica, at least for the moment: the US Geological Survey in 2016 showed great deposits of Permian coal.[58]

Gondwanaland's massive and flexible symbolic capital offsets the effects of this extraction of its material coal-capital. This is part of what the cultural history of modern Gondwanaland is and does. The deepest time world-making of Gondwanaland serves environmentalist ethics and culture well. It is a useful instance, a substantive case, for a critical geography and environmental humanities that inquire into ways of comprehending "earth-without-humans."[59] At the same time, part of the fascination, the urgency, the cultural politics, and the sheer modernity of Gondwanaland is that its global myth-making often displaces the Gond people at its geohistory center. More than that, in at least one context, Gond *Adivasi* have been *re*placed in relation to Gondwanaland by a vague connection with other Aboriginal people. In this settler popular culture, Gondwanaland aesthetics and poetics are plain settler nationalism, in desperate need of deep-time belonging to land. In Anthropocene cultural colonizing of Gondwanaland, the ancient megacontinent is being recast as homeland once again—not of the Gond but of the human species who have lost Nature.

AFTERWORD

Alison Bashford, Emily M. Kern, and Adam Bobbette

The cosmologies, theologies, and temporalities described in *New Earth Histories* are not merely the remnants of a premodern, more enchanted world, one now lost to us in the twenty-first century. They are still present, as the progenitors and shapers of contemporary earth cosmologies. Nor are these ways of thinking about the earth absent from even the most modern of geo-environmental preoccupations. The Anthropocene is in part a new geo-mythology, a story about society and culture, origins and time, where politics and history are inscribed into the earth itself. And yet, the new earth histories in this book show us how the seemingly innovative nature of the Anthropocene is hardly unique to our own times: ancestral deities, supernatural entities, and even powerful mortal rulers or revolutionary heroes have long been incorporated into physical landscapes and explanations of how the earth has taken shape. The idea that scientific modernity disenchanted the world, and that the solution to the Anthropocene crisis is a re-enchantment of nature, is in fact old earth history. *New Earth Histories*, instead, reminds us just how profoundly the geological sciences have been enchanted all along.

In his foreword, Dipesh Chakrabarty signals that this book advances the debate about the Anthropocene. The first generation of Anthropocene scholarship often held scientists and humanists to account for attachments to undifferentiated conceptions of the human, or for insufficiently grappling with the *who* and *where* of responsibility for environmental crises. New earth histories raise new problems for the next generation of Anthropocene scholars. For instance, contemporary earth scientists and humanists have yet to take seriously what it would mean to acknowledge that the material world is ancestral—that stones, sand, rivers, and ice make moral claims upon us and that we are indebted to them. Yet, the conditions exist to understand this reality: melting glaciers, forest fires, rising sea levels, and heat waves all now bristle with human history. The human combustion of fossil fuels over centuries is now present to us in a broken climate. This is social history made natural. One reason to think about new earth histories now is to learn from traditions that already know how to listen to ancestral matter. *New Earth Histories* raises a related problem for Anthropocene scholars: how to inherit other people's ancestors. If we are to take seriously that the geological record is social, how are we to make sense of an asphalt road made of mined gravel from stolen Indigenous lands, for example? Do the commodity chains of global capitalism not put us in contact with people's displaced ancestors, histories to whom we owe a duty to learn how to inherit? Such questions raise novel trajectories for thinking about the material world.

New Earth Histories multiplies the voices that shaped our understanding of nature. The decolonization of the environmental sciences is more familiar in histories of biology but has yet to touch the earth sciences as profoundly as it might. This book is one attempt to address this. More than decentering the significance of the West, we have also attended to the erased voices that made these sciences. The orthodox conception of the Earth System that many of us take for granted—even for our understanding of the Anthropocene and the climate crisis—was made by a much more cosmopolitan world of thinkers, researchers, and visionaries than has been previously imagined.

The original New Earth Histories conference was convened in December 2019 during Australia's catastrophic Black Summer, when more than twenty-four million hectares of land burned across the country. In Sydney, the city was covered by a layer of smoke so thick that it set off smoke alarms indoors. Three months later, during the first months of COVID-19 lockdown in New South Wales, we began work on the book. In the subsequent two years, this volume came together during periods of disease, quarantine, and isolation worldwide, as well as a war in Europe and accelerating global climate breakdown.

As the chapters in this volume make clear, this is not the first time that studies of the earth's deep history have coincided with periods of crisis. Grappling with both literal and metaphorical instability has been one of the key drivers of earth-knowledge-making. Thinking on a geological or a planetary scale can be both comforting and alarming. On one hand, considered against the vastness of planetary or universal time, two or three years is irrelevant. On the other, we are sobered by the scale of the many and variable disasters engulfing our world. We might ask, what kinds of new geo-poetics and geo-theologies can arise from this rupture? What new—or old—cosmologies might form or return to prominence as individuals, communities, and nations grapple with the rapidly changing world in which we find ourselves and wrestle with our individual, societal, or species-wide complicity? While rethinking the past, *New Earth Histories* is one guide to the present.

Notes

Foreword

1. Nigel Clark and Bronislaw Szerszynski, *Planetary Social Thought: The Anthropocene Challenge to the Social Sciences* (Cambridge: Polity, 2021). Parenthetically, I should note that the authors also suggest that social should in turn be geologized. I am putting this comment to one side as it is not an exercise this volume undertakes.

2. Introduction to this volume.

Introduction

1. "What Is Country," Australian Institute of Aboriginal and Torres Strait Islander Studies, last modified May 25, 2022, https://aiatsis.gov.au/explore/welcome-country#toc-what-is-country.

2. Roy Porter, *The Making of Geology: Earth Science in Britain, 1660–1815* (Cambridge: Cambridge University Press, 1977); James A. Secord, *Controversy in Victorian Geology: The Cambrian-Silurian Dispute* (Princeton, NJ: Princeton University Press, 1986); Gabriel Gohau, *A History of Geology* (New Brunswick: Rutgers University Press, 1990); David Oldroyd, *Thinking about the Earth: A History of Ideas in Geology* (Cambridge, MA: Harvard University Press, 1996); Naomi Oreskes, *The Rejection of Continental Drift: Theory and Method in American Earth Science* (Oxford: Oxford University Press, 1999); Peter Bowler, *The Earth Encompassed: A History of the Environmental Sciences* (New York: Norton, 2000).

3. Adam Bobbette and Amy Donovan, eds., *Political Geology: Active Stratigraphies and the Making of Life* (Basingstoke: Palgrave, 2017).

4. Martin Rudwick, *Bursting the Limits of Time: The Reconstruction of Geohistory in the Age of Revolution* (Chicago: University of Chicago Press, 2005).

5. Satpal Sangwan, "Reordering the Earth: The Emergence of Geology as a Scientific Discipline in Colonial India," *Earth Sciences History* 12, no. 2 (1993): 224–33; Suzanne Zeller, "The Colonial World as Geological Metaphor: Strata(gems) of Empire in Victorian Canada," *Osiris* 15, no. 1 (2000): 85–107; Robert A. Stafford, "Annexing the Landscapes of the Past: British Imperial Geology in the Nineteenth Century World," in *Imperialism and the Natural World*, ed. John A. Mackenzie (Manchester: Manchester University Press, 1990); Pratik Chakrabarti, *Inscriptions of Nature: Geology and the Naturalization of Antiquity* (Baltimore: Johns Hopkins University Press, 2020).

6. James A. Secord, "Global Geology and the Tectonics of Empire," in *Worlds of Natural History*, ed. Helen Anne Curry et al. (Cambridge: Cambridge University Press, 2018).

7. Simon Schaffer et al., eds., *The Brokered World: Go-Betweens and Global Intelligence, 1770–1820* (Sagamore Beach, MA: Science History Publications, 2009); John Gascoigne, "Cross-Cultural Knowledge Exchange in the Age of the Enlightenment," in *Indigenous Intermediaries*, ed. Shino Konishi, Maria Nugent, and Tiffany Shellam (Canberra: ANU Press, 2015), 131–46.

8. Pratik Chakrabarti and Joydeep Sen, "'The World Rests on the Back of a Tortoise': Science and Mythology in Indian History," *Modern Asian Studies* 50, no. 2 (2016): 808–40.

9. Amy J. Elias and Christian Moraru, eds., *The Planetary Turn* (Evanston, IL: Northwestern University Press, 2015); Nigel Clark and Bronislaw Szerszynski, *Planetary Social Thought: The Anthropocene Challenge to the Social Sciences* (Cambridge: Polity, 2021).

10. Tim Ingold, "Globes and Spheres: The Topology of Environmentalism," in *Environmentalism*, ed. Kay Milton (London: Routledge, 1993), 31–42; Dennis Cosgrove, *Apollo's Eye* (Baltimore: Johns Hopkins University Press, 2003).

11. Sumathi Ramaswamy, *Terrestrial Lessons: The Conquest of the World as Globe* (Chicago: University of Chicago Press, 2017).

12. Alison Bashford, "Terraqueous Histories," *Historical Journal* 60 (2017): 253–72; Emilie Savage-Smith, *Islamicate Celestial Globes: Their History, Construction, and Use* (Washington, DC: Smithsonian Institution Press, 1985), 3; Sreeramula Rajeswara Sarma, "From al-Kura to Bhagola: The Dissemination of the Celestial Globe in India," *Studies in the History of Medicine and Sciences* 13 (1994): 69–85; Jerry Brotton, "Terrestrial Globalism: Mapping the Globe in Early Modern Europe," in *Mappings*, ed. Denis Cosgrove (London: Reaktion, 1999).

13. I. W. Mabbett, "The Symbolism of Mount Meru," *History of Religions* 23, no. 1 (1983): 64–83.

14. Elizabeth A. Povinelli, *Geontologies: A Requiem to Late Liberalism* (Durham, NC: Duke University Press, 2016).

15. P. Chu, "The Debate over the Sphericity of the Earth in China," *Bull. Inst. Hist. and Philology* 69, no. 3 (1998): 589–670; T. H. Yang, "The Development of Geology in Republican China," in *Philosophy and Conceptual History of Science in Taiwan*, ed. Cheng-hung Lin and Daiwie Fu (Dordrecht: Kluwer Academic Publishers, 1993), 221–24; A. Wang, *Cosmology and Political Culture in Early China* (Cambridge: Cambridge University Press, 2000); Benjamin Elman, *On Their Own Terms: Science in China, 1550–1900* (Cambridge, MA: Harvard University Press, 2005).

16. Adam Bobbette, "Priests on the Shore: Climate Change and the Anglican Church in Melanesia," *GeoHumanities* 5, no. 2 (2019), 554–69.

17. Jill Ker Conway, Kenneth Keniston, and Leo Marx, eds., *Earth, Fire, Air, Water* (Amherst: University of Massachusetts Press, 1999).

18. Alison Bashford and Sarah W. Tracy, eds., "Modern Airs, Waters, and Places," special issue, *Bulletin of the History of Medicine* 86, no. 4 (Winter 2012).

19. J. W. Gregory, *Dead Heart of Australia* (London: John Murray, 1906).

20. Dipesh Chakrabarty, *The Climate of History in a Planetary Age* (Chicago: University of Chicago Press, 2021).

21. Alison Bashford, "The Anthropocene is Modern History," *Australian Historical Studies* 44 (2013): 341–49. See also Libby Robin and Will Steffen, "History for the Anthropocene," *History Compass* 5, no. 5 (2007): 1694–1719.

22. Ian J. Barrow, *Making History, Drawing Territory: British Mapping in India, 1756–1905* (New Delhi: Oxford University Press, 2003); Ian J. Barrow, *Surveying and Mapping in Colonial Sri Lanka: 1800–1900* (New Delhi: Oxford University Press, 2008); Matthew H. Edney, *Mapping*

an Empire: The Geographical Construction of British India, 1765–1843 (Chicago: University of Chicago Press, 1997).

23. Deborah Bird Rose, "Hard Times: An Australian Study," in *Quicksands: Foundational Histories in Australia and Aotearoa New Zealand*, ed. Klaus Neumann, Nicholas Thomas, and Hilary Ericksen (Sydney: UNSW Press, 1999), 219; Ann McGrath and Mary Anne Jebb, *Long History, Deep Time: Deepening Histories of Place* (Canberra: ANU Press, 2015).

24. William Walter Skeat, *Malay Magic* (London: Macmillan, 1900), 250–77. See also Teren Sevea, *Miracles and Material Life: Rice, Ore, Traps and Guns in Islamic Malaya* (Cambridge: Cambridge University Press, 2020). For a consideration of the material agency of spirits, see Diana Espirito Santo and Ruy Blane, "Introduction: On the Agency of Intangibles," in *The Social Life of Spirits* (Chicago: University of Chicago Press, 2014), 1–32.

25. Rudwick, *Bursting the Limits of Time*.

26. For a consideration of temporalities and deep history in the Pacific context, see Alison Bashford, "World History and the Tasman Sea," *American Historical Review* 126, no. 3 (September 2021), 922–48.

27. Sumathi Ramaswamy, *The Lost Land of Lemuria: Fabulous Geographies, Catastrophic Histories* (Berkeley: University of California Press, 2004).

28. For example, Nigel Clark, *Inhuman Nature: Sociable Life on a Dynamic Planet* (London: Sage, 2011); Jeffrey Jerome Cohen, *Stone: An Ecology of the Inhuman* (Minneapolis: University of Minnesota Press, 2015).

Chapter One

1. Sheila Jasanoff, "Heaven and Earth: The Politics of Environmental Images," in *Earthly Politics: Local and Global in Environmental Governance*, ed. Sheila Jasanoff and Marybeth Long Martello (Cambridge, MA: MIT Press, 2004), 49.

2. James Lovelock, "Gaian Pilgrimage," *Encyclopedia of Religion and Nature*, vol. 1, ed. Bron Taylor (London: Bloomsbury, 2005), 685.

3. Peter Sloterdijk, *Globes: Macrosphereology*, trans. Wieland Hoban (Cambridge, MA: MIT Press, 2014), 45–46. For an alternate reading on the distinction between "globe" and "sphere," see Tim Ingold, "Globes and Spheres: The Topology of Environmentalism," in *Environmentalism: The View from Anthropology*, ed. Kay Milton (London: Routledge, 1993), 31–42.

4. Sloterdijk, *Globes*, 4.

5. Sloterdijk, *Globes*, 774.

6. For example, Dennis Cosgrove, *Apollo's Eye: A Cartographic Genealogy of the Earth in Western Imagination* (Baltimore: Johns Hopkins University Press, 2001); David Woodward, "The Image of the Spherical Earth," *Perspecta* 25 (1989): 2–15. For a recent analysis of how such an image of a smooth and perfect sphere facilitates the free and frictionless flow of capital, see Samuel Ferdinand, Irene Villaescusa-Illán, and Esther Pereen, "Introduction," in *Other Globes: Past and Peripheral Imaginations of Globalization*, ed. Samuel Ferdinand et al. (Cham, Switzerland: Palgrave Macmillan, 2019), 1–39.

7. Jennifer Wenzel, "Planet vs. Globe," *English Language Notes* 52, no. 1 (2014): 19–30.

8. Sloterdijk, *Globes*, 927.

9. Edward Said, *Culture and Imperialism* (New York: Vintage, 1994), 7.

10. For the growing realization from the seventeenth century in Europe that our planet is neither perfectly nor smoothly spherical, see Woodward, "The Image of the Spherical Earth," 10–11. See also Kenneth R. Olwig, "The Earth Is Not a Globe: Landscape versus the 'Global-

ist' Agenda," *Landscape Research* 36, no. 4 (2011): 401–15. For wide-ranging essays that track the growing ambivalence, even outright hostility, to "anthropomorphism"—a term apparently coined in 1753—in the age of the scientific modern in Europe, see "Notes from the Field: Anthropomorphism," *Art Bulletin* XCIV, no. 1 (2012): 10–31. Of course, it is worth recalling that in Cesare Ripa's *Iconologia* (1593), the European Renaissance's influential manual of iconography, Earth ("Terra") is a matronly woman seated on a globe, holding a cornucopia in one hand and a scepter in the other. Such a feminized figure has also to contend with the enchantment with the spherical under scientific modernity.

11. For the artist's biography, see *American Debut: Shalinee Kumari, June 18–July 19, 2009* (San Francisco: Frey Norris Gallery, 2009). See also "Shalinee Kumari," produced by Tula Goenka and Susan Wadley, video, 4:00, June 13, 2013, https://vimeo.com/68330811. On Mithila art, see especially Carolyn B. Heinz, "Documenting the Image in Mithila Art," *Visual Anthropology Review* 22, no. 2 (2006): 5–33.

12. David Kinsley, *Hindu Goddesses: Visions of the Divine Feminine in the Hindu Religious Tradition* (Delhi: Motilal Banarsidass, 1987), 9.

13. Minoru Hara, "The King as a Husband of the Earth (*Mahī-Pati*)," *Asiatische Studien Etudes Asiatiques* 27, no. 2 (1973): 113–14.

14. O. P. Dwivedi, "Dharmic Ecology," in *Hinduism and Ecology: The Intersection of Earth, Sky and Water*, ed. Christopher Key Chapple and Mary Evelyn Tucker (Cambridge, MA: Harvard University Press, 2000), 10.

15. Maurice Bloomfield, trans., *Hymns of the Atharva-Veda, Together with Extracts from the Ritual Books and the Commentaries* (Oxford: Clarendon Press, 1897), 199.

16. Archana Venkatesan, "Bhūdevi," in *Brill's Encyclopedia of Hinduism*, vol. I, *Regions, Pilgrimage, Deities*, ed. Knut A. Jacobsen, Helene Basu, and Angelika Malinar (Leiden, The Netherlands: Brill, 2009), 491–98.

17. Hara, "The King as a Husband of the Earth," 109.

18. Sumathi Ramaswamy, "Of Gods and Globes: The Territorialisation of Hindu Deities in Popular Visual Culture," in *India's Popular Culture: Iconic Spaces and Fluid Images*, ed. Jyotindra Jain (Mumbai: Marg Publications, 2007), 19–31.

19. Martin W. Lewis and Kären Wigen, *Myth of the Continents: A Critique of Metageography* (Berkeley: University of California Press, 1997).

20. Denis Wood and John Fels, *The Power of Maps* (New York: Guilford Press, 1992), 5.

21. Sumathi Ramaswamy, "Conceit of the Globe in Mughal India," *Comparative Studies in Society and History* 49, no. 4 (2007): 751–82. See also Sumathi Ramaswamy, "Going Global in Mughal India," accessed March 21, 2022, http://sites.duke.edu/globalinmughalindia.

22. Thongchai Winichakul, *Siam Mapped: A History of the Geo-Body of a Nation* (Honolulu: University of Hawai'i Press, 1994), 37–61.

23. Sumathi Ramaswamy, *Terrestrial Lessons: The Conquest of the World as Globe* (Chicago: University of Chicago Press, 2017).

24. Quoted in Ramaswamy, *Terrestrial Lessons*, 26.

25. Quoted in Ramaswamy, *Terrestrial Lessons*, 27.

26. Simon Digby, "The Bhugola of Kshema Karna: A Dated Sixteenth Century Piece of Indian Metalware," *Art and Archaeology Research Papers* 4 (1973): 10–31.

27. Quoted in Ramaswamy, *Terrestrial Lessons*, 11.

28. Ramaswamy, *Terrestrial Lessons*. The Sanskrit word *gola* means both round (as in a coin or disk) as well as sphere (as in a ball).

29. Krishna Kumar, "'People's Science' and Development Theory," *Economic and Political Weekly* 19, no. 28 (1984): 1082-84.

30. Gauhar Raza et al., *Confluence of Science and Peoples' Knowledge at the Sangam* (New Delhi: NISTADS, 1996), 30–33.

31. Sumathi Ramaswamy, "Visualizing India's Geo-Body: Globes, Maps, Bodyscapes," in *Beyond Appearances? Visual Practices and Ideologies in Modern India*, ed. Sumathi Ramaswamy (New Delhi: Sage, 2003), 157–95. See also Sumathi Ramaswamy, *The Goddess and the Nation: Mapping Mother India* (Durham, NC: Duke University Press, 2010).

32. Cosgrove, *Apollo's Eye*, 15–16.

33. Ramaswamy, "Of Gods and Globes," 25.

34. Christopher Pinney, *Photos of the Gods: The Printed Image and Political Struggle in India* (New Delhi: Oxford University Press, 2004), 67. See also Raja Ravi Varma, *Sita Bhumipravesh*, 1880, oil on canvas, 56 × 40", Vadodara, Gujarat, India, https://artsandculture.google.com/asset/sita-bhumipravesh-raja-ravi-varma/jwHVOcEP79vPhg?hl=en. (I thank Arnav Adhikari for bringing this site to my attention.) See also https://artsandculture.google.com/asset/sita-bhumi-pravesh-raja-ravi-varma/jwEyTCqmNqpCsQ.

35. Pinney, *Photos of the Gods*, 67.

36. National Archives of India, "Innovations in the Ramlila Processions," Government of India, Home, Political (Confidential), December 1911, No. 14 (Deposit).

37. Pinney, *Photos of the Gods*, 109.

38. In the coexistence thus of the spherical and the anthropomorphic, Ramaiah's Bhoodevi resonates, curiously, with Ripa's Terra from several centuries earlier (see note 10).

39. On the illustrator Pratap Mulick, who produced the images for this comic book, declared by one scholar to be a "bumper hit," see Nandini Chandra, *The Classic Popular: Amar Chitra Katha, 1967–2007* (New Delhi: Yoda Press, 2008), 126–29.

40. Author's interview with Shalinee Kumari, June 3, 2021.

41. *American Debut: Shalinee Kumari*, plates 2, 3, 7, 8, and 11.

42. For these images, see Ramaswamy, "Of Gods and Globes," 28–29.

43. Ramaswamy, *The Goddess and the Nation.*

44. Cosgrove, *Apollo's Eye*, 105–6.

45. Max Horkheimer and Theodor W. Adorno, *Dialectic of Enlightenment*, trans. John Cumming (New York: Seabury, 1972), 3.

46. Bruno Latour, "What Is Iconoclash? Or Is There a World Beyond the Image Wars?" in *Iconoclash*, ed. Bruno Latour and Peter Weibel (Cambridge, MA: ZKM and MIT Press, 2002), 14–15; emphasis in original.

47. Vijaya R. Nagarajan, "Embedded Ecologies and the Earth Goddess," in *Feeding a Thousand Souls: Women, Ritual, and Ecology in India—An Exploration of Kolam* (New York: Oxford University Press, 2019), 204–24. For a recent Indian activist project for women's rights that has taken recourse to Bhumata, see https://en.wikipedia.org/wiki/Bhumata_Brigade and "Bhumata Brigade a.k.a. The Angry Goddesses," Blush Originals, S1 E8, August 27, 2016, YouTube video, 6:00, https://www.youtube.com/watch?v=jsNhPQthwJ4. For a feminist critique of the Whole Earth image popularized by NASA's photographs of *Earthrise* (1968) and *Blue Marble* (1972), which shows a wariness about returning to feminine imagery of our planet, see Yaakov Jerome Garb, "Perspective or Escape? Ecofeminist Musings on Contemporary Earth Imagery," in *Reweaving the World: The Emergence of Ecofeminism*, ed. Irene Diamond and Gloria Feman Orenstein, (San Francisco: Sierra Club Books, 1990), 264–78.

48. See especially Miriam Tola, "Composing with Gaia: Isabelle Stengers and the Feminist Politics of the Earth," *PhoenEx* 11, 1 (2016): 1–21; Miriam Tola, "Between Pachamama and Mother Earth: Gender, Political Ontology and the Rights of Nature in Contemporary Bolivia," *Feminist Review* 118 (2018): 25–40; and Miriam Tola, "Pachamama," in *An Ecotopian Lexicon*, ed. Matthew Schneider-Mayerson and Brent Ryan Bellamy (Minneapolis: University of Minnesota Press, 2019), 194–203.

49. Bruno Latour, "Waiting for Gaia: Composing the Common World through Arts and Politics" (lecture, French Institute, London, November 2011, http://www.bruno-latour.fr/sites/default/files/124-GAIA-LONDON-SPEAP_0.pdf.

50. Bruno Latour, "Some Experiments in Art and Politics," *E-flux* 23 (2011): 1.

51. Latour, "Waiting for Gaia," 12. For Latour's more extended later discussion on Gaia, see Bruno Latour, *Facing Gaia: Eight Lectures on the New Climatic Regime* (Cambridge, MA: Polity Press, 2017).

52. Rohini Swamy, "BJP's 'Antics' Raja in Tamil Nadu Gets a Lot of Attention but Not the Kind His Party Needs," *The Print*, June 19, 2019, https://theprint.in/politics/bjps-antics-raja-in-tamil-nadu-gets-a-lot-of-attention-but-not-the-kind-his-party-needs/251643/.

Chapter Two

1. Đặng Xuân Bảng, *Sử Học Bị Khảo [Historical Reference]* (Hà Nội: Viện Sự Học & NXB Văn Hoá—Thông Tin, 1997), 17. The publication date of the text is unknown, although it is likely to have been in the mid-nineteenth century, before Đặng Xuân Bảng's retirement in 1888.

2. This same theory is also outlined in a text published by Phạm Phục Trai in 1853, *A Recitation of the Essentials for Enlightening Children* (*Khải Đồng Thuyết Ước*, 啓童說約), National Library of Vietnam, Manuscript No. NLVNPF-0617.

3. Elite members of the precolonial Nguyễn court, from 1802 until French Indochina was established in 1887, were introduced to classical Chinese culture through their education. They read Chinese books and sat public examinations that tested them on their knowledge of Chinese history and the traditional Confucian corpus, the *Four Books and Five Classics* (四書五經). See Alexander Woodside, *Vietnam and the Chinese Model: A Comparative Study of Nguyễn and Ch'ing Civil Government in the First Half of the Nineteenth Century* (Cambridge, MA: Harvard University Press, [1971] 1988), 187.

4. One of the earliest and clearest descriptions of the celestial sphere emerges in the writings of the astronomer Chang Heng. In his "Commentary on the Armillary Sphere" (117 CE), Chang Heng clearly outlines the notion of an egg-like cosmos, a hard spherical shell containing a yolk-like earth: "the heavens are like a hen's egg . . . the earth is like the yolk of an egg and lies alone in the centre." Translation in Joseph Needham, *Science and Civilisation in China*, vol. 3 (Cambridge: Cambridge University Press, 1954), 217. The original celestial sphere theory included the idea that the bottom half of the sky-sphere is filled with water, upon which the earth floats. Also see Shigeru Nakayama, *A History of Japanese Astronomy: Chinese Background and Western Impact* (Cambridge, MA: Harvard University Press, 1969), 35–37.

5. Chu-tzu yu-lei 朱子語類 [Classified Conversations of Master Chu] compiled in 1270, YL2.10a0 translated and cited in Yung Sik Kim, *The Natural Philosophy of Chu Hsi (1130–1200)* (Philadelphia: American Philosophical Society, 2000), 139.

6. It is worth noting that the cosmic egg did not explain the actual physical shape of the heavens and earth. The sky-sphere was conceived of as perfectly round, rather than oblong, and up until the seventeenth century, Chinese philosophers subscribed to the view that the earth

was square. The terraqueous globe was integrated into celestial sphere theory through Chinese philosophers' interactions with Jesuit missionaries. For an excellent discussion of this process, see Qiong Zhang, *Making the New World Their Own: Chinese Encounters with Jesuit Science in the Age of Discovery* (Leiden, The Netherlands: Brill, 2015), especially chapter four, "The Introduction and Refashioning of the Terraqueous Globe," 148–202.

7. Chu Hsi's theory of the cosmos, which Đặng Xuân Bảng draws upon, is described in detail by Yim Sim Kim. See Yung Sik Kim, *The Natural Philosophy of Chu Hsi (1130–1200)* (Philadelphia: American Philosophical Society, 2000), 135–36.

8. "[He] established coordinates of a myriad leagues, delineated the wild, [and] demarcated the administrative districts." See "Treatise on Earth's Patterns," *History of Han*, translated in Douglas Howland, *Borders of Chinese Civilization: Geography and History at Empire's End* (Durham, NC: Duke University Press, 1996), 1.

9. I have borrowed Liam Kelley's translation of "Domain of Manifest Civility." See Liam Kelley, *Beyond the Bronze Pillars: Envoy Poetry and the Sino-Vietnamese Relationship* (Honolulu: University of Hawai'i Press, 2005), 28–36.

10. John Fairbank, ed., *The Chinese World Order: Traditional China's Foreign Relations* (Cambridge, MA: Harvard University Press, 1968). The "Sinosphere" was coined by James Matisoff in his article "On Megalocomparison," *Language*, 66 (1990): 106–20.

11. Caroline Dodds Pennock and Amanda Power, "Globalizing Cosmologies," *Past & Present* 238, no. S13 (2018): 88–115.

12. Pennock and Power, "Globalizing Cosmologies," 88–115.

13. John K. Fairbank, "Tributary Trade and China's Relations with the West," *The Journal of Asian Studies* 1, no. 2 (1942): 129–49.

14. Qiong Zhang, *Making the New World Their Own*, 188–94.

15. Tim Ingold, "Globes and Spheres: The Topology of Environmentalism," in *The Perception of the Environment: Essays on Livelihood, Dwelling and Skill* (London: Routledge, 2011), 209–18.

16. Ingold, "Globes and Spheres," 210.

17. Kathlene Baldanza, *Ming China and Vietnam: Negotiating Borders in Early Modern Asia* (Cambridge: Cambridge University Press, 2018), 6.

18. *Khâm Định Đại Nam Hội Điển Sự Lệ* [Official Compendium of Institutions and Usages of Imperial Vietnam], vol. 8 (Huế: Nhà Xuất Bản Giáo Dục, 2004), 527, 529.

19. *Khâm Định Đại Nam Hội Điển Sự Lệ*, vol. 8, 527, 529.

20. *Khâm Định Đại Nam Hội Điển Sự Lệ*, vol. 8, 530. These results were to be submitted to the emperor to peruse.

21. *Đại Nam Thực Lục* [Veritable Records of Đại Nam] (translated into Vietnamese), vol. 7 (Hà Nội: Nhà Xuất Bản Giáo Dục, 2002), 483. In an attempt to lift standards, Tự Đức offered a reward of a certificate and three ounces of silver to teachers for every student who successfully passed the three-year course and punished them with a fine equivalent to six months' salary if there were no graduates.

22. *Đại Nam Thực Lục*, vol. 7, 483.

23. *Khâm Định Đại Nam Hội Điển Sự Lệ*, vol. 8, 530–531.

24. I am grateful to Catherine Churchman for assistance translating these titles.

25. George Dutton, "The Nguyễn State and the Book Collecting Project," unpublished paper, Conference on Nguyễn Vietnam, Harvard University, Cambridge, MA, May 11–12, 2013; Philippe Langlet, *L'ancienne historiographie d'État au Viêtnam*, tome 1, *Raisons d'être, conditions d'élaboration et caractères au siècle des Nguyễn* (Paris: École Française d'Extrême-Orient, 1990), 277–78.

26. *Đại Nam Thực Lục*, vol. 2, 642, 694; 大南寔錄 [Veritable Records of Đại Nam] (original Chinese) (Tokyo: Institute of Cultural and Linguistic Studies, Keio University, 1961–1974), vol. 6, 2017(187); vol. 6, 2046(216).

27. In 1810, for instance, the mandarin Nguyễn Hữu Thận presented Emperor Gia Long with the book *The Calendar of the Great Qing Dynasty (Đại Thanh Lịch Tương Khảo Thánh)* on his return trip from China, with a request that it be immediately handed over to the Observatory to update their collection; *Đại Nam Thực Lục*, vol. 1, 785.

28. There is excellent scholarly literature written on this subject. See especially, Catherine Jami, *The Emperor's New Mathematics: Western Learning and Imperial Authority during the Kangxi Reign (1662–1722)* (New York: Oxford University Press, 2011).

29. According to Shi Yunli, *Later Volumes on the Thorough Investigation of Calendrical Astronomy* was compiled in 1742 in Beijing under the direction of Ignatius Kogler (1680–1746) and Andre Pereira (1689–1743). See Shi Yunli, "The Yuzhi Lixiang Kaocheng Houbian in Korea," in *History of Mathematical Sciences: Portugal and East Asia*, ed. Luís Saraiva (Singapore: World Scientific, 2012), 205. The *Record on the Newly Built Astronomical Instruments for the Beijing Observatory* was compiled by the Flemish Jesuit missionary Ferdinand Verbiest (1623–1688). See Joachim Kurtz, "Framing European Technology in Seventeenth-Century China: Rhetorical Strategies in Jesuit Paratexts," in *Cultures of Knowledge: Technology in Chinese History*, ed. Dagmar Schäfer (Leiden, The Netherlands: Brill, 2011), 210.

30. Qiong Zhang, *Making the New World Their Own*, 150.

31. *Đại Nam Thực Lục*, vol. 2, 534.

32. *Đại Nam Thực Lục*, vol. 8, 337–38; 大南寔錄 [Veritable Records of Đại Nam] (original Chinese), vol. 18, 7015(190). For more on Jesuit missionaries in Vietnam in the seventeenth century, see Alexi Volkov, "Evangelization, Politics and Technology Transfer in 17th Century Cochinchina: The Case of Joao Da Cruz," in *Europe and China: Science and Arts in the 17th and 18th Centuries*, ed. Luís Saraiva (Singapore: World Scientific, 2012), 31–70.

33. Other books sent to the Hải Dương printing press included *Coal Mines and Mining (Khai Môi Yếu Pháp)*, authored by Warington Wilkinson Smyth in 1867 and translated into Chinese by John Fryer in 1871. Also, *Elements of International Law (Vạn Quốc Công Pháp)*, first published in 1836 by the American Henry Wheaton.

34. *Đại Nam Thực Lục*, vol. 8, 337–338; 大南寔錄 [Veritable Records of Đại Nam] (original Chinese), vol. 18, 7015(190).

35. See Kelley, *Beyond the Bronze Pillars*, 28–36, for further historical context on theories that the celestial and terrestrial realms were patterned and how writing enabled humans to communicate with this realm.

36. Sun Xiaochun and Jacob Kistemaker, *The Chinese Sky during the Han: Constellating Stars and Society* (Leiden, The Netherlands: Brill, 1997); Ho Peng Yoke, *Chinese Mathematical Astrology: Reaching Out of the Stars* (London: Routledge, 2004).

37. Yung Sik Kim, *The Natural Philosophy of Chu Hsi*, 147. The North Pole/Emperor star was believed to be at the center of the sky itself. Unlike other stars that were swept along in an orbit of perpetual cosmic motion, the North Star was thought to be fixed in space and a reference point around which all the heavens revolved.

38. *Đại Nam Thực Lục*, vol. 7, 722; 大南寔錄 [Veritable Records of Đại Nam] (original Chinese), vol. 16, 6208(134)–6209(135).

39. *Đại Nam Thực Lục*, vol. 7, 722; 大南寔錄 [Veritable Records of Đại Nam] (original Chinese), vol. 16, 6208(134)–6209(135).

40. Astral-terrestrial correspondences were developed during the late Spring and Autumn

period (722–481 BCE), but it wasn't until the early millennium that the ancient Nine Provinces (later twelve states) of north and central China were assigned different astral fields in relation to the Yellow River. See David Pankenier, "Astronomy," in *Routledge Handbook of Early Chinese History*, ed. Paul R. Goldin (London: Routledge, 2018). Also see David W. Pankenier, "On Chinese Astrology's Imperviousness to External Influences," in *Astrology in Time and Place: Cross-Cultural Questions in the History of Astrology*, ed. Nicholas Campion and Dorian Gieseler Greenbaum (Newcastle upon Tyne: Cambridge Scholars Publishing, 2015), 3–26.

41. Such connections between the stars and the ancient geographical space of China occurred frequently in Nguyễn reading of the skies. For example, in searching for the meaning of Venus rising, Trương Quốc Dụng and Trương Đăng Quế pondered the connection of the Dipper Mansion to Yong Province (雍州) (one of the legendary Nine Provinces) during the Qin dynasty (221 to 206 BCE). *Đại Nam Thực Lục*, vol. 7, 656; 大南寔錄 [Veritable Records of Đại Nam] (original Chinese), vol. 16, 6159(85)–6160(86).

42. Chinese history was a cornerstone of formal education in Vietnam right up until the late nineteenth century largely because—as discussed by Philippe Langlet—of the "neutral ground" it provided in approaching the past. Langlet suggests that the teaching of Vietnamese history, by contrast, was more dangerous because of the contested legitimacy of the Nguyễn state. See Langlet, *L'ancienne historiographie d'État au Vietnam*, 135.

43. See Trịnh Hoài Đức, *Gia Định Thành Thông Chí* [Gia Dinh Citadel Records] (Hanoi: Nhà Văn Hóa, 1972), 3.

44. Trịnh Hoài Đức, *Gia Định Thành Thông Chí*, 4.

45. Trịnh Hoài Đức, *Gia Định Thành Thông Chí*, 4.

46. *Đại Nam Nhất Thống Chí* [Unified Gazetteer of Đại Nam], 5 vols. (Huế: Nhà Xuất Bản Thuận Hoá, 1993).

47. *Đại Nam Nhất Thống Chí*, vol. 5, 199.

48. *Đại Nam Nhất Thống Chí*, vol. 4, 184.

49. *Đại Nam Nhất Thống Chí*, vol. 4, 199. Đặng Xuân Bảng also searches for Vietnam's governing star bodies in his text *Sử Học Bị Khảo [Historical Reference]*, 146–150. After citing a number of ancient texts, he concludes that Trịnh Hoài Đức's theory (i.e., that Vietnam is governed by the first and second asterism of the Ox Mansion) is correct.

50. In 1715, Nishikawa Joken not only questioned why portents should be connected to the geographical framework of China, but the system itself, in a book entitled *Wakan Hensho Kai Bendan* (Defense and Refutation of Theories of Climatic Anomalies): "Heaven is too vast to be monopolised by one country. How is it reasonable to discuss the relation between the nature of portents and the fate of one specific country when the phenomena are observable from any part of the world?" See Nakayama, *A History of Japanese Astronomy*, 54.

51. *Đại Nam Thực Lục*, vol. 4, 506–7; 大南寔錄 [Veritable Records of Đại Nam] (original Chinese), vol. 10, 3523(23)–3524(24).

52. The *Veritable Records* documents thunder within numerous sites in the imperial palace, including the East Gate Tower (1841) and the Northern Gate and Southern Pavilion (1869). In 1872, it is recorded that thunder "shook the flagpole in the capital citadel." See *Đại Nam Thực Lục*, vol. 7, 244, 256, 1369. Within official discourse, these episodes of thunder are often interpreted as pointed heavenly assaults on the court, causing emperors to quiver with fear and show reverence.

53. *Khán Thiên Tượng* (看天象) [Observing Astronomy], Viện Hán Nôm (Sino-Vietnamese Institute), Hà Nội, Manuscript No. VHV.474.

54. *Châu Bản Triều Nguyễn* [Vermilion Records of the Nguyễn Dynasty], National Archives No. 1, Hà Nội. No. 111/155 (12/12/TD16).

55. 一蹟天文家傳 [Nhất Tích Thiên Văn Gia Truyền, Astronomical Vestiges], Manuscript No. VHV.1382, Hán-Nôm Institute, Hanoi, 2a.

56. *Đại Nam Thực Lục*, vol. 3, 457; 大南寔錄 [Veritable Records of Đại Nam] (original Chinese), vol. 8, 2644(4)–2665(5).

57. *Đại Nam Thực Lục*, vol. 5, 180; 大南寔錄 [Veritable Records of Đại Nam], vol. 11, 4152(238).

58. *Đại Nam Thực Lục*, vol. 4, 514–15; 大南寔錄 [Veritable Records of Đại Nam], vol. 10, 3530(30).

59. Kathryn Dyt, *The Nature of Kingship and the Nguyễn Weather-World in Nineteenth-Century Vietnam* (forthcoming with University of Hawai'i Press).

60. Sarah Strauss, "Weather Wise: Speaking Folklore to Science in Leukerbad," in *Weather, Climate, Culture*, ed. Sarah Struss and B. S. Orlove (Oxford: Berg Publishers, 2003), 48–49.

61. *Đại Nam Thực Lục*, vol. 5, 560; 大南寔錄 [Veritable Records of Đại Nam] (original Chinese), vol. 11, 4475(145).

62. *Đại Nam Thực Lục*, vol. 3, 276; 大南寔錄 [Veritable Records of Đại Nam] (original Chinese), vol. 7, 2520(272).

63. *Đại Nam Nhất Thống Chí*, vol. 2, 12–13.

64. *Đại Nam Nhất Thống Chí*, vol. 2, 12–13.

65. Sources such as the *Veritable Records of Đại Nam* (*Đại Nam Thực Lục*) and the *Unified Gazetteer of Dai Nam* (*Đại Nam Nhất Thống Chí*) contain numerous such portrayals. See also Philip Taylor, *The Khmer Lands of Vietnam: Environment, Cosmology and Sovereignty* (Singapore: National University of Singapore Press, 2014), 7–9.

Chapter Three

1. I have chosen to spell the mountain's name "Chomolungma," as this is the most common spelling used by the Sherpa community and within the climbing world. As far as I can ascertain, the convention of spelling the mountain's name "Chomolungma" developed in early climbing accounts, and the Sherpa community adopted this usage. However, there are multiple other ways to spell the mountain's name. When the name is transliterated from Tibetan (the Sherpas use the Tibetan script), it is written *Jo mo glang ma*. Its most phonetically accurate transcription in English is *Jomolangma*. The initiation *J* can often sound like a *Ch*, so Chomolungma is also used. As this chapter will show, the mountain's naming conventions have been and continue to be a site of contestation.

2. Ngag dbang bstan 'dzin nor bu, *Rnam thar 'chi med bdud rtsi'i rol mtsho* (n.p., n.d.), 291a. Sa hib is a transliteration of a British Indian term. Sherry Ortner, *Life and Death on Mt. Everest: Sherpas and Himalayan Mountaineering* (Princeton, NJ: Princeton University Press, 2020), 5–18.

3. Ngag dbang bstan 'dzin nor bu, "Rdza rong phu gnas yig," in *Gsung 'bum* (Kathmandu: Ngagyur Dongak Choling Monastery, 2004), 401–26.

4. Ngag dbang bstan 'dzin nor bu, *Rnam thar 'chi med bdud rtsi'i rol mtsho*, 291b.

5. Francis Younghusband, *Everest: The Challenge* (London: Thomas Nelson & Sons Ltd, 1936), 56.

6. S. G. Burrard and H. H. Hayden, *A Sketch of the Geography and Geology of the Himalaya Mountains and Tibet* (Calcutta: Superintendent Government Printing, 1908), 21.

7. 珠穆朗玛. The Chinese government also represents the name in the Tibetan Pinyin system as Qomolangma. Many Tibetans view this spelling as a symbol of China's disputed sovereignty and refuse to use it.

8. M. C. Meyer et al., "Landscape Dynamics and Human-Environment Interactions in the Northern Foothills of Cho Oyu and Mount Everest (Southern Tibet) during the Late Pleistocene and Holocene," *Quaternary Science Reviews* 229 (2020): 106127. See also Jade d'Alpoim Guedes and Mark Aldenderfer, "The Archaeology of the Early Tibetan Plateau: New Research on the Initial Peopling through the Early Bronze Age," *Journal of Archaeological Research* 28 (2019): 1–54; Pasang Wangdu et al., *Shel dkar chos 'byung: History of the "White Crystal," Religion and Politics of Southern La Stod* (Wien: Verlag der Österreichischen Akademie der Wissenschaften, 1996), 2–5.

9. Toni Huber, "Traditional Environmental Protectionism in Tibet Reconsidered," *The Tibet Journal* 16, no. 3 (1991): 64–66. Martin Mills outlined the Tibetans' approach to mining in "King Srong btsan sGam po's Mines: Wealth Accumulation and Religious Asceticism in Buddhist Tibet," *The Tibet Journal* 31, no. 4 (2006): 89–106.

10. Tibetan: *Brtan ma bcu gnyis*, *Tshe ring mched lnga*, and *zhing skyong 'gro ma*. I discussed these goddesses in Ruth Gamble, *Reincarnation in Tibetan Buddhism: The Third Karmapa and the Invention of a Tradition* (New York: Oxford University Press, 2018), 141–48.

11. Tibetan: *jo mo* and *gangs skar*.

12. For an overview of the multiplicity allowed by this theory of "two truths" see Jay L. Garfield, "Taking Conventional Truth Seriously: Authority regarding Deceptive Reality," in *Moonshadows: Conventional Truth in Buddhist Philosophy*, ed. The Cowherds (New York: Oxford University Press, 2011), 23–37. The allowance of various descriptions for relative truth had limits based on doctrines such as cause and effect and rebirth, which proscribed killing, for example.

13. Jan Nattier, *Once upon a Future Time: Studies in a Buddhist Prophecy of Decline* (New York: Asian Humanities Press, 1991), 8.

14. Lewis Doney, *The Zangs gling ma: The First Padmasambhava Biography; Two Exemplars of the Earliest Attested Recension* (Andiast, Switzerland: International Institute for Tibetan and Buddhist Studies, 2015).

15. This traditional story is outlined in Ngag dbang bstan 'dzin nor bu, "Rdza rong phu gnas yig," 401–26.

16. Andrew Quintman, "Toward a Geographic Biography: Mi la ras pa in the Tibetan Landscape," *Numen* 55, no. 4 (2008): 363–410.

17. René de Nebesky-Wojkowitz, *Oracles and Demons of Tibet: The Cult and Iconography of the Tibetan Protective Deities* (The Hague: Mouton & Co., 1956): 177–82.

18. Ngag dbang bstan 'dzin nor bu, "Rdza rong phu gnas yig," 402.

19. Geoffrey Samuel, "Hidden Lands of Tibet in Myth and History," in *Hidden Lands in Himalayan Myth and History: Transformations of* sbas yul *through Time*, ed. Frances Garrett, Elizabeth McDougal, and Geoffrey Samuel (Leiden, The Netherlands: Brill, 2020), 51–91.

20. Ruth Gamble, John Powers, and Paul Hackett, "Famines of the Early Little Ice Age (1260–1360): The Impacts of Pre-modern Climate Change in Central Tibet," *Bulletin of the School of African and Oriental Studies* 85, no. 2 (2022), 1–19.

21. Luciano Petech, *China and Tibet in the Early XVIIIth Century: History of the Establishment of Chinese Protectorate in Tibet* (Leiden, The Netherlands: Brill 1997), 51–66.

22. Sangs rgyas bstan 'dzin, *Shar pa'i chos 'byung dang mes rabs* (Lhasa: Bod ljongs bod yig dpe rnying dpe skrun khang, 2003), 12–14.

23. Stanley F. Stevens, *Claiming the High Ground: Sherpas, Subsistence, and Environmental Change in the Highest Himalaya* (Berkeley: University of California Press, 1993), 95–211.

24. Lindsay A. Skog, "Khumbi yullha and the Beyul: Sacred Space and the Cultural Politics

of Religion in Khumbu, Nepal," *Annals of the American Association of Geographers* 107, no. 2 (2017): 546–54.

25. Ngag dbang bstan 'dzin nor bu, "Rdza rong phu gnas yig," 404.

26. Sherry B. Ortner, *High Religion: A Cultural and Political History of Sherpa Buddhism* (Princeton, NJ: Princeton University Press, 1989), 70–90.

27. In Nepali histories of this time, Nyalam is called Kuti. Prem Uprety, "Treaties between Nepal and her Neighbors: A Historical Perspective," *Tribhuvan University Journal* 19, no. 1 (1996): 15–24.

28. Nigel Harris, "The Geological Exploration of Tibet and the Himalaya," *The Alpine Journal* 96 (1992): 66–74.

29. Matthew H. Edney, *Mapping an Empire: The Geographical Construction of British India, 1765–1843* (Chicago: University of Chicago Press, 2009), 199–234.

30. Derek J. Waller, *The Pundits: British Exploration of Tibet and Central Asia* (Louisville: University Press of Kentucky, 2004), chap. 6.

31. Waller, *The Pundits*, chap. 6.

32. Edney, *Mapping an Empire*, 262–70.

33. There have been many academic articles written about the mountain's name, but none, it seems, by anyone who knows Tibetan, Sherpa, Nepali, or Hindi. Douglas W. Freshfield, "Tibet I. Notes from Tibet," *The Geographical Journal* 23, no. 3 (1904): 361–66. During the 1930s, there were a series of debates about the mountain's name in a Tibetan-language newspaper: *Melong* vol. 7, no. 3, March 1933; *Melong* vol. 7, no. 5, May 1933; *Melong* vol. 7, no. 4, April 1933.

34. Paper read by Andrew Scott Waugh to the Royal Geographical Society on May 12, 1857, reported in *Illustrated London News*, August 15, 1857, 170.

35. "Proceedings of the Geographical Section of the British Association. Aberdeen Meeting, 1885," *Proceedings of the Royal Geographical Society and Monthly Record of Geography* 7, no. 11 (1885): 748–52.

36. Rasoul Sorkhabi, "Historical Development of Himalayan Geology," *Journal of Geological Society of India* 49, no. 1 (1997): 95.

37. James Herbert, "Report of the Mineralogical Survey of the Region Lying between the Rivers Satlaj and Kali, with a Geological Map," *Journal of the Asiatic Society of Bengal* 11, vol. I (1842): 734.

38. Sorkhabi, "Historical Development of Himalayan Geology," 97.

39. Sorkhabi, "Historical Development of Himalayan Geology," 97.

40. Edward Suess, "Are Ocean Depths Permanent?," *Natural Science: A Monthly Review of Scientific Progress*, 2 (March 1893): 180–87.

41. Suess, "Are Ocean Depths Permanent?," 183.

42. Edward Suess, *The Face of the Earth: Das Antlitz der Erde*, trans. Hertha Sollas and William Sollas (Oxford: Clarendon Press, 1904–1908).

43. Suess, *The Face of the Earth*, vol. 2, 18.

44. John Auden, "Traverses in the Himalaya," *Geological Survey of India*, 69 (1935): 123–67.

45. Martin Rudwick, *Earth's Deep History: How It Was Discovered and Why It Matters* (Chicago: University of Chicago Press, 2021), 217.

46. Sherry Ortner, *Life and Death on Mt. Everest: Sherpas and Himalayan Mountaineering* (Princeton, NJ: Princeton University Press, 2020), 26–54.

47. M. P. Ward, "Mapping Everest," *The Cartographic Journal* 31, no. 1 (1994): 33–44.

48. Alex McKay, *Tibet and the British Raj: The Frontier Cadre, 1904–1947* (London: Psychology Press, 1997), 1–29.

49. Burrard and Hayden, *Geography and Geology of the Himalaya*, 44.

50. Charles H. D. Ryder, "Exploration and Survey with the Tibet Frontier Commission, and from Gyangtse to Simla Via Gartok," *The Geographical Journal* 26, no. 4 (1905): 370.

51. Burrard and Hayden, *Geography and Geology of the Himalaya*, 21.

52. Hugh Ruttledge, "The Mount Everest Expedition, 1933," *The Geographical Journal* 83, no. 1 (1934): 1–9.

53. Charles Bell, *Portrait of a Dalai Lama: The Life and Times of the Great Thirteenth* (Boston: Wisdom Publications, 1987), 274–75.

54. This passport has not been analyzed yet. I thank Chung Tsering for explaining that the term *rgya'i mthun lam gang che zhu* implies an official treaty and allyship rather than mere friendship, as it has been more usually translated. See Charles K. Howard-Bury, *Mount Everest: The Reconnaissance, 1921* (London: Edward Arnold & Co., 1922), 24.

55. Tib: bya ma lung. Bell translated this name as "the southern district where birds are kept." Bell, *Portrait of a Dalai Lama*, 277.

56. Howard-Bury, *Mount Everest*, 84.

57. Alexander Heron, "Appendix III: A Note on the Geological Results of the Expedition," in Howard-Bury, *Mount Everest*, 338.

58. A. M. Heron, "The Rocks of Mount Everest," *Geographical Journal* 60 (1922): 219–20.

59. Heron, "Appendix III," 338–40.

60. This exchange is described in Wade Davis, *Into the Silence: The Great War, Mallory, and the Conquest of Everest* (London: Vintage, 2011), 368. Davis drew his account from Charles Bell's papers.

61. Royal Geographic Society Archives, box 11, file 4, and box 27, file 3. Reported in Wade Davis, *Into the Silence*, 367.

62. Ngag dbang bstan 'dzin nor bu, *Bdud rtsi'i rol mtsho*, 291a.

63. Ngag dbang bstan 'dzin nor bu, "Rdza rong phu gnas yig," 403.

64. Ngag dbang bstan 'dzin nor bu, *Bdud rtsi'i rol mtsho*, 291b.

65. Ngag dbang bstan 'dzin nor bu, *Bdud rtsi'i rol mtsho*, 291a–291b.

66. Noel Odell, "Observations on the Rocks and Glaciers of Mount Everest," *The Geographical Journal*, 66, no. 4 (1925), 289.

67. Odell, "Observations," 313.

68. This scandal has been analyzed in Peter H. Hansen, "The Dancing Lamas of Everest: Cinema, Orientalism, and Anglo-Tibetan Relations in the 1920s," *The American Historical Review* 101, no. 3 (1996): 712–47.

69. Lawrence Wager, "A Review of the Geology and Some New Observations," in *Everest 1933*, ed. Hugh Ruttledge (London: Hodder and Stroughton, 1934), 312–37.

70. Wager, "A Review of the Geology," 312.

71. Wager, "A Review of the Geology," 319–24.

72. Wager, "A Review of the Geology," 336.

73. At the end of August Gansser and Arnold Heim, *Central Himalaya, Geological Observations of the Swiss Expedition 1936* (Zürich: Gebrüder Fretz, 1939).

74. They called it the "great thrust fold of Darjeeling." Gansser and Heim, *Central Himalaya*, 225.

75. Paldin and Kirken's stories are told in August Gansser and Arnold Heim, *The Throne of the Gods: An Account of the First Swiss Expedition to the Himalayas*, trans. Eden and Cedar Paul (New York: Macmillan Company, 1939).

76. Gansser and Heim, *Central Himalaya*, 16.

77. Sorkhabi, "Historical Development of Himalayan Geology," 100.

78. Ortner, *Life and Death*, 248–79.

79. For an overview of this time, see Tsering Shakya, *The Dragon in the Land of Snows: A History of Modern Tibet Since 1947* (New York: Columbia University Press, 1999). For the influence of the Cold War on this history, see Carole McGranahan, "Tibet's Cold War: The CIA and the Chushi Gangdrug Resistance, 1956–1974," *Journal of Cold War Studies* 8, no. 3 (2006): 102–30.

80. Luca Dal Zilio et al., "Building the Himalaya from Tectonic to Earthquake Scales," *Nature Reviews Earth & Environment* 2, no. 4 (2021): 251–68.

81. Jingfang Bo et al., "Upper Triassic Reef Coral Fauna in the Renacuo Area, Northern Tibet, and Its Implications for Palaeobiogeography," *Journal of Asian Earth Sciences* 146 (2017): 114–33.

82. Rodolfo Carosi et al., "20 Years of Geological Mapping of the Metamorphic Core across Central and Eastern Himalayas," *Earth-Science Reviews* 177 (2018): 124–38.

83. Sorkhabi, "Historical Development of Himalayan Geology," 100–103. In Nepal, the Geological Survey and the Department of Mines (1942) were combined to form the Department of Mines and Geology in 1977. Its website includes a historical overview: https://www.dmgnepal.gov.np/. Grace Shen Yen describes geology's role in Chinese nationalist imaginings. She notes instances of geologists claiming Tibet cartographically in 1934–35 but notes no geological surveys on the Plateau. Grace Yen Shen, *Unearthing the Nation: Modern Geology and Nationalism in Republican China* (University of Chicago Press: Chicago, 2014), 136.

84. Sorkhabi, "Historical Development of Himalayan Geology," 100–103.

85. Zhang et al., *A Scientific Guidebook to South Xizang* (Tibet) (Peking: s.n., 1980).

86. Department of Mines and Geology, *Annual Report No. 11 DMG* (Kathmandu: Government of Nepal, 2019); Department of Mines and Geology, *Annual Report No. 12 DMG* (Kathmandu: Government of Nepal, 2020). Chinese support for surveyance and scientific analysis is acknowledged throughout both reports.

87. "地质学家,中国工程院首位藏族院士多吉驻守高原'寻宝'" [Dorje, a Geologist and the First Tibetan Academic Appointed to the Chinese Academy of Engineering, Is on a "Treasure Hunt" on the Plateau], *China Science Communication*, November 12, 2017.

88. Many social and educational structures make it difficult for Tibetans to study science. See Lubei Zhang and Linda Tsung, "Exploring Sustainable Multilingual Language Policy in Minority Higher Education in China: A Case Study of the Tibetan Language," *Sustainability* 12, no. 18 (2020): 7267.

89. Ortner, *Life and Death*, 248–79.

90. Jeremy Spoon, "The Heterogeneity of Khumbu Sherpa Ecological Knowledge and Understanding in Sagarmatha (Mount Everest) National Park and Buffer Zone, Nepal." *Human Ecology* 39, no. 5 (2011): 670.

91. Geological Society of America, "Grant Announcement: Tshering Lama Sherpa Wins Microscopy Grant for Innovative Geoscience Research," *Eureka Alert*, December 9, 2020, https://www.eurekalert.org/news-releases/736253; Arley Titzler, "GLOF Risk Perception in Nepal Himalaya," *GlacierHub*, January 30, 2019, https://glacierhub.org/2019/01/30/glof-risk-perception-in-nepal/.

Chapter Four

1. Tui Atua Tupua Tamasese Ta'isi Efi, "Climate Change and the Perspective of the Fish," in *Pacific Climate Cultures: Living Climate Change in Oceania*, ed. Tony Crook and Peter Rudiak-Gould (Warsaw: De Gruyter, 2018), 155–59.

2. Epeli Hau'ofa, "Our Sea of Islands," in *A New Oceania: Rediscovering Our Sea of Islands*, ed. Eric Waddell, Vijay Naidu, and Epeli Hau'ofa (Suva: University of the South Pacific in association with Beake House, 1993), 147–61.

3. Hau'ofa, "Our Sea of Islands," 153.

4. United Nations, *The Second World Ocean Assessment* (New York: United Nations, 2021), 5.

5. Teuira Henry, "Tahitian Astronomy: Birth of the Heavenly Bodies," *Journal of the Polynesian Society* 16, no. 2 (1907): 101–4.

6. See Anne Salmond, *The Trial of the Cannibal Dog: Captain Cook in the South Seas* (London: Penguin, 2004), 1–33.

7. For a detailed account of Tupaia and his role in the *Endeavour* voyage, see Anne Salmond, "Their Body Is Different, Our Body Is Different: European Navigators in the Eighteenth Century," *History and Anthropology* 16, no. 2 (June 2005): 170–80.

8. For a detailed account of the 'Arioi cult, see Anne Salmond, *Aphrodite's Island: The European Discovery of Tahiti* (Auckland: Penguin Viking, 2009), 25–32.

9. For a more detailed account of ancestral Tahitian ideas about the ocean, see Anne Salmond, "Star Canoes, Voyaging Worlds," *Interdisciplinary Science Reviews* 46, no. 3 (2021): 267–68.

10. Teuira Henry, *Ancient Tahiti, Based on Material Recorded by J. M. Orsmond*, Bernice P. Bishop Museum Bulletin No. 48 (Honolulu: Bernice P. Bishop Museum, 1928), 433–43.

11. Henry, *Ancient Tahiti*, 389.

12. For a remarkable reconstruction of Tahitian understandings of the cosmos and navigational methods, based on an analysis of early European and Tahitian accounts and linguistic evidence, see Claude Teriierooiterai, "Mythes, astronomie, découpage du temps et navigation traditionnelle: l'héritage océanien contenu dans les mots de la langue tahitienne" (PhD diss., l'Université de la Polynésie française, 2013).

13. James Cook, *The Journals of Captain James Cook on His Voyages of Discovery*, ed. J. C. Beaglehole, vol. 1, *The Voyage of the Endeavour, 1768–1771* (Cambridge: Published for the Hakluyt Society at the University Press, 1955–1974), 54.

14. Cook, *Journals of Captain James Cook*, 139n.

15. For discussions of Tupaia's Map, see David Turnbull, "Cook and Tupaia, a Tale of Cartographic 'Méconnaissance,'" in *Science and Exploration in the Pacific: European Voyages to the Southern Oceans in the 18th Century*, ed. Margarette Lincoln (Woodbridge, Suffolk: Boydell Press, 1998), 117–31; Anne Di Piazza and Erik Pearthree, "A New Reading of Tupaia's Chart," *Journal of the Polynesian Society* 116, no. 3 (2007): 321–40; Lars Eckstein and Anja Schwarz, "The Making of Tupaia's Map: A Story of the Extent and Mastery of Polynesian Navigation, Competing Systems of Wayfinding on James Cook's Endeavour, and the Invention of an Ingenious Cartographic System," *The Journal of Pacific History* 54, no. 1 (2007): 1–95; Anne Salmond, "Hidden Hazards: Reconstructing Tupaia's Chart. Forum on Tupaia's Chart," *Journal of Pacific History* 54, no. 4 (October 2019): 534–37.

16. Mytea (Meheti'a), Imao (Mo'orea), Tapooamanu (Maiao), Tethuroa (Teti'aroa), Huiheine (Huahine), Ulietea (Ra'iatea), Otaha (Tahaa), Bolabola (Borabora), Tubai (Tupai), and Maurua (Maupiti).

17. Island list transcribed from the second draft of Tupaia's Chart (T2), recorded in James Cook's "Journal" (T2/C), here as copied by his clerk Richard Orton, [Mitchell MS], State Library of New South Wales, Sydney, Safe 1/71.

18. Charles Frake, "Cognitive Maps of Time and Tide among Mediaeval Seafarers," *Man* 20, no. 2 (June 1985): 254–70.

19. Anne Salmond, "Their Body Is Different," 166–67.

20. Jordan Branch, "Mapping the Sovereign State: Cartographic Technology, Political Authority, and Systemic Change" (PhD diss., University of California, Berkeley, 2011).

21. M. H. Edney, "British Military Education, Mapmaking, and Military 'Map-Mindedness' in the Later Enlightenment," *The Cartographic Journal* 31, no. 1 (1994): 14–20; Nicholas Blomley, "Law, Property and the Geography of Violence: The Frontier, the Survey and the Grid," *Annals of the Association of American Geographers* 93, no. 1 (March 2003): 121–41; Tore Frängsmyr, J. L. Heilbron, and Robin E. Rider, eds., *The Quantifying Spirit in the 18th Century* (Berkeley: University of California Press, 1990).

22. Joseph Banks, *The Endeavour Journal of Joseph Banks: 1768–1771, I*, ed. J. C. Beaglehole (Sydney: The Trustees of the Public Library of New South Wales in Association with Angus and Robertson, 1962), 435.

23. Version of the English text of the Treaty of Waitangi taken from the first schedule to the Treaty of Waitangi Act 1975, Waitangi Tribunal / Te Rōpū Whakamana i te Tiriti o Waitangi, "The Treaty of Waitangi / Te Tiriti O Waitangi," updated September 19, 2016, https://waitangitribunal.govt.nz/treaty-of-waitangi/english-version/.

24. Ngahuia Murphy, text message to author, January 4, 2021.

25. Anaru Reedy, *Ngā Kōrero a Mohi Ruatapu: The Writings of Mohi Ruatapu*, translated, edited, and annotated by Anaru Reedy (Christchurch: Canterbury University Press, 1993), 124.

26. Pauline Harris et al., "A Review of Māori Astronomy in Aotearoa–New Zealand," *Journal of Astronomical History and Heritage* 16, no. 3 (2013): 329.

27. Apirana Ngata, "Rauru-nui-a-Toi and Ngati Kahungunu Origins" Lecture 2, the Porourangi Maori Cultural School Rauru-nui-a-Toi Course Lectures 1–7, delivered in 1944 at The Pavilion, Whakarua Park, Ruatoria (Wellington: Victoria University of Wellington, Multilith Printing Department, 1972): 2.

28. GNS Science Te Pū Ao, "Te Riu-a-Māui Our Continent," https://www.gns.cri.nz/our-science/land-and-marine-geoscience/te-riu-a-maui-our-continent/.

29. Bradford Haami, "Te Whānau Puha—Whales: Whales in Māori Tradition," Te Ara: The Encyclopaedia of New Zealand, June 12, 2006, https://teara.govt.nz/en/te-whanau-puha-whales/print.

30. Billie Lythberg and Wayne Ngata, "Heeding the Call of Paikea: A Whakapapa Approach to Whaling and Whale People in Aotearoa–New Zealand," in *Across Species and Cultures: New Histories of Pacific Whaling*, ed. Ryan Tucker Jones and Angela Wanhalla (Honolulu: University of Hawai'i Press, 2022), 319.

31. Reedy, *Ngā Kōrero a Mohi Ruatapu*, 96–98, 200–204.

32. Lythberg and Ngata, "Heeding the Call of Paikea," 319.

33. Tiimi Waata Rimini, George Davies, and E. Tregear, "Te puna kahawai i Motu," *The Journal of the Polynesian Society*, 10, no. 4 (December 1901): 186.

34. Rimini et al., "Te puna kahawai i Motu," 186.

35. Rimini et al., "Te puna kahawai i Motu," 186–87.

36. Tūtere Wi Repa, "Maoris of East Coast: Research by State Ethnographical Party," *The Gisborne Times*, April 12, 1923, 5.

37. Wiremu Kaa, conversation with the author, August 18, 2016, Rangitukia.

38. Kaa, conversation with the author.

39. Tate Pēwhairangi, WAI 272. *Affidavit in the Matter of the Treaty of Waitangi Act 1975 and a Claim by Apirana Tuahae Mahuika for and on Behalf of Te Runanga o Ngāti Porou*, 2000, 13, https://www.parliament.nz/resource/en-NZ/49SCMA_EVI_00DBHOH_BILL10537_1_A194245/fdc852689e7cfcceb44b1e74695ce8a2775c4de2.

40. Tautini Glover, WAI 272, *Affidavit in the Matter of the Treaty of Waitangi Act 1975 and a Claim by Apirana Tuahae Mahuika for and on Behalf of Te Runanga o Ngāti Porou*, 2000, 3, https://www.parliament.nz/resource/en-NZ/49SCMA_EVI_00DBHOH_BILL10537_1_A194225/d3fa5ab48f3841af16c00920cbb4a8ff52470 72a.

41. Daniel Hikuroa, "Mātauranga Māori—the Ūkaipō of Knowledge in New Zealand," *Journal of the Royal Society of New Zealand* 47, no. 1 (2017): 5.

42. "Claim to the Pacific Dismissed" (from the *New Zealand Herald*), Notes and Queries, *Journal of the Polynesian Society* 64, no. 1 (1955): 162–63.

43. Mare liberum is a doctrine articulated by Hugo Grotius in defense of the right of the Dutch to sail to the East Indies, as against the Portuguese claim to exclusive control over those waters: "The sea is common to all, because it is so limitless that it cannot become a possession of any one, and because it is adapted for the use of all, whether we consider it from the point of view of navigation or of fisheries," Hugo Grotius, *The Freedom of the Seas: Or, The Right Which Belongs to the Dutch to Take Part in the East Indian Trade*, trans. Ralph Magoffin (New York: Oxford University Press, 1916), 28.

44. Prue Taylor, "The Common Heritage of Mankind: Expanding the Oceanic Circle," in *The Future of Ocean Governance and Capacity Development: Essays in Honor of Elisabeth Mann Borgese (1918–2002)*, ed. Dirk Werle et al. (Leiden, The Netherlands: Brill Nijhoff, 2018), 142.

45. Taylor, "Common Heritage of Mankind," 142–43.

46. Matthew Rout et al., *Māori Marine Economy: A Literature Review: Whai Rawa, Whai Mana, Whai Oranga: Creating a World-Leading Indigenous Blue Marine Economy* (Wellington, New Zealand: Sustainable Seas National Science Challenge, 2019), 29.

47. Rout et al., *Māori Marine Economy*, 29.

48. Department of Justice, Waitangi Tribunal, *Report of the Waitangi Tribunal on the Muriwhenua Fishing Claim (WAI22)/Muriwhenua Fisheries Report* (Wellington: Waitangi Tribunal, 1988, 1989, 1996), https://forms.justice.govt.nz/search/Documents/WT/wt_DOC_68478237/Muriwhenua%20Fishing%20Report%201988.compressed.pdf.

49. Department of Justice, *Report of the Waitangi Tribunal.*

50. Peter Garven, Marty Nepia, and Harold Ashwell (from information supplied by Kai Tahu Whanau Whanui), *Te whakatau kaupapa o Murihiku: Ngai Tahu Resource Management Strategy for the Southland*, ed. Maarire Goodall (Wellington: Aoraki Press, 1997).

51. This provoked a flurry of legal debates, see for instance F. M. Brookfield, "Maori Customary Title in Foreshore and Seabed," *New Zealand Law Journal* 1 (2004): 34–38.

52. Sian Elias, CJ, in Ngati Apa v. Attorney-General, [2003] NZCA 117, Court of Appeal of New Zealand (CA 173/01).

53. *Report on the Crown's Foreshore and Seabed Policy*, WAI 1071, Waitangi Tribunal, 2004.

54. See Department of Conservation, Te Papa Atawhai, "Revitalising the Gulf: Government Action on the Sea Change Plan," accessed December 2, 2022, https://www.doc.govt.nz/our-work/sea-change-hauraki-gulf-marine-spatial-plan/.

55. The Nobel Prize, "Our Planet, Our Future," Nobel Prize Summit, April 26–28, 2021, virtual event, accessed January 17, 2022, https://www.nobelprize.org/events/nobel-prize-summit/2021.

56. For instance, Turama Thomas Hawira, 2007, brief of evidence for the Whanganui District Inquiry (do B28), 11.

57. For related papers, see Anne Salmond, "Tears of Rangi: People, Water and Power in New Zealand," *HAU: Journal of Ethnographic Theory* 4, no. 3 (2014): 285–309; Gary Brierley et al., "A Geomorphic Perspective on the Rights of the River in Aotearoa New Zealand," *River Research Applications* 35, no. 10 (2019), 1640–51; Anne Salmond, Gary Brierley, and Dan Hikuroa, "Let the Rivers Speak: Thinking about Waterways in Aotearoa New Zealand," *Policy Quarterly* 15, no. 3 (August 2019): 45–54.

Chapter Five

1. William Branwhite Clarke, *Transactions of the Royal Society of New South Wales* (1867), vol. 1, 1868: 1, 26–27.

2. James Secord, "Global Geology and the Tectonics of Empire," in *Worlds of Natural History*, ed. H. A. Curry et al. (Cambridge: Cambridge University Press, 2018), 401–2; see also Daniel Zizzamia, "Restoring the Paleo West: Fossils, Coal, and Climate in Late Nineteenth-Century America," *Environmental History* 24, no. 1 (January 2019): 141–45.

3. Geology and ethnography were also pursued simultaneously in other colonial sites like the Cape Colony and in India. Saul Dubow, "Earth History, Natural History, and Prehistory at the Cape, 1860–1875," *Comparative Studies in Society and History* 46, no. 1 (January 2004): 132–33; Pratik Chakrabarti, "Gondwana and the Politics of the Deep Past," *Past and Present*, 242 (2019): 151–53; Ian Keen, "The Anthropologist as Geologist: Howitt in Colonial Gippsland," *Australian Journal of Anthropology* 11, no. 1 (April 2000): 78–97.

4. Charles Gillespie, *Genesis and Geology: A Study in the Relations of Scientific Thought, Natural Theology, and Social Opinion in Great Britain, 1790–1850* (Cambridge, MA: Harvard University Press, 1951, 1996), 3–40. Citations refer to the 1996 edition.

5. Approaching colonial geology in this way builds on Josh Gibson and Helen Gardner's approach to Alfred William Howitt's late nineteenth-century anthropological insights. Jason Gibson and Helen Gardner, "Conversations on the Frontier: Finding the Dialogic in Nineteenth-Century Anthropological Archives," *History Workshop Journal* 88 (2019): 4–5. See also Cameron Strang, "Indian Storytelling, Scientific Knowledge, and Power in the Florida Borderlands," *William & Mary Quarterly* 70, no. 4 (2013): 673.

6. Pratik Chakrabarti, *Inscriptions of Nature: Geology and the Naturalization of Antiquity* (Baltimore: Johns Hopkins University Press, 2020), 87–88; see also Martin Rudwick, *Earth's Deep History: How It Was Discovered and Why It Matters* (Chicago: University of Chicago Press, 2014), 155–63.

7. William Branwhite Clarke to Adam Sedgwick, August 13, 1840, Cambridge University Library, Correspondence of Adam Sedgwick, Add 7652, 1/F/82.

8. William Branwhite Clarke to Adam Sedgwick, June 29, 1841, Cambridge University Library, Correspondence of Adam Sedgwick, Add 7652, 1/D/155.

9. For the utility of and reliance on Indigenous spatial and environmental knowledge in colonial Australia, see Shino Konishi, Maria Nugent, and Tiffany Shellam, eds., *Indigenous Intermediaries: New Perspectives on Exploration Archives* (Canberra: ANU Press, 2015); Tiffany Shellam, *Meeting the Waylo: Aboriginal Encounters in the Archipelago* (Perth: UWA Press, 2019); Penny Olsen and Lynette Russell, *Australia's First Naturalists: Indigenous Peoples' Contribution to Early Zoology* (Canberra: National Library of Australia Publishing, 2019); Grace Karskens,

"Floods and Flood-Mindedness in Early Colonial Australia," *Environmental History* 21, no. 2 (April 2016): 319–26; for an American equivalent, see Cameron Strang, *Frontiers of Science: Imperialism and Natural Knowledge in the Gulf South Borderlands, 1500–1850* (Chapel Hill: University of North Carolina Press, 2018); on Mitchell, see D. W. A. Baker, *The Civilised Surveyor: Thomas Mitchell and Australian Aborigines* (Melbourne: Melbourne University Press, 1997).

10. A corroboree is a ritual conducted across the Australian continent, during which Aboriginal people commune with each other and the Dreaming. Corroborees take place during periods of meeting, assembly, or gathering; during particular times of the year; or as part of initiation and other rites. See William Branwhite Clarke to Adam Sedgwick, November 18, 1843, Cambridge University Library, Correspondence of Adam Sedgwick, Add7652, 1/F/95a; Adam Sedgwick to William Branwhite Clarke, October 9, 1844, W. B. Clarke Papers and Notebooks, 1827–1951, ML (Mitchell Library) MSS 139/47, 81–88; copy MSS 139/109; Ann Mosley, "James Dwight Dana in New South Wales, 1839–1840," *Journal of the Proceedings of the Royal Society of New South Wales* 97, 6A (1964): 185–91.

11. Account in this paragraph described in Charles Wilton, "Scientific Intelligence," *Australian Quarterly Journal of Theology, Literature & Science* 4 (1828), 380–85. Aboriginal guides often became uncomfortable when they were too far from their country; see Annemarie McLaren, "No Fish, No House, No Melons: The Earliest Aboriginal Guides in Colonial New South Wales," *Aboriginal History* 43 (December 2019): 33–55.

12. *Sydney Gazette*, August 18, 1828. Reprinted in *Australian Quarterly Journal of Theology, Literature & Science* 1 (1828): viii–ix.

13. Charles Pleydell Wilton, "The Burning Mountain of Australia," *The Sydney Gazette and New South Wales Advertiser*, March 14, 1829, 2. Wilton first visited the mountain in February of 1829 and took another trip there in late 1831: Charles Pleydell Wilton, "Mount Wingen—The Burning Mountain of Australia," original correspondence, *The Sydney Gazette and New South Wales Advertiser*, November 12, 1831, 3.

14. Michael Gladwin, "The Journalist in the Rectory: Anglican Clergymen and Australian Intellectual Life, 1788–1850," *History Australia* 7, no. 3 (2010): 10; Gladwin quotes from Clarke in *The Sydney Morning Herald*, October 3, 1846.

15. Lancelot Threlkeld, *An Australian Language as Spoken by the Awabakal, the People of Awaba or Lake Macquarie (near Newcastle, New South Wales) Being an Account of Their Language, Traditions and Customs*, ed. John Fraser (Sydney: Charles Potter, Government Printer, 1892), 51.

16. See Greg Blyton, Deirdre Heitmeyer, and John Maynard, *Wannin Thanbarran: A History of Aboriginal and European Contact in Muswellbrook and the Upper Hunter Valley* (Muswellbrook, Australia: Muswellbrook Shire Aboriginal Reconciliation Committee, 2004).

17. Lancelot Threlkeld, *An Australian Grammar: Comprehending the Principles and Natural Rules of the Language, as Spoken by the Aborigines in the Vicinity of Hunter's River, Lake Macquarie, &c* (Sydney: Stephens & Stokes, 1834), 131.

18. "Lake Macquarie Coals," *The Sydney Gazette and New South Wales Advertiser*, April 22, 1841, 3; "Important Discovery of Coal," *The Australian*, June 13, 1827, 2.

19. Roger Pierson, "McCoy and Clarke: Their Dispute over the Age of Australia's Black Coal," *Victorian Naturalist* 118, no. 5 (2001): 219–25.

20. "Australian Geology," *The Sydney Morning Herald*, August 4, 1845, 2. This abstract was read before the Geological Society in 1843 by Adam Sedgwick, who received the abstract in August 1842.

21. William Branwhite Clarke, "On a Fossil Pine Forest at Kurrur-Kurran, in the Inlet of

Awaaba [Lake Macquarie], East Coast of Australia," *Annual Report of the Department of Mines of N.S.W. for the Year 1884* (Sydney, 1885), 156–59. Original article signed and dated "W. B. Clarke. Parramatta, 29th August 1842."

22. Adam Bobbette and Amy Donovan, "Political Geology: An Introduction," in *Political Geology: Active Stratigraphies and the Making of Life*, ed. Adam Bobbette and Amy Donovan (London: Palgrave Macmillan, 2019), 1–34.

23. Rudwick, *Earth's Deep History*, 161–63.

24. The majority of Ridley's insights came from the Kamilaroi in northwestern New South Wales, but he drew on information about neighboring groups. William Ridley, "The Aborigines of Australia," *The Empire* (published in three parts), September 17, 19, and 21, 1864, 2, 2, and 8. Ridley also wrote a travel account of his trip and an ethnography: *Report . . . of a Journey along the Condamine, Barwan and Namoi Rivers* (1855) and *Gurre Kamilaroi: Or Kamilaroi Sayings* (1856).

25. Lydia Barnett, *After the Flood: Imagining the Global Environment in Early Modern Europe* (Baltimore: Johns Hopkins University Press, 2019), 93–98. By the early nineteenth century, naturalists, following Jean André Deluc and Georges Cuvier, had drawn on an array of "multicultural records," including biblical history and stories from China, Egypt, and the Middle East to nominate a period of inundation as *the* "boundary-event" between "the present and the former world" and between "human history and prehistory." Rudwick, *Earth's Deep History*, 110–14; Martin Rudwick, *Bursting the Limits of Time: The Reconstruction of Geohistory in the Age of Revolution* (Chicago: University of Chicago Press, 2005), 158, 454.

26. Charles Pleydell Wilton, "On the Connection between Science and Religion," *Australian Quarterly Journal of Theology, Literature and Science* 1 (1828), 1–6; Charles Pleydell Wilton, "Geology—No. 1," *Australian Quarterly Journal of Theology, Literature and Science* 2 (1828), 191–98.

27. Ann Moyal, *The Web of Science: The Scientific Correspondence of the Rev. W.B. Clarke, Australia's Pioneer Geologist*, 2 vols. (Melbourne: Australian Scholarly Publishing, 2003), 5–6; Paul Carter, *The Road to Botany Bay: An Exploration of Landscape and History* (Chicago: University of Chicago Press, 1987), 58, 114. For how this played out globally, see Simon Naylor and Simon Schaffer, "Nineteenth-Century Survey Sciences: Enterprises, Expeditions and Exhibitions," *Notes and Records of the Royal Society* 73, no. 2 (2019): 135–47.

28. William Branwhite Clarke, "Review: *Remarks on the Probable Origin and Antiquity of the Aboriginal Natives of New South Wales, Deduced from Certain of Their Customs, Superstitions, and Existing Caves and Drawings, in Connexion with Those of the Nations of Antiquity*," *The Sydney Morning Herald*, October 3, 1846, 2–4. At the time of Clarke's training, Sedgwick was a late advocate of the correspondence between the geological deluge and the biblical Flood, a conviction that he "recanted" later on in the nineteenth century. Rudwick, *Earth's Deep History*, 172–73; see also Martin Rudwick, *Worlds Before Adam: The Reconstruction of Geohistory in the Age of Reform* (Chicago: University of Chicago Press, 2008), 177–91, 291–93.

29. Tom Griffiths, *Hunters and Collectors: The Antiquarian Imagination in Australia* (Melbourne: Cambridge University Press, 1996); Tom Griffiths, "Environmental History, Australian Style," *Australian Historical Studies* 46, no. 2 (2015): 157–73.

30. Clarke, "*Review: Remarks on the Probable Origin*," 4.

31. James Turner, *Philology: The Forgotten Origins of the Modern Humanities* (Princeton, NJ: Princeton University Press, 2000), 131, 242–50.

32. Shellam, *Meeting the Waylo*, 3–17, 89–104, 138–57. Australian contexts challenged naturalists with a wide array of new evidence, but their forebears had been reading the Bible and other

ancient texts in ways that sought to link them to geological events since at least the eighteenth century. Rudwick, *Earth's Deep History*, 208; Rudwick, *Bursting the Limits of Time*, 250–58.

33. Charles Pleydell Wilton, "*Review of New Publications: Narrative of a Survey of the Intertropical and Western Coasts of Australia, Performed between the Years 1818 and 1822, by Captain Phillip P. King*," *Australian Quarterly Journal of Theology, Literature & Science* 1 (1828): 20–24.

34. Ridley, "The Aborigines of Australia," September 19, 1864, 2.

35. Ridley, "The Aborigines of Australia," September 17, 1864, 2.

36. Clarke, "*Review: Remarks on the Probable Origin*," 4.

37. Emily Kern, "Bodies, Cultures, Tongues: Race between Philology and Biology in the Nineteenth Century," in *A Cultural History of Race in the Age of Revolution, Empire and Nation State (1760–1920)*, ed. Marina Mogilner (London: Bloomsbury Academic, 2021), 37–55.

38. William Branwhite Clarke to Adam Sedgwick, November 18, 1843, Cambridge University Library, Correspondence of Adam Sedgwick, Add7652, 1/F/95a.

39. William Branwhite Clarke to Adam Sedgwick, April 20, 1844, Cambridge University Library, Correspondence of Adam Sedgwick, Add 7652, 1/F/95d. This letter was heavily underlined by Sedgwick. The omission of these stories or etymologies matches up with other major publications and reports from the mid-1830s. For example, while Robert Montgomery Martin mentioned Wingen in the geological portion of his 1837 *History of Austral-Asia*, he avoided any reference to Aboriginal etymologies of cosmologies. A series of articles in *The Australian* (probably from Clarke) in March 1844 also failed to mention this etymology, even though one did refer to similar phenomena in the Hunter and elsewhere as "Wingans" or "Wingens." William Branwhite Clarke, "Mount Wingen," *The Australian*, March 13, 1844, 2.

40. Adam Sedgwick to William Branwhite Clarke, October 9, 1844, MLMSS 139/47, 81–88; copy MSS 139/109.

41. Rudwick, *Earth's Deep History*, 85–86.

42. William Branwhite Clarke to Adam Sedgwick, May 30, 1845, Cambridge University Library, Correspondence of Adam Sedgwick, Add 7652, 1/F/95e and 95f.

43. I thank Adam Bobbette for this insight.

44. Rudwick, *Earth's Deep History*, 207–12; Rudwick, *Bursting the Limits of Time*, 55–58.

45. For Hutton's natural theology, see Mott Greene, *Geology in the Nineteenth Century: Changing Views of a Changing World* (Ithaca, NY: Cornell University Press, 1982), 22. According to Greene, Hutton's theory of the earth was "a classic of deistic natural theology and natural philosophy in which geological phenomena were evidence for the conclusion that no system so balanced and purposeful could ever have emerged without intelligent design." For Dawson, see Nanna Katrine Lüders Kaalund, "Of Rocks and 'Men': The Cosmogeny of John William Dawson," in *Historicizing Humans: Deep Time, Evolution, and Race in Nineteenth-Century British Sciences*, ed. Efram Sera Schriar (Pittsburgh: The University of Pittsburgh Press, 2018), 50, 54.

46. Adam Sedgwick quoted in James Secord, *Controversy in Victorian Geology: The Cambrian-Silurian Dispute* (Princeton, NJ: Princeton University Press, 1986, 2014), 227. Citations refer to the 2014 edition.

47. William Branwhite Clarke to Adam Sedgwick, April 20, 1844, Cambridge University Library, Correspondence of Adam Sedgwick, Add 7652, 1/F/95d. This letter heavily underlined by Sedgwick.

48. Moyal, *The Web of Science*, vol. 1, 52–53.

49. Alison Bashford and Joyce Chaplin, *The New Worlds of Thomas Robert Malthus: Rereading the Principle of Population* (Princeton, NJ: Princeton University Press, 2016), 17–23.

50. Wilton, "On the Connection between Science and Religion," 3.

51. Wilton, "Geology—No. 1," 193; See also, Jarrod Hore, "Settlers in Earthquake Country: Apprehending Instability in New Zealand and California," *Pacific Historical Review* 91, no. 1 (2022): 1–32.

52. William Branwhite Clarke, "Review: The History of Transmuted Rocks, or Bible Geology. By E.W. Rudder, 1854," *The Sydney Morning Herald*, February 5, 1855, 3.

53. Rudwick, *Earth's Deep History*, 3–5, 29; Michael Gladwin, "The Journalist in the Rectory: Anglican Clergymen and Australian Intellectual Life, 1788–1850," *History Australia* 7, no. 3 (2010): 10.

54. Adam Sedgwick to William Branwhite Clarke, October 9, 1844, MLMSS 139/47, 81–88; copy MSS 139/109. This echoes the kind of writing that geologists did in the American West; Zizzamia, "Restoring the Paleo-West," 130–56.

Chapter Six

1. Martin Rudwick, *Earth's Deep History* (Chicago: University of Chicago Press, 2014), 258.

2. Henry R. Frankel, *Continental Drift Controversy: Paleomagnetism and Confirmation of Drift* (Cambridge: Cambridge University Press, 2012), xi–xviii.

3. Naomi Oreskes, *Science on a Mission: How Military Funding Shaped What We Do and Don't Know about the Ocean* (Chicago: University of Chicago Press, 2020).

4. Volcanological surveys and observatories were relatively new, and volcanology was a novel science linked to meteorology and seismology. The observatory at Vesuvius is often regarded as the first permanent volcano observatory; it was opened in 1841, though it operated intermittently into the early twentieth century. The observatory at Kilauea, built in 1911, and often recognized as the second permanent volcano observatory, was modeled on the operations at Vesuvius.

5. Some *babad* ran to thousands of pages, many of them written in *Kawi*, a language which few people read today. A small number of *babad* have been translated into Javanese or Indonesian, and only one is widely available in Indonesian, though in severely abbreviated form, the *Babad Tanah Jawi*—or *Chronicle of the Land of Java*. It was translated into Dutch in 1941 by W. L. Olthof and much later into Indonesian. Much of the extensive geological and geographical knowledge contained in the *babad* remains unknown and unstudied beyond a very small group of scholars.

6. P. J. Veth, *Java, Geographisch, Ethnologisch, Historisch* (Haarlem, The Netherlands: De Erven F. Bohn, 1882), 377.

7. See, for instance, H. R. Sumarsono, *Babad Tanah Jawi: Mulai dari Nabi Adam sampai tahun 1647* (Yogyakarta, Indonesia: Narasi, 2014), 93–98. This edition is an Indonesian translation of W. L. Olthof's *Poenika serat Babad Tanah Djawi wiwit saking Nabi Adam doemoegi ing taoen 1647* ('s-Gravenhage, The Netherlands: Nijhoff, 1941).

8. R. Soedjana Tirtakoesoema, "De Verjaring van den Verheffingsdag van Z. H. den Sultan van Jogjakarta (Tingalan Pandjĕnĕngan)," *Djåwå* 6, 13 (1933): 372–88.

9. Tirtakoesoema's text is ambiguous on the relationship between the sultan and Nyai Ratu Kidul and if the throne is for Senopati's descendant or Nyai Ratu Kidul. "Over den Indischen oceaan heerscht de machtige godin Ratoe Kidoel, die Panembahan Sénapati de mededeeling zou hebben gedaan, dat op den troon steeds één zijner nazaten zou zetelen, met als patih Njahi Lara Kidoel." Tirtakoesoema, "De Verjaring van den Verheffingsdag," 381.

10. See Tirtakoesoema, "De Verjaring van den Verheffingsdag." An English translation ap-

pears as R. Soedjana Tirtakoesoema, "The Anniversary of the Accession of His Highness the Sultan of Yogyakarta (Tingalan Panjenengan)," in *The Kraton: Selected Essays on Javanese Courts*, ed. Stuart Robson (Leiden, The Netherlands: KITLV Press, 2003), 156.

11. M. C. Rickleffs, *Jogjakarta under Sultan Mangkubumi 1749–1792* (London: Oxford University Press, 1974), 376–408.

12. Theodore Pigeaud, "Alexander, Sakèndèr en Sénapati," *Djåwå* 1, 7 (1927): 321–61.

13. Lucien Adam, "Eenige historische en legendarische plaatsnamen in Jogjakarta," *Djåwå* 4–5, 10(1930): 150–52.

14. R. D. M. Verbeek, *Oudheden van Java* (Batavia: Bataviaasch Genootschap van Kunsten en Wetenschappen, M. Nijhoff, 1891), 1–18.

15. Vening Meinesz, "Gravity Anomalies in the East Indian Archipelago," *Geographical Journal* 77, 4 (April 1931), 329.

16. F. A. Vening Meinesz, J. H. F. Umbgrove, and P. H. Kuenen, *Gravity Expeditions at Sea 1923–1932*, vol. II, *Report on the Gravity Expeditions in the Atlantic of 1932 and the Interpretation of the Results*, 2nd ed. (Delft, The Netherlands: Rijkscommissie voor Geodesie, 1964), 163–82.

17. P. H. Kuenen, "The Negative Isostatic Anomalies in the East Indies (with Experiments)," *Leidsche Geologische Mededeelingen* 8, 2 (1936): 169–214.

18. G. A. F. Molengraaff, "Modern Deep-Sea Research in the East Indian Archipelago," *The Geographical Journal* 57, 2 (February 1921): 107.

19. Meinesz, "Gravity Anomalies," 329.

20. Dirk van Hinloopen Labberton, "Oud-Javaansche gegevens omtrent de vulkanlogie van Java," *Natuurkundig Tijdschrift voor Nederlandsch-Indië* 81 (1921): 150.

21. Labberton, "Oud-Javaansche gegevens," 143.

22. S. W. Visser, *Natuurkundig Tijdschrift voor Nederlandsch-Indië* 81 (1921): 159–65.

23. R. W. van Bemmelen, "Merapi, No. 41," *Bulletin of the East Indian Volcanological Survey for the Year 1941* (Bulletin Nrs. 95–98) (1943): 69–72.

24. Reinout W. van Bemmelen, *Mountain Building: A Study Primarily Based on Indonesia Region of the World's Most Active Crustal Deformations* (The Hague: Martinus Nijhoff, 1954), 2.

25. Johannes Umbgrove to his mother, May 15, 1928, Brief 109 JHFU Java; from personal collection of Frederik van Veen, Groningen, Netherlands.

26. A. C. de Jongh, "On the Valency of the Chemical Atoms in Connection with Theosophical Conceptions Concerning Their Exterior Form," *The Theosophist* 35, 10 (July 1914): 535–71.

27. G. Kemmerling, "Vulkanologische Berichten," *Natuurkundig Tijdschrift voor Nederlandsch-Indië* 82, 2 (1920): 188–96.

28. Kemmerling, "Vulkanologische Berichten," 188.

29. F. A. Vening Meinesz and F. E. Wright, "The Gravity Measuring Cruise of the U.S. Submarine S-21," *Publications of the U.S. Naval Observatory Second Series* 13 (1933): 49.

30. Harry Hess, "Interpretation of Geological and Geophysical Observations (with Figures)," in *The Navy-Princeton Gravity Expedition to the West Indies in 1932* (Washington, DC: United States Government Printing Office, 1933), 28.

31. J. Lamar Worzel, *Pendulum Gravity Measurements at Sea, 1936–1959* (New York: J. Wiley & Sons, 1965).

32. Oreskes, *Science on a Mission*, 164–211.

33. Xavier Le Pichon, "Fifty Years of Plate Tectonics: Afterthoughts of a Witness," *Tectonics* 38, 8 (2019): 2924.

34. Warren Hamilton, "Tectonics of the Indonesian Region," *Bulletin of the Geological Society of Malaysia* 6 (July 1973): 3–10.

35. Surono et.al. "The 2010 Explosive Eruption of Java's Merapi Volcano—A '100-year' Event," *Journal of Volcanology and Geothermal Research* 241–242 (October 2012): 121–35.

Chapter Seven

1. For example, Pope Francis, *Laudato Si: On Care for Our Common Home* (Rome: Our Sunday Visitor, 2015). See also Simone Kotva, "Attention in the Anthropocene: On the Spiritual Exercises of Any Future Science," in *Political Geology: Active Stratigraphies and the Making of Life*, ed. Adam Bobbette and Amy Donovan (Cham, Switzerland: Palgrave Macmillan, 2019), 239–61.

2. Mary Evelyn Tucker and John Grim, eds. *Religions of the World and Ecology*, 9 vols. (Cambridge, MA: Harvard University Press, 1997–2004). See also Kotva, "Attention in the Anthropocene."

3. Martin S. J. Rudwick, "Biblical Flood and Geological Deluge: The Amicable Dissociation of Geology and Genesis," in *Geology and Religion: A History of Harmony and Hostility*, ed. Martina Kölbl-Ebert (London: Geological Society of London, 2009), 103–10. For more general accounts on the intersections of religion and science, see Martin S. J. Rudwick, *Bursting the Limits of Time: The Reconstruction of Geohistory in the Age of Revolution* (Chicago: University of Chicago Press, 2005); Stephen Jay Gould, *Time's Arrow, Time's Cycle: Myth and Metaphor in the Discovery of Geological Time* (Cambridge, MA: Harvard University Press, 1987).

4. Lydia Barnett, *After the Flood: Imagining the Global Environment in Early Modern Europe* (Baltimore: Johns Hopkins Press, 2019), 3.

5. Bronislaw Szerszynski, "Gods of the Anthropocene: Geo-Spiritual Formations in the Earth's New Epoch," *Theory, Culture & Society* 34, nos. 2–3 (2017): 253–75; Bronislaw Szerszynski, "Praise Be to You, Earth-Beings," *Environmental Humanities* 8, no. 2 (2016): 291–97.

6. Andrew R. George, *The Epic of Gilgamesh* (Harmondsworth, UK: Penguin, 1999), 88–95.

7. Gen. 8:4.

8. Quran 11:44.

9. Gen. 8:4.

10. Irving Finkel, *The Ark before Noah: Decoding the Story of the Flood* (London: Hachette UK, 2014).

11. Andrew Collins, *Gobekli Tepe: Genesis of the Gods: The Temple of the Watchers and the Discovery of Eden* (New York: Simon and Schuster, 2014), 263.

12. Gertrude L. Bell, *Amurath to Amurath* (London: William Heinemann, 1911), 706–7.

13. Samar Abbas, "The Glorious Gutians: Historic Kurdistan as Ancient Gutium," Iranian.com, March 24, 2005, https://iranian.com/History/2005/March/Gutians/index.html; Mehrdad Izady, "In Guti We Trust," *Kurdish Life*, no. 14 (1995), https://kurdistanica.com/285/in-guti-we-trust/.

14. H. Akın Ünver, "Schrödinger's Kurds: Transnational Kurdish Geopolitics in the Age of Shifting Borders," *Journal of International Affairs* 69, no. 2 (2016): 65.

15. Michael M. Gunter, "The Kurdish Question in Perspective," *World Affairs* 166, no. 4 (2004): 197–205.

16. KJA Peoples and Faith Commission, "Press Statement on the Destruction of Our Peoples' Graveyards," January 2016, https://peaceinkurdistancampaign.com/wp-content/uploads/2016/01/kja-destruction-of-our-peoples-graveyards-1.pdf.

17. Mahmut Bozarslan, "Why Have PKK cemeteries Become a Target?" *Al-Monitor*, September 30, 2015.

18. Jacob van Etten et al., "Environmental Destruction as a Counterinsurgency Strategy in the Kurdistan Region of Turkey," *Geoforum* 39, no. 5 (2008): 1786–97.

19. van Etten et al., "Environmental Destruction," 1796.

20. Zozan Pehlivan, "Wildfires in Mount Cudi and the Ecological, Ideological, Political, and Historical Dimensions of Forest Fires: Turkey's Destruction of the Kurdish Environment," interview by Anıl Olcan, *Jadaliyya*, September 30, 2020.

21. Pehlivan, "Wildfires in Mount Cudi."

22. Pehlivan, "Wildfires in Mount Cudi."

23. Zeynep Oguz, "Cavernous Politics: Geopower, Territory, and the Kurdish Question in Turkey," *Political Geography* 85 (March 2021).

24. Banu Bargu, "Another Necropolitics," *Theory & Event* 19, no. 1 (2016), https://muse.jhu.edu/article/610222.

25. Bargu, "Another Necropolitics," 5.

26. Hişyar Özsoy, "Between Gift and Taboo: Death and the Negotiation of National Identity and Sovereignty in the Kurdish Conflict in Turkey" (PhD diss., University of Texas, 2010), 1.

27. "Teröristlerin mezarlığa çevirdiği türbe Mehmetçik tarafından yeniden ziyarete açıldı," *Sabah*, April 6, 2018, https://www.sabah.com.tr/gundem/2018/04/06/teroristlerin-mezarliga-cevirdigi-turbe-mehmetcik-tarafindan-yeniden-ziyarete-acildi.

28. Fecri Barlik, "Terörden temizlenen Herekol'da petrol aranıyor," *Anadolu Ajansi*, August 31, 2018, https://www.aa.com.tr/tr/turkiye/terorden-temizlenen-herekolda-petrol-araniyor/1243039.

29. Barlik, "Terörden temizlenen Herekol'da petrol aranıyor."

30. Şırnak Üniversitesi, "2. Uluslararası Nuh Tufanı ve Cudi Dağı Sempozyumu" [The Second International Symposium of the Flood, Noah's Ark, and Mount Cudi], streamed live on May 25, 2021, YouTube video, 1:36:47, https://www.youtube.com/watch?v=g9Q6U60gRnE.

31. Şırnak Üniversitesi, "2. Uluslararası Nuh Tufanı ve Cudi Dağı Sempozyumu."

32. "Noah's Ark? Boatlike Form Is Seen near Ararat," *Life*, September 5, 1960, 112, 114.

33. Rene Noorbergen, *The Ark File* (London: Hodder & Stoughton, 1980), 128.

34. Lorence Gene Collins and David Franklin Fasold, "Bogus 'Noah's Ark' from Turkey Exposed as a Common Geologic Structure," *Journal of Geoscience Education* 44, no. 4 (1996): 439–44.

35. Harun Yahya, *The Atlas of Creation*, vol. 1 (Istanbul: Global Yayincilik, 2006).

36. John Larsen and Andrew Jones, "The Results of the Subsurface Imaging Project of Noah's Ark," accessed August 1, 2021, https://noahsarkscans.nz/.

37. Larsen and Jones, "Subsurface Imaging Project of Noah's Ark."

38. AICUNI2007, "6. Uluslararası Ağrı Dağı ve Nuh'un Gemisi Sempozyumu Açılış Programı" [The Sixth International Symposium of Mount Ararat and Noah's Ark: Opening Program], October 26, 2021, YouTube video, 1:51:49, https://www.youtube.com/watch?v=BNIXm5ceDqU.

Chapter Eight

I am grateful to Alexis Rider, Etienne Benson, Anthony Irwin, Harun Küçük, Justin McDaniel, Simon Schaffer, and Pamela H. Smith for their comments and suggestions for this chapter.

1. Richard Burnaby at Ayutthaya to Francis Bowyear and Council at Bantam, December 10, 1680, no. 326 in *The English Factory in Siam, 1612–1685*, ed. Anthony Farrington and Dhiravat Na Pombejra, vol. 1 (London: British Library, 2007), 555. Henceforth all EIC sources cited from this

collection will be referred to using the following format: EFS i.326, 555. Narai would later appoint Burnaby as governor of the port of Mergui in 1683.

2. Emphasis added. Memorandum by Richard Burnaby at Ayutthaya, October 11, 1680, EFS i.317, 543. This chapter also takes inspiration from the themes of contestation over authority and replicability in Simon Schaffer, "Glass Works: Newton's Prisms and the Uses of Experiment," in *The Uses of Experiment: Studies in the Natural Sciences*, ed. David Gooding, Trevor Pinch, and Simon Schaffer (Cambridge: Cambridge University Press, 1989), 67–104. Thanks to Simon Schaffer for first turning my attention toward the English factory records from Siam.

3. The East India Company in London to Robert Parker and Council at Bantam, October 5, 1677, EFS i.243, 426.

4. Hamon Gibbon, William Ramsden, and Benjamin Sangar at Ayutthaya to the East India Company in London, December 22, 1675, EFS i.216, 396. There was a long-standing tradition in Southeast Asia of relocating artisans. For example, see the request made by King Mangrai of Lanna to the Lord of Ava [Burma] for smiths and jewelers c. 1290, recorded in *Chiang Mai Chronicle*, trans. David K. Wyatt and Aroonrut Wichienkeeo, 2nd ed. (Chiang Mai, Thailand: Silkworm Books, 1998), 37–38; for a complex chronology of this early modern text, see xxx–xxxii.

5. Bhawan Ruangsilp, *Dutch East India Company Merchants at the Court of Ayutthaya: Dutch Perceptions of the Thai Kingdom c. 1604–1765* (Leiden, The Netherlands: Brill, 2007), 111–48; Giorgio Riello, "'With Great Pomp and Magnificence': Royal Gifts and the Embassies between Siam and France in the Late Seventeenth Century," in *Global Gifts: The Material Culture of Diplomacy in Early Modern Eurasia*, ed. Zoltán Biedermann, Anne Gerritsen, and Giorgio Riello, (Cambridge: Cambridge University Press, 2018), 235–65; Sarah Benson, "European Wonders at the Court of Siam," in *Collecting Across Cultures: Material Exchanges in the Early Modern Atlantic World*, ed. Daniela Bleichmar and Peter C. Mancall (Philadelphia: University of Pennsylvania Press, 2011), 155–74. On lesser-known European relations with Siam, see Stefan Halikowski Smith, *Creolization and Diaspora in the Portuguese Indies: The Social World of Ayutthaya, 1640–1720* (Leiden, The Netherlands: Brill, 2011); Michael Smithies and Luigi Bressan, *Siam and the Vatican in the Seventeenth Century* (Bangkok: River Books, 2001); and Isabel Leonor da Silva Diaz de Seabra, *The Embassy of Pero Vaz de Siqueira to Siam (1684–1686)*, trans. Custodio Cavaco Martins, Mario Pinharanda Nunes, and Alan Norman Baxter (Macao: University of Macau, 2005).

6. Lydia Barnett, "Showing and Hiding: The Flickering Visibility of Earth Workers in the Archives of Earth Science," *History of Science* 58, no. 3 (2020): 245–74; H. S. Torrens, *The Practice of British Geology, 1750–1850* (Aldershot, UK: Ashgate, 2002); Allison Margaret Bigelow, *Mining Language: Racial Thinking, Indigenous Knowledge, and Colonial Metallurgy in the Early Modern Iberian World* (Chapel Hill, NC: Omohundro Institute of Early American History & Culture, 2020). On practical science, see also B. Harun Küçük, *Science without Leisure: Practical Naturalism in Istanbul, 1660–1732* (Pittsburgh: University of Pittsburgh Press, 2019).

7. Ann Blair, "Mosaic Physics and the Search for a Pious Natural Philosophy in the Late Renaissance," *Isis* 91, no. 1 (March 2000): 32–58. Ivano dal Prete's important recent study has shown that European belief in an extremely old or eternal earth was much more widespread than previously thought. Ivano Dal Prete, *On the Edge of Eternity: The Antiquity of the Earth in Medieval and Early Modern Europe* (Oxford: Oxford University Press, 2022).

8. Tina Asmussen, "Spirited Metals and the Oeconomy of Resources in Early Modern European Mining," *Earth Sciences History* 39, no. 2 (November 12, 2020): 371–88; Pablo F. Gómez, "Caribbean Stones and the Creation of Early-Modern Worlds," *History and Technology* 34, no. 1 (January 2, 2018): 11–20; Orlando Bentancor, *The Matter of Empire: Metaphysics and Mining*

in Colonial Peru (Pittsburgh: University of Pittsburgh Press, 2017); Robyn d'Avignon, "Spirited Geobodies: Producing Subterranean Property in Nineteenth-Century Bambuk, West Africa," *Technology and Culture* 61, no. 2S (2020): S20–48; Teren Sevea, *Miracles and Material Life: Rice, Ore, Traps and Guns in Islamic Malaya* (Cambridge: Cambridge University Press, 2020).

9. Matthew H. Edney, *Mapping an Empire: The Geographical Construction of British India, 1765–1843* (Chicago: University of Chicago Press, 1997); Sumathi Ramaswamy, *Terrestrial Lessons: The Conquest of the World as Globe* (Chicago: University of Chicago Press, 2017); Thongchai Winichakul, *Siam Mapped: A History of the Geo-Body of a Nation* (Honolulu: University of Hawai'i Press, 1994); Donald S. Lopez Jr., *Buddhism and Science: A Guide for the Perplexed* (Chicago: University of Chicago Press, 2008), chap. 1.

10. John Tresch, "Cosmologies Materialized: History of Science and History of Ideas," in *Rethinking Modern European Intellectual History*, ed. Darrin M. McMahon and Samuel Moyn (Oxford: Oxford University Press, 2014), 155.

11. For examples of precious stones as material metaphors, see Quincy Ngan, "Collecting Azurite Blue and Malachite Green as Curios and Medicines in Late Imperial China," *Ming Qing Yanjiu* 24, no. 1 (May 15, 2020): 67–102; Patrick R. Crowley, "Crystalline Aesthetics and the Classical Concept of the Medium," *West 86th: A Journal of Decorative Arts, Design History, and Material Culture* 23, no. 2 (September 1, 2016): 220–51; Richard Oosterhoff, "Ingenious Materials: Salts as Material Metaphor," in *Secrets of Craft and Nature in Renaissance France: A Digital Critical Edition and English Translation of BnF Ms. Fr. 640*, ed. Pamela H. Smith et al. (New York: Making and Knowing Project, 2020), https://www.doi.org/10.7916/4vf4-2867.

12. Phyllis Granoff, "Maitreya's Jewelled World: Some Remarks on Gems and Visions in Buddhist Texts," *Journal of Indian Philosophy* 26, no. 4 (1998): 347–71, 352.

13. Mattia Silvani, "Ratna: A Buddhist World of Precious Things," in *Soulless Matter, Seats of Energy: Metals, Gems, and Minerals in South Asian Traditions*, ed. Fabrizio M. Ferrari and Thomas W. P. Dähnhardt (Sheffield, UK: Equinox Publishing): 219–54; Vanessa R. Sasson, ed., *Jewels, Jewelry, and Other Shiny Things in the Buddhist Imaginary* (Honolulu: University of Hawai'i Press, 2021).

14. Phyllis Granoff, "Relics, Rubies and Ritual: Some Comments on the Distinctiveness of the Buddhist Relic Cult," *Rivista Degli Studi Orientali* 81, no. 1/4 (2008): 59–72.

15. The idea that minerals grew out of some kind of seed had widespread purchase in Renaissance European philosophy as well. See Hiro Hirai, "Seed Concept," in *Encyclopedia of Renaissance Philosophy*, ed. Marco Sgarbi (Cham, Switzerland: Springer International Publishing, 2015), 1–4.

16. For an excellent overview, see James McHugh, "Gemstones," *Brill's Encyclopedia of Hinduism Online*, May 29, 2018, http://dx.doi.org/10.1163/2212-5019_BEH_COM_000372.

17. See, for example, Hugo Miguel Crespo, "Rock Crystal Carving in Portuguese Asia: An Archaeometric Analysis," in *The Global City: On the Streets of Renaissance Lisbon* (London: Paul Holberton Publishing, 2015), 186–212; Anna Contadini, "Facets of Light: The Case of Rock Crystals," in *God Is the Light of the Heavens and the Earth: Light in Islamic Art and Culture* (New Haven, CT: Yale University Press, 2015), 121–55; Cynthia Hahn and Avinoam Shalem, eds. *Seeking Transparency: Rock Crystals across the Medieval Mediterranean* (Berlin: Gebruder Mann Verlag, 2020).

18. Stefania Gerevini, "Christus Crystallus: Rock Crystal, Theology and Materiality in the Medieval West," in *Matter of Faith: An Interdisciplinary Study of Relics and Relic Veneration in the Medieval Period*, British Museum Research Publication 195 (London: British Museum Press, 2014), 92–99. On Buddhist relics that have also been found in caskets of rock crystal, see John S. Strong, "Beads and Bones: The Case of the Piprahwa Gems," in *Jewels, Jewelry, and Other Shiny*

Things in the Buddhist Imaginary, ed. Vanessa R. Sasson (Honolulu: University of Hawai'i Press, 2021), 185–207; on crystals formed from Buddhist bodily remains, see John S. Strong, *Relics of the Buddha*, (Princeton, NJ: Princeton University Press, 2004), 10–12.

19. On the cosmological significance of personal adornment, see Christoph Emmrich, "'I Don't Want a Wife without Ear Cuffs': Jewels, Gender, and the Market among the Newars of Nepal," in *Jewels, Jewelry, and Other Shiny Things in the Buddhist Imaginary*, ed. Vanessa R. Sasson (Honolulu: University of Hawai'i Press, 2021), 112–53, esp. 118.

20. Kris E. Lane, *The Colour of Paradise: The Emerald in the Age of Gunpowder Empires* (New Haven, CT: Yale University Press, 2010), on Indigenous mining of emeralds, 36–43; on Middle Eastern emeralds, 249, n. 11.

21. Sven Dupré, "The Art of Glassmaking and the Nature of Stones," in *Steinformen: Materialität, Qualität, Imitation*, ed. Isabella Augart et al. (Boston: De Gruyter, 2019), 207–20.

22. For a comprehensive account of European writings on Asia in the seventeenth century, see Donald Lach and Edwin J. Van Kley, *Asia in the Making of Europe*, vol. 3, *A Century of Advance*, books 1–3 (Chicago: University of Chicago Press, 1993).

23. Banarasidas, *Ardhakathanak: A Half Story*, trans. Rohini Chowdhury (New Delhi: Penguin Classics, 2010).

24. The classic contrasting accounts of this phenomenon are Anthony Reid, *Southeast Asia in the Age of Commerce, 1450–1680*, 2 vols. (New Haven, CT: Yale University Press, 1993), and Victor B. Lieberman, *Strange Parallels: Southeast Asia in Global Context, c. 800—1830*, 2 vols. (Cambridge: Cambridge University Press, 2003).

25. Chris Baker and Phongpaichit Pasuk, *A History of Ayutthaya: Siam in the Early Modern World* (Cambridge: Cambridge University Press, 2017), 120–21. David K. Wyatt, *Thailand: A Short History*, 2nd ed. (New Haven, CT: Yale University Press), 95–96; Bhawan Ruangsilp and Pimmanus Wibulsilp, "Ayutthaya and the Indian Ocean in the 17th and 18th Centuries: International Trade, Cosmopolitan Politics, and Transnational Networks," *Journal of the Siam Society* 105 (2017): 97–114.

26. Baker and Pasuk, *Ayutthaya*, 119–72.

27. Baker and Pasuk, *Ayutthaya*, 140; see also the account of the 1682 Safavid embassy to Siam, Moḥammad Rabi' ibn Muhammad Ibrahim, *The Ship of Sulaiman*, trans. John O'Kane, (London: Routledge, 1972).

28. For example, William Eaton at Hirado wrote to Sir Thomas Smythe in London, December 20, 1617, noting that many precious stones could be profitably purchased in Siam. EFS i.75, 232.

29. Baker and Pasuk, *Ayutthaya*, 139.

30. Alan Strathern, "Sacred Kingship under King Narai of Ayutthaya (1656–88): Divinisation and Righteousness," *Journal of the Siam Society* 107, no. 1 (2019): 72.

31. Alan Strathern, "Tensions and Experimentations of Kingship: King Narai and His Response to Missionary Overtures in the 1680s," *Journal of the Siam Society* 107, pt. 2 (2019): 17–41; see also the blending of Buddhist and Hindu honorifics in the "Royal Title" bestowed upon Narai during his coronation, in Richard D. Cushman and David K. Wyatt, trans., *The Royal Chronicles of Ayutthaya* (Bangkok: The Siam Society, 2000), 232–33.

32. On the prominent role of the Persian community, see Ruangsilp and Wibulsilp, "Ayutthaya and the Indian Ocean," 103-5.

33. *Kamsuan Samut*, see Baker and Pasuk, *Ayutthaya*, 133–34.

34. Bhawan Ruangsilp quotes an unpublished English translation by Han ten Brummelhuis *Dutch East India Company Merchants at the Court of Ayutthaya* (Leiden, The Netherlands: Brill,

2007), 55. The Dutch original was published as Cornelis van Nijenrode, "Remonstrantie en verthoninge der gelegentheyt des coninck- rijx van Siam mitsgaders haeren handel ende wandel ende waar de negotie meest in bestaet etc.," *Kroniek van het Historisch Genootschap Gevestigd te Utrecht*, 10 (1854), 176-91; this quote appears on 178.

35. Assadullah Souren Melikian-Chirvani, "The Jewelled Objects of Hindustan," and Jennifer Scarce, "A Splendid Harmony: Mughal Jewelry and Dress," in *Jewelled Arts of Mughal India*, ed. Beatriz Chadour-Sampson and Nigel Israel, *Jewellery Studies*, vol. 10 (London: Society of Jewellery Historians, 2004), 9–32, 33–40.

36. Rudolf Distelberger, "The Castrucci and the Miseroni: Prague, Florence, Milan," in *Art of the Royal Court: Treasures in Pietre Dure from the Palaces of Europe*, ed. Wolfram Koeppe and Anna Maria Giusti (New York: Metropolitan Museum of Art, 2008), 29.

37. Sher Banu A.L. Khan, "The Jewel Affair: The Sultana, Her Orang Kaya and the Dutch Foreign Envoys," in *Mapping the Acehnese Past*, ed. R. Michael Feener, Patrick Daly, and Anthony Reed (Leiden: Brill, 2011), 141–62.

38. Hugo Miguel Crespo, "Rock Crystal Carving in Portuguese Asia: An Archaeometric Analysis"; see also Molly A. Warsh, *American Baroque: Pearls and the Nature of Empire, 1492–1700* (Chapel Hill: University of North Carolina Press, 2018).

39. Hiram Bingham, *Elihu Yale: The American Nabob of Queen Square* (New York: Dodd, Mead & Company, 1939), 120–33.

40. Ibn Muhammad Ibrahim, *Ship of Sulaiman*, 156.

41. The first French Jesuits arrived in Siam in 1662. For a comparative account of Catholic missionaries in early modern Southeast Asia, see Tara Alberts, *Conflict and Conversion: Catholicism in Southeast Asia, 1500–1700* (Oxford: Oxford University Press, 2013).

42. Traiphum cosmography dated from the medieval Sukothai era. For its modern history, see Thongchai, *Siam Mapped*, chap. 1. Thai cosmological texts in the Traiphum tradition do discuss gem classifications to some extent—e.g., the fourteenth-century *Traiphum Phra Ruang* describes eighty-four thousand different gems and discusses their different sizes and colors. See Frank E. Reynolds and Mani B. Reynolds, eds., *Three Worlds According to King Ruang: A Thai Buddhist Cosmology* (Berkeley, CA: Asian Humanities Press / Motila Banarsidass, 1982), 163–64. On the Brahmanical influence in early modern Ayutthaya, see Baker and Pasuk, *Ayutthaya*, 146–47; Strathern, "Tensions and Experimentations," 28. On the role of astrology and cosmology, see Ian Hodges, "Western Science in Siam: A Tale of Two Kings," *Osiris* 13 (1998): 80–95, and Ian Hodges, "Time in Transition: King Narai and the Luang Prasoet Chronicle of Ayutthaya," *Journal of the Siam Society* 87, nos. 1 & 2 (1999): 33–44. On Wat San Paulo, see Wayne Orchiston, Martin George, and Boonrucksar Soonthornthum, "Exploring the First Scientific Observations of Lunar Eclipses Made in Siam," *Journal of Astronomical History and Heritage* 19, no. 1 (2016): 25–45.

43. Ian Hodges, "Western Science in Siam," 84.

44. Sarah Benson, "European Wonders at the Court of Siam."

45. Strathern, "Tensions and Experimentations," 32.

46. Dedo von Kerssenbrock-Krosigk, "Glass for the King of Siam: Bernard Perrot's Portrait Plaque of King Louis XIV and Its Trip to Asia," *Journal of Glass Studies* 49 (2007): 73–74.

47. Tara Alberts, "Missionaries as Merchants and Mercenaries: Religious Controversies over Commerce in Southeast Asia," in *Trade and Finance in Global Missions in the 16th–18th Centuries*, ed. Hélène Vu Thanh and Ines G. Županov (Leiden, The Netherlands: Brill, 2020), 237–68; Ronald S. Love, "Monarchs, Merchants, and Missionaries in Early Modern Asia: The Missions Étrangères in Siam, 1662–1684," *The International History Review* 21, no. 1 (March 1, 1999): 1–27.

48. This fact has, in turn, inspired a creative historiography focusing on the diplomatic em-

bassies of Europe and Persia with Ayutthaya, especially the two voyages from each side between Versailles and Ayutthaya that have attracted extended commentary from historians of material culture. See Meredith Martin, "Mirror Reflections: Louis XIV, Phra Narai, and the Material Culture of Kingship," *Art History* 38, no. 4 (2015): 652–67; Ronald S. Love, "Rituals of Majesty: France, Siam, and Court Spectacle in Royal Image-Building at Versailles in 1685 and 1686," *Canadian Journal of History* 31, no. 2 (August 1, 1996): 171–98.

49. Michael Smithies, *A Siamese Embassy Lost in Africa, 1686: The Odyssey of Ok-Khun Chamnan* (Chiang Mai, Thailand: Silkworm Books, 1999).

50. Michael Smithies and Dhiravat na Pombejra, "Instructions Given to the Siamese Envoys Sent to Portugal, 1684," *Journal of the Siam Society* 90.1 & 2 (2002), 125–35.

51. Ruangsilp, *Dutch East India Company Merchants at the Court of Ayutthaya*, 142.

52. Ruangsilp, *Dutch East India Company Merchants at the Court of Ayutthaya*, 153 (goldsmith), 138 (pyrotechnics expert).

53. For another physician in Narai's employ, see Tara Alberts, "Making Drinkable Gold for the King of Siam," *The Recipes Project* (blog), March 2, 2015, https://recipes.hypotheses.org/5162.

54. Simon de La Loubère, *The Kingdom of Siam* [French original 1691; English translation 1693], with introduction by David K. Wyatt (Oxford: Oxford University Press, 1986), 14.

55. Ruangsilp, *Dutch East India Company Merchants at the Court of Ayutthaya*, 142.

56. La Loubère, *The Kingdom of Siam*, 14. These "diamonds" are likely to have been colorless sapphires, or rock crystal, although Myanmar and southern Thailand are known to have small alluvial diamond deposits.

57. For this early history, see Dhiravat na Pombejra, "The Voyages of the 'Sea Adventure' to Ayutthaya, 1615–1618: The English East India Company and Its Siam-Japan Trade," *Itinerario* 37, no. 03 (December 2013): 49–69.

58. For the development of the tin industry, see D. K. Bassett, "British 'Country' Trade and Local Trade Networks in the Thai and Malay States, c. 1680–1770," *Modern Asian Studies* 23, no. 4 (January 1, 1989): 625–43.

59. Richard Burnaby at Ayutthaya to Robert Parker and Council at Bantam, October 28, 1678, EFS i.259, 444–46; Samuel Pots at Songkhla to Richard Burnaby at Ayutthaya, December 19, 1678, EFS i.267, 455–57.

60. Ruangsilp, *Dutch East India Company Merchants at the Court of Ayutthaya*, 141.

61. The East India Company in London to Robert Parker and Council at Bantam, October 5, 1677, EFS i.243, 426.

62. Michael Noble, "The 'Invencon' of Lead Crystal—or Was It Flint Glass?" *Journal of Glass Studies* 58 (2016): 185–95. See also the use of "christalline metal" in the English translation of Antonio Neri, *The Art of Glass* (1612), trans. Christopher Merrett (1662), ed. Michael Cable (Sheffield, UK: Society of Glass Technology, 2000).

63. Significant rock-crystal deposits exist in modern-day Myanmar, Indonesia, China, Sri Lanka, India, Nepal, Iran, Afghanistan, Pakistan, and China. For locality information, see "Rock Crystal," Mindat.org, accessed March 14, 2022, https://www.mindat.org/min-6128.html.

64. Cushman and Wyatt, *Royal Chronicles*, 249.

65. The Setangkhamani, or White Crystal Buddha, carved from rock crystal, was the subject of the fifteenth-century rivalry between Ayutthaya and Chiang Mai, and it remains one of the most important Buddha images in Northern Thailand. Angela Chiu, *The Buddha in Lanna: Art, Lineage, Power and Place in Northern Thailand* (Honolulu: University of Hawai'i Press, 2017), 24–25; for a valuable discussion of the role of copying in Buddhist art, see 164–84. For discussion

of Siamese rock-crystal carvings of the Buddha from the fifteenth to the eighteenth centuries, see A. B. Griswold, "A Glass Image and Its Associations," *Journal of Glass Studies* 5 (1963): 75–104.

66. On Borneo diamonds, see Peter Borschberg, "Batu Sawar Johor: A Regional Centre of Trade in the Early Seventeenth Century," in *Early Modern Southeast Asia, 1350–1800*, ed. Ooi Keat Gin and Hoang Anh Tuan (New York: Routledge, 2015), 136–53; Jack Ogden, *Diamonds: An Early History of the King of Gems* (London: Yale University Press, 2018), 327–29. For background on English engagements with Sukadana, see F. Andrew Smith, "Misfortunes in English Trade with Sukadana at the End of the Seventeenth Century, with an Appendix on Thomas Gullock, a Particularly Unlucky Trader," *Borneo Research Bulletin* 46 (January 1, 2015): 75–93.

67. Henry Dacres, Arnold White, and Council at Bantam to Hamon Gibbon and factors at Ayutthaya, July 30, 1676, EFS i.229, 416.

68. It should be noted that this incident with Sukadana is not recorded in the Ayutthaya chronicles, which may render it either apocryphal or even more valuable; it nevertheless records something of importance to both parties in these negotiations. VOC records do, however, note that an embassy from Sukadana in 1662 brought Narai royal letters and gifts, including rough diamonds. Ruangsilp, *Dutch East India Company Merchants at the Court of Ayutthaya*, 131.

69. This arrangement remained in place until the VOC negotiated control of Sukadana and Landak from Bantam in 1778. Tijl Vanneste, *Blood, Sweat and Earth: The Struggle for Control over the World's Diamonds throughout History* (London: Reaktion Books, 2021), 65.

70. Claire Sabel, "The Impact of European Trade with Southeast Asia on the Mineralogical Studies of Robert Boyle," in *Gems in the Early Modern World*, ed. Michael Bycroft and Sven Dupré (Cham, Switzerland: Springer, 2019), 87–116. On Boyle's broader gemological studies, see Michael Bycroft, "Robert Boyle's Restless Gems," in *Ingenuity in the Making: Materials and Technique in Early Modern Art and Science*, ed. Richard J. Oosterhoff, José Ramon Marcaida, and Alexander Marr (Pittsburgh: University of Pittsburgh Press, 2021), 36–50. South and Southeast Asian sources of corundum continued to be relevant through the eighteenth century, see Stephen T. Irish, "The Corundum Stone and Crystallographic Chemistry," *Ambix* 64, no. 4 (October 2017): 301–25.

71. Anna Winterbottom, *Hybrid Knowledge in the Early East India Company World* (Basingstoke, UK: Palgrave Macmillan, 2016), 54–81.

72. *Robert Boyle Papers*, Royal Society Archives, BP 1/39/76-7. Michael Hunter has identified this informant as Antoine Pascot (c. 1646-1689), a French missionary who accompanied the Siamese embassy to France in 1684 and then traveled onward to England. Michael Hunter, *Boyle Studies: Aspects of the Life and Thought of Robert Boyle (1627-91)* (Routledge, 2016), 222n3.

73. The Siamese revolution of 1688, which led to the assassination of Narai and the expulsion of all European merchants besides the Dutch, has been analyzed in detail elsewhere and is beyond the scope of this chapter. See Bhawan, *Dutch East India Company Merchants*, 149–55; Wyatt, *Thailand*, 117; Dhiravat na Pombejra, "Ayutthaya at the End of the Seventeenth Century: Was There a Shift to Isolation?," in *Southeast Asia in the Early Modern Era: Trade, Power, and Belief*, ed. Anthony Reid (Ithaca, NY: Cornell University Press), 250–72.

74. Bhawan, *Dutch East India Company Merchants*, 159.

75. The VOC also definitively left Ayutthaya in the wake of the Burmese invasion; see Bhawan, *Dutch East India Company Merchants*, 25.

76. Robert J. Stern et al., "Plate Tectonic Gemstones," *Geology* 41, no. 7 (July 1, 2013): 723–26; Katie A. Smart et al., "Early Archaean Tectonics and Mantle Redox Recorded in Witwatersrand Diamonds," *Nature Geoscience* 9, no. 3 (March 2016): 255–59; R. J. Stern, M. I. Leybourne, and

Tatsuki Tsujimori, "Kimberlites and the Start of Plate Tectonics," *Geology* 44, no. 10 (October 1, 2016): 799–802.

Chapter Nine

1. Michael Bravo discusses ice visibility in "A Cryopolitics to Reclaim Our Frozen Material States," in *Cryopolitics: Frozen Life in a Melting World*, ed. Joanna Radin and Emma Kowal (Cambridge, MA: MIT Press, 2017), 27–57.

2. Sverker Sörlin, "Cryo-History: Narratives of Ice and the Emerging Arctic Humanities," in *The New Arctic*, ed. Birgitta Evengård, Joan Nymand Larsen, and Øyvind Paasche (Heidelberg: Springer, 2015), 327–41.

3. Henri Bader, "United States Polar Ice and Snow Studies in the International Geophysical Year," in *Geophysical Monograph Series*, ed. Hugh Odishaw and Stanley Ruttenberg (Washington, DC: American Geophysical Union, 2013): 177. Originally published January 1, 1958.

4. Bravo, "Material States," 31.

5. See Noah Heringman, "Deep Time at the Dawn of the Anthropocene," *Representations* 129, no. 1 (2015): 56–85.

6. For an account of precious gems extraction and trade in the early modern period, see Claire Conklin Sabel, "Glass Worke," chapter 8 in this volume.

7. There are valid and important critiques about assigning the Anthropocene a start date and even naming a new epoch. See, for example, Heather Davis and Zoe Todd, "On the Importance of a Date, or Decolonizing the Anthropocene," *ACME: An International Journal of Critical Geographies* 16, no. 4 (2017): 761–80; Eileen Crist, "On the Poverty of Our Nomenclature," *Environmental Humanities* 3, no. 1 (2013): 129–47.

8. The theory certainly did not originate with Agassiz. See Tobias Krüger, *Discovering the Ice Ages: International Reception and Consequences for a Historical Understanding of Climate* (Boston: Brill, 2013).

9. Archibald Geikie, *On the Phenomena of the Glacial Drift of Scotland* (Glasgow: John Grey, 1863), 214.

10. James Croll, "Autobiographical Sketch," in James Irons, *Autobiographical Sketch of James Croll* (London: Edward Stanford, 1896), 33.

11. Croll, "Autobiographical Sketch," 2.

12. Croll left each of these jobs because of bodily ailments—including ceaseless headaches, painful corns, and an ossified elbow joint.

13. Croll, "Autobiographical Sketch," 14.

14. James Croll, "On the Physical Cause of the Change of Climate during Geological Epochs," *The London, Edinburgh, and Dublin Philosophical Magazine and Journal of Science* 28, no. 187 (August 1864): 129.

15. Croll cites Poisson as the originator of the theory that space varied in temperature. See Croll, "On the Physical Causes," 123.

16. Croll, "Physical Causes," 129.

17. James Croll, *Climate and Time in Their Geological Relations: A Theory of Secular Changes of the Earth's Climate* (London: Daldy, Isbister, & Co., 1875), 19–20. Both Emily M. Kern, "Earth Time, Ice Time, Species Time," and Perrin Selcer, "Holocene Time Perspective," chapters 11 and 13, respectively, in this volume, explore historical episodes where researchers attempted to grasp the absolute measure of geological time using ice; the former focuses on archaeological research

and glacial till, the latter on varves—sediment records produced by receding and expanding glaciers—and dating the Holocene.

18. Croll, *Climate and Time*, 13.

19. Croll, *Climate and Time*, 13.

20. See Rebecca Woods, "Nature and the Refrigerating Machine: The Politics and Production of Cold in the Nineteenth Century," in *Cryopolitics: Frozen Life in a Melting World*, ed. Joanna Radin and Emma Kowal (Cambridge, MA: MIT Press, 2017), 89–116.

21. Cara New Daggett, *The Birth of Energy: Fossil Fuels, Thermodynamics, and the Politics of Work* (Durham, NC: Duke University Press, 2019), 69.

22. William Thomson, "On the Age of the Sun's Heat," *Macmillan's Magazine* 5 (March 1862): 388–93.

23. Kathryn Yusoff writes: "Locked into a belatedness in becoming human enough in relation to the ideal (white) humanist subject, the spatializing of time along a vertical line is used as a mechanism to deny juridical rights, wherein Whiteness becomes the achievement of one's temporal identity in geologic time." Kathryn Yusoff, *A Billion Black Anthropocenes or None* (Minneapolis: University of Minnesota Press, 2018), 77.

24. As Daggett makes clear, both the notions of energy and entropy were (and still are) abstract concepts that are historically specific "concept-things." See chapter 2, "A Steampunk Production," in *The Birth of Energy*.

25. Daggett, *Birth of Energy*, 47.

26. Camille Flammarion, *Astronomie Populaire: Description Générale du Ciel* (Paris: C. Marpon and E. Flammarion, 1881), 101.

27. Gillian Beer, *Open Fields: Science in Cultural Encounter* (New York: Oxford University Press, 1996), 225.

28. In chapter 13 of this volume, "Holocene Time Perspective," Perrin Selcer explores the central position of Scandinavia—the site of much glacial action and a launching spot for Arctic exploration—in the development of paleoclimatology in the twentieth century.

29. See Emily M. Kern, "Earth Time, Ice Time, Species Time," chapter 11 in this volume, for a discussion of the significance of interglacials in twentieth-century chronologies of prehistoric people and for arguments around human civilization.

30. This was also true of American culture. See Michael Robinson, *The Coldest Crucible: Arctic Exploration and American Culture* (Chicago: University of Chicago Press, 2006).

31. The notion of the Antarctic having never been visited prior to the nineteenth century is very much Eurocentric: Pacific peoples had visited the region as early as the seventh century. See Priscilla M. Wehi et al., "A Short Scan of Māori Journeys to Antarctica," *Journal of the Royal Society of New Zealand* 52, no. 5 (2022): 587–98, https://doi.org/10.1080/03036758.2021.1917633.

32. Michael Bravo, *North Pole: Nature and Culture* (London: Reaktion Books, 2019), 114.

33. Russell Potter, *Arctic Spectacles: The Frozen North in Visual Culture, 1818–1875* (Seattle: University of Washington Press, 2007), 38. Chauncy Loomis further argues that, as well as searching for the Northwest Passage, these Arctic expeditions had "practical, scientific, strategic, and even commercial purposes [while] the press, and the public came to believe that somehow British manhood and British power were on the line"; see Chauncy C. Loomis, "The Arctic Sublime," in *Nature and the Victorian Imagination*, ed. U. C. Knoepflmacher and G. B. Tennyson (Berkeley: University of California Press, 1977), 95.

34. Loomis, "The Arctic Sublime," 110.

35. Potter tallies that "between 1818 and 1883, no fewer than 60 Arctic shows—including 22

moving panoramas, 3 fixed panoramas, 12 lantern exhibitions, 4 mechanical automata theatres, and 4 exhibitions of 'Esquimaux' or Arctic natives—were presented to the public"; Potter, *Arctic Spectacles*, 12.

36. Letter from Croll to Mr. Bennie dated January 28, 1868, printed in Irons, *Autobiographical Sketch*, 179.

37. Such authors included American naval officer Charles Wilkes (1798–1877), British naval officer James Clarke Ross (1800–1862), Finnish-Swede explorer Adolf Nordenskiöld (1832–1901), American explorer Isaac Israel Hayes (1832–1881), and Danish explorer Hinrich Johannes Rink (1819–1893).

38. Isaac Israel Hayes, *The Open Polar Sea: A Narrative of a Voyage of Discovery towards the North Pole, in the Schooner United States* (London: Sampson Low, Son, and Marston, 1867) quoted in Croll, *Climate and Time*, 380.

39. David Sugden, "James Croll (1821–1890): Ice, Ice Ages, and the Antarctic Connection," *Antarctic Science* 26, no. 06 (December 2014): 604–13.

40. Croll, *Climate and Time*, 377.

41. James Croll, "On the Thickness of the Antarctic Ice Sheet and Its Relation to the Glacial Epoch," *The Journal of Science* 1 (1879): 2.

42. "Arctic and Antarctic Lands," *All the Year Round: A Weekly Journal*, April 13, 1861, 55.

43. Archibald Geikie, "The Geological Influences Which Have Affected British History," *Macmillan's Magazine*, March 1882, 365. Newspapers, for example *The Glasgow Herald*, published lectures by geologists explaining in detail the effects of the ice age.

44. Gillen D'Arcy Wood, "Interglacial Victorians," in *Victorian Sustainability in Literature and Culture*, ed. Wendy Parkins (New York: Routledge, 2018), 220.

45. "Is Another Ice Age Approaching?" *The Blackburn Standard and Weekly Express*, May 23, 1891.

46. Robert Ball, *The Cause of an Ice Age* (New York: D. Appleton and Company, 1891), 25.

47. Ball, *Ice Age*, 25; emphasis added.

48. Ball, *Ice Age*, 170.

49. Ball, *Ice Age*, 168.

50. J. Horner, "Time and Change," *Our Corner* 3, no. 1 (1884): 28.

51. "Is an Ice Age Periodic?" *Chambers's Journal of Popular Literature, Science, and Art*, December 9, 1893, 782.

52. Robert Macfarlane, *Mountains of the Mind: How Desolate and Forbidding Heights Were Transformed into Experiences of Indomitable Spirit* (New York: Pantheon Books, 2003).

53. "Prehistoric Man," *Chambers's Journal*, July 3, 1886, 420.

54. See Robert Dingley, "The Ruins of the Future: Macaulay's New Zealander and the Spirit of the Age," in *Histories of the Future: Studies in Fact, Fantasy, and Science Fiction*, ed. Alan Sandison and Robert Dingley (New York: Palgrave, 2000), 15–33.

55. John C. Elliot, "Our Next Ice-Age," *The Arena*, August 1907, 149.

56. Elliot, "Our Next Ice-Age," 158.

Chapter Ten

1. Alexander du Toit, "The Problem of the Great Australian Artesian Basin," *Journal and Proceedings of the Royal Society of New South Wales* 51 (1917): 135–208.

2. Du Toit, "The Problem," 206.

3. Edward Fisher Pittman, "Note on the Great Australian Artesian Basin," *Journal and Proceedings of the Royal Society of New South Wales* 51 (1917): 431–34.

4. John Walter Gregory, "The Machinery of the Earth," *Nature* 126, no. 3190 (1930): 959–63.

5. Carolyn Strange and Alison Bashford, *Griffith Taylor: Visionary, Environmentalist, Explorer* (Toronto: University of Toronto Press, 2008); Joseph M. Powell, *Griffith Taylor and "Australia Unlimited"* (St. Lucia: University of Queensland Press, 1993); Sarah Mirams, "'The Attractions of Australia': E. J. Brady and the Making of *Australia Unlimited*," *Australian Historical Studies* 42, no. 2 (2012): 270–86.

6. Alison Bashford, "Nation, Empire, Globe: The Spaces of Population Debate in the Interwar Years," *Comparative Studies in Society and History* 49, no. 1 (2007): 170–201; Joseph M. Powell, "The Bowman, Huntington and Taylor Correspondence, 1928," *Australian Geographer* 14, no. 2 (1978): 123–25; Russell McGregor, "A Dog in the Manger: White Australia and Its Vast Empty Spaces," *Australian Historical Studies* 32, no. 2 (2012): 157–73.

7. Donald Meinig, *On the Margins of the Good Earth: The South Australian Wheat Frontier* (Adelaide: Rigby, 1962); Des Heathcote, *Back of Bourke: A Study of Land Appraisal and Settlement in Semi-arid Australia* (Melbourne: Melbourne University Press, 1965); Tim Sherratt, Tom Griffiths, and Libby Robin, eds., *A Change in the Weather: Climate and Culture in Australia* (Canberra: National Museum of Australia Press, 2005); Donald Garden, *Droughts, Floods and Cyclones: El Niños That Shaped Our Colonial Past* (Melbourne: Australian Scholarly Publishing, 2009); Emily O'Gorman, *Flood Country: An Environmental History of the Murray-Darling Basin* (Collingwood, Australia: CSIRO Publishing, 2012); Deb Anderson, *Endurance: Australian Stories of Drought* (Collingwood, Australia: CSIRO Publishing, 2014); Grace Karskens, "Floods and Flood-Mindedness in Early Colonial Australia," *Environmental History* 21, no. 2 (2016): 315–42; Rebecca Jones, *Slow Catastrophes: Living with Drought in Australia* (Clayton, Australia: Monash University Publishing, 2017).

8. Mike Davis, *Late Victorian Holocausts: El Niño Famines and the Making of the Third World* (London: Verso Books, 2002); Katharine Anderson, *Predicting the Weather: Victorians and the Science of Meteorology* (Chicago: University of Chicago Press, 2005); Anya Zilberstein, *A Temperate Empire: Making Climate Change in Early America* (New York: Oxford University Press, 2016); Richard Grove and George Adamson, *El Niño in World History* (Basingstoke, UK: Palgrave Macmillan, 2018); Ruth Morgan, "Natural Worlds: Cultures of Climate Concern in the Age of Empires," in *A Cultural History of Western Empires in the Age of Empires*, ed. Kirsten McKenzie (London: Bloomsbury, 2019), 67–86; Ruth Morgan, "Prophecy and Prediction: Forecasting Drought and Famine in British India and the Australian Colonies," *Global Environment* 13 (2020): 96–133.

9. See, for example, Meredith McKittrick, "Theories of 'Reprecipitation' and Climate Change in the Settler Colonial World," *History of Meteorology* 8 (2017): 74–94; Philipp Lehmann, "Average Rainfall and the Play of Colors: Colonial Experience and Global Climate Data," *Studies in History and Philosophy of Science Part A* 70 (2018): 38–49; Meredith McKittrick, "Talking about the Weather: Settler Vernaculars and Climate Anxieties in Early Twentieth-Century South Africa," *Environmental History* 23, no. 1 (2018): 3–27.

10. Rachael Squire and Klaus Dodds, "Subterranean Geopolitics," *Geopolitics* 25, no. 1 (2020): 4–16.

11. Bruce Braun, "Producing Vertical Territory: Geology and Governmentality in Late Victorian Canada," *Ecumene* 7, no. 1 (2000): 24.

12. Notable exceptions are Joseph Powell, *Plains of Promise, Rivers of Destiny: Water Man-*

agement and the Development of Queensland, 1824–1990 (Brisbane: Boolarong, 1991); Heather Goodall, "Digging Deeper: Ground Tanks and the Elusive 'Indian Archipelago,'" in *Beyond the Black Stump: Rethinking Rural Histories in Australia*, ed. Alan Mayne (Adelaide: Wakefield Press, 2008), 129–60; Kim de Rijke, Paul Munro, and Maria de Lourdes Melo Zurita, "The Great Artesian Basin: A Contested Resource Environment of Subterranean Water and Coal Seam Gas," *Society and Natural Resources* 29, no. 6 (2016): 696–710.

13. Marie Habermehl, "The Evolving Understanding of the Great Artesian Basin (Australia), from Discovery to Current Hydrogeological Interpretations," *Hydrogeology Journal* 28 (2020): 13–36.

14. Rod Fensham, "Artesian Springs," in *Desert Channels: The Impulse to Conserve*, ed. Libby Robin, Chris Dickman, and Mandy Martin (Collingwood, Australia: CSIRO Publishing, 2010), 146.

15. Mike Smith, *The Archaeology of Australia's Deserts* (Melbourne: Cambridge University Press, 2013), 70; Russell Fairfax and Rod Fensham, "In the Footsteps of J. Alfred Griffiths: A Cataclysmic History of Great Artesian Basin Springs in Queensland, Australia," *Australian Geographical Studies* 40, no. 2 (2002): 210–30.

16. Clem Lloyd, *Either Drought or Plenty: Water Development and Management in New South Wales* (Sydney: Department of Water Resources, 1988), 104.

17. O. C. Powell, "Song of the Artesian Water: Aridity, Drought and Disputation along Queensland's Pastoral Frontier in Australia," *Rangeland Journal* 34, no. 3 (2012): 308.

18. Thomas E. Rawlinson, "Subterranean Water Supply in the Interior," abridged, *Philosophical Society of Adelaide* (1878): 124–26.

19. For example, Walter Gibbons Cox, *Artesian Wells as a Means of Supply* (Brisbane: Sapsford & Co., 1895); Robert Logan Jack, *Artesian Water in the Western Interior of Queensland* (Brisbane: Government Printer, 1895).

20. Thom Blake and Margaret Cook, *Great Artesian Basin: Historical Overview* (Brisbane: Department of Natural Resources and Mines, 2006), 16.

21. Lloyd, *Either Drought or Plenty*, 112.

22. John Walter Gregory, *The Dead Heart of Australia: A Journey around Lake Eyre in the Summer of 1901–1902, with Some Account of the Lake Eyre Basin and the Flowing Wells of Central Australia* (London: John Murray, 1906).

23. For example, Henry Chamberlain Russell, "The River Darling—the Water Which Should Pass through It," *Journal of the Royal Society of New South Wales* 13 (1879): 169–70; Ralph Tate, "Geology in Its Relation to Mining and Subterranean Water Supply in South Australia," *Transactions and Proceedings of the Royal Society of South Australia* 4 (1880): 128–34; T. W. Edgeworth David, "Artesian Water in New South Wales (Preliminary Notes)," *Journal and Proceedings of the Royal Society of New South Wales* 25 (1891): 286–96.

24. Gregory, *The Dead Heart*, 286–87. See also Powell, "Song of the Artesian Water," 305–17.

25. Gregory, *The Dead Heart*, 340.

26. Gregory, *The Dead Heart*, 335.

27. Blake and Cook, *Great Artesian Basin*, 45–56.

28. Ernest Favenc, *History of Australian Exploration from 1788 to 1888* (Sydney: Turner and Henderson, 1888), 289–90.

29. Edward F. Pittman, "Problems of the Artesian Water Supply of Australia, with Special Reference to Professor Gregory's Theory," *Journal and Proceedings of the Royal Society of New South Wales* 41 (1907): 100–39.

30. Edward F. Pittman and T. W. Edgeworth David, "Irrigation Geologically Considered

with Special Reference to the Artesian Area of New South Wales," *Journal and Proceedings of the Royal Society of New South Wales* 37 (1903): CIII–CLIII.

31. Pittman, "Problems of the Artesian Water Supply," 105.

32. Pittman, "Problems of the Artesian Water Supply," 138.

33. David Branagan and Elaine Lim, "J.W. Gregory, Traveller in the Dead Heart," *Historical Records of Australian Science* 6, no. 1 (1984): 71–84; A. E. Faggion, "The Australian Groundwater Controversy 1870–1910," *Historical Records of Australian Science* 10, no. 4 (1995): 337–48; Bernard Leake, *The Life and Work of Professor J.W. Gregory FRS (1864–1932): Geologist, Writer and Explorer* (London: Geological Society, 2011), 90.

34. See Oscar Meinzer, "The History and Development of Ground-Water Hydrology," *Journal of the Washington Academy of Sciences* 24, no. 1 (1934): 6–32; John Kemp et al., *Artesian Water Supplies in Queensland* (Brisbane: Department of the Co-ordinator-General of Public Works, 1954); Powell, "Song of the Artesian Water," 305–17.

35. John W. Gregory, "Artesian Water in Victoria: The Prospect of Tapping It," *The Argus*, January 8, 1903, 7.

36. Eduard Suess, "Ueber Heisse Quellen," *Naturwissenschaftliche Rundschau* 17, no. 47 (1902): 585–611; Eduard Suess, "Hot Springs," trans. D. H. Newland, *Engineering and Mining Journal* 76, no. 1 (1903): 8–9, 76; no. 2 (1903): 51–52.

37. Eduard Suess, *The Face of the Earth (Das Antlitz der Erde)*, vol. 4, trans. Hertha B. C. Sollas (Oxford: Clarendon Press, 1909), 548–49.

38. Christopher J. Duffy, "The Terrestrial Hydrologic Cycle: An Historical Sense of Balance," *WIREs Water* 4 (2017): e1216.

39. Meinzer, "The History and Development of Ground-Water Hydrology," 12.

40. Otto Volger, "Die Wissenschaftliche Lösung der Wasser-, insbesondere der Quellenfrage," *Zeitschrift Ver Deutscher Ingenieure* 21 (1877): 482–50, cited in Meinzer, "The History and Development of Ground-Water Hydrology," 13.

41. John W. Gregory, "Suess's Theories of Geographical Evolution," *Natural Science* 12, no. 72 (1898): 120.

42. Eduard Suess, "Die Brüche des Östlichen-Afrika," in *Beiträge zur Geologischen Kenntniss des Östlichen Afrika*, ed. L. R. Hohnell et al. (Vienna: F. Tempsky, 1891), 447–84; John W. Gregory, "Eduard Suess Memorial Tablet," *Quarterly Journal of the Geological Society* 85 (1929): cxxxvi–cxli.

43. Gregory, "Eduard Suess Memorial Tablet," cxxxix; Leake, *The Life and Work*, 19–20.

44. John W. Gregory, *The Great Rift Valley: Being the Narrative of a Journey to Mount Kenya and Lake Baringo* (London: John Murray, 1896), 94.

45. Leake, *The Life and Work*, 30.

46. Gregory, *The Dead Heart*, 3–16.

47. Alfred W. Howitt, "The Dieri and Other Kindred Tribes of Central Australia," *Journal of the Anthropological Institute of Great Britain and Ireland* 20 (1891): 30–104; Alfred W. Howitt and Otto Siebert, "Two Legends of the Lake Eyre Tribe, Central Australia," *Report of the Australasian Association for the Advancement of Science*, 9 (1902): 525–32; Alfred W. Howitt and Otto Siebert, "Legends of the Dieri and Kindred Tribes of Central Australia," *Journal of the Anthropological Institute of Great Britain and Ireland* 34 (1904): 100–29; Alfred W. Howitt, *The Native Tribes of South-East Australia* (London: Macmillan, 1904). See also Mike Smith, "The Historiography of *Kardimarkara*: Reading a Desert Tradition as Cultural Memory of the Remote Past," *Journal of Social Archaeology* 19, no. 1 (2019): 47–66.

48. Howitt and Siebert, "Two Legends of the Lake Eyre Tribe," 525–32.

49. Gregory, *The Dead Heart*, 74.

50. Ralph Tate, "Post-Miocene Climate in South Australia (Being in Part a Rejoinder to Mr. Scoular's Paper)," *Transactions and Proceedings and Report of the Royal Society of South Australia* 8 (1884–85): 53; Mike Smith, "How the Desert Got a Past: A History of Quaternary Research in Australia's Deserts," *Historical Records of Australian Science* 25 (2014): 172–85.

51. Gregory, *The Dead Heart*, 183.

52. Gregory, "Suess's Theories," 119; John W. Gregory, "The Face of the Earth," *Nature* 72, no. 1861 (1905): 193–94.

53. Eduard Suess, *Das Antlitz der Erde Bd 1* (Vienna: Tempsky, 1885); Eduard Suess, *The Face of the Earth*, vol. 1, trans. Hertha B. C. Sollas (Oxford: Clarendon Press, 1904).

54. Suess, *The Face of the Earth*, 17.

55. Vybarr Cregan-Reid, "The Gilgamesh Controversy: The Ancient Epic and Late-Victorian Geology," *Journal of Victorian Culture* 14 (2009): 224–37.

56. Suess, *The Face of the Earth*, 18, 30–31, 39.

57. Suess, *The Face of the Earth*, 31; emphasis in original.

58. Suess, *The Face of the Earth*, 33; Mott Greene, *Geology in the Nineteenth Century: Changing Views of a Changing World* (Ithaca, NY: Cornell University Press, 1982), 160–91.

59. Deborah Coen, *The Earthquake Observers: Disaster Science from Lisbon to Richter* (Chicago: University of Chicago Press, 2013), 170. See Eduard Suess, *Der Boden der Stadt Wien nach seiner Bildungsweise, Beschaffenheit und seinen Beziehungen zum Bürgerlichen Leben: Eine Geologische Studie* (Vienna: W. Braumüller, 1862).

60. Leake, *The Life and Work*, 54.

61. Gregory, "Eduard Suess Memorial Tablet," cxxxvi–cxli.

62. Gregory, "Eduard Suess Memorial Tablet," cxxxvii.

63. Coen, *The Earthquake Observers*, 21.

64. Percy G. H. Boswell, "John Walter Gregory—1864–1932," *Obituary Notices of the Royal Society* 1 (1936): 53–59.

65. Boswell, "John Walter Gregory," 55.

66. Powell, "Song of the Artesian Water," 310–11; Lloyd, *Either Drought or Plenty*, 113.

67. Garald G. Parker Sr., "Early Stage of Hydrogeology in the United States, 1776 to 1912," *Water Resources Bulletin* 22, no. 5 (1986): 705.

68. Pittman and David, "Irrigation," cxxxi.

69. Robert Tierney, Kevin Parton, and Deanna Duffy, "Three Raindrops and Some Dust: Combining Archival and GIS Analysis to Map the Spatial Distribution of the Impact of the Federation Drought of 1895–1903 in Rural New South Wales," *Journal of Historical Geography* 55 (2017): 1–16.

70. Donald Garden, "The Federation Drought of 1895–1903, El Niño and Society in Australia," in *Common Ground: Integrating the Social and Environmental in History*, ed. Genevieve Massard-Guilbaud and Stephen Mosley (Newcastle upon Tyne: Cambridge Scholars, 2010), 270–92.

71. Percy Allan, "The Drought Antidote for the North-West, or the Utilisation of the Artesian Resources of New South Wales," *Journal and Abstract of Proceedings of the Sydney University Engineering Society, New South Wales* 11 (1906): 1.

72. John W. Gregory, "The Geographical Factors that Control the Development of Australia," *Geographical Journal* 35 (1910): 658–82; John W. Gregory, "The Flowing Wells of Central Australia," *Geographical Journal* 38, no. 1 (1911): 34–59; John W. Gregory, "The Flowing Wells of Central Australia (continued)," *Geographical Journal* 38, no. 2 (1911): 157–81.

73. Alfred Harker, *The Natural History of Igneous Rocks* (London: Methuen, 1909), 47–48.

74. Edward F. Pittman et al., *Report of the Interstate Conference on Artesian Water* (Sydney: Government Printer, 1913), x.

75. Pittman et al., *Report of the Interstate Conference*, xiii.

76. Edward F. Pittman et al., *Report of the Second Interstate Conference on Artesian Water, Brisbane, 1914* (Brisbane: Government Printer, 1914).

77. Glynn Connolly, cited in *Report of the Second Interstate Conference*, 68.

78. Pittman, *Report of the Second Interstate Conference*, xi–xii.

79. Edward F. Pittman, *The Great Australian Artesian Basin and the Source of Its Water* (Sydney: Government Printer, 1914), 41.

80. In a subsequent report, Pittman included a similar plate to that featured in figure 10.3. Now the caption read for emphasis, "The Most Porous Intake-Beds of the Artesian Basin." See Edward F. Pittman, *The Composition and Porosity of the Intake Beds of the Great Australian Artesian Basin* (Sydney: Government Printer, 1915), plate 1.

81. Leake, *The Life and Work*, 148; David Branagan, *T.W. Edgeworth David: A Life* (Canberra: National Library of Australia, 2005).

82. Charles P. Lucas, "Presidential Address: Man as a Geographical Agency," in *Report of the British Association for the Advancement of Science, Australia, July 28–August 31, 1914* (London: John Murray, 1915), 435.

83. John W. Gregory, "The Flowing Wells of Western Queensland," *Queensland Geographical Journal* XXX–XXXI (1916): 1–29.

84. For example, G. E. Bunning, cited in Pittman et al., *Report of the Second Interstate Conference*, 90; Powell, "Song of the Artesian Water," 311; Powell, *Plains of Promise*, 142.

85. Kemp et al., *Artesian Water Supplies*, 7.

86. John W. Gregory, "The Machinery of the Earth," *Nature* 126 (1930): 961–62.

87. John W. Gregory, *The Menace of Colour: A Study of the Difficulties Due to the Association of White and Coloured Races* (London: Seeley Service & Co., 1925), 152. See also John W. Gregory, "The Principles of Migration Restriction," in *Proceedings of the World Population Conference*, ed. Margaret Sanger (London: Edward Arnold, 1927), 302–5.

88. John W. Gregory, *Australia* (Cambridge: Cambridge University Press, 1916), 146.

89. T. W. Edgeworth David, *Explanatory Notes to Accompany a New Geological Map of the Commonwealth of Australia* (Sydney: Australasian Medical Publishing, 1932), 175.

90. See Tom Griffiths, *Hunters and Collectors: The Antiquarian Imagination in Australia* (Melbourne: Cambridge University Press, 1996), 26.

91. David, *Explanatory Notes*, 176.

92. Patrick Nunn and Nicholas Reid, "Aboriginal Memories of Inundation of the Australian Coast Dating from More than 7000 Years Ago," *Australian Geographer*, 47, no. 1 (2016): 11–47; Patrick Nunn, *The Edge of Memory: Ancient Stories, Oral Tradition and the Post-glacial World* (London: Bloomsbury, 2018); Tom Griffiths, "The Planet Is Alive: Radical Histories for Uncanny Times," *Griffith Review* 63 (2019): 61–72.

93. Keith McConnochie, "Desert Departures: Isolation, Innovation and Introversion in Ice-Age Australia," in *Departures: How Australia Reinvents Itself*, ed. Xavier Pons (Melbourne: Melbourne University Press, 2002), 27–28.

Chapter Eleven

1. Martin Rudwick, *Earth's Deep History: How It Was Discovered and Why It Matters* (Chicago: University of Chicago Press, 2014); Joe D. Burchfield, *Lord Kelvin and the Age of the Earth*, new ed. (1975; Chicago: University of Chicago Press, 1990).

2. On the history of European prehistory, anthropology, and archaeology in the nineteenth century, see Marianne Sommer, *Bones and Ochre: The Curious Afterlife of the Red Lady of Paviland* (Cambridge, MA: Harvard University Press, 2007); Marianne Sommer, *History Within: The Science, Culture, and Politics of Bones, Organisms, and Molecules* (Chicago: University of Chicago Press, 2016); Chris Manias, *Race, Science, and the Nation: Reconstructing the Ancient Past in Britain, France, and Germany* (New York: Routledge, 2013); Efram Sera-Shriar, ed. *Historicizing Humans: Deep Time, Evolution, and Race in Nineteenth-Century British Sciences* (Pittsburgh: University of Pittsburgh Press, 2018); Maria Stavrinaki, *Transfixed by Prehistory: An Inquiry into Modern Art and Time*, trans. Jane Marie Todd (New York: Zone Books, 2022); Brent Maner, *Germany's Ancient Pasts: Archaeology and Historical Interpretation since 1700* (Chicago: University of Chicago Press, 2018); Alice Conklin, *In the Museum of Man: Race, Anthropology, and Empire in France, 1850–1950* (Ithaca, NY: Cornell University Press, 2013); Cécile Debray, Rémi Labrusse, and Maria Stavrinaki, *Préhistoire: Une énigme moderne* (Paris: Éditions du Centre Pompidou, 2019); Bruce Trigger, *A History of Archaeological Thought*, 2nd ed. (Cambridge: Cambridge University Press, 2006), 121–65.

3. Natalie J. Whitcomb and William Miller, "Lyell's Proposal of the Term 'Pleistocene,'" *Tulane Studies in Geology and Paleontology* 18, no. 2 (1984), 77–81.

4. Rudwick, *Earth's Deep History*, 146.

5. Burchfield, *Lord Kelvin*, 97.

6. F. G. Houtermans, "History of the K/Ar-Method of Geochronology," in *Potassium Argon Dating*, comp. O. A. Schaeffer and J. Zähringer (Heidelberg: Springer-Verlag, 1966), 1–6; R. E. Taylor and Ofer Bar-Yosef, *Radiocarbon Dating: An Archaeological Perspective*, 2nd ed. (Walnut Creek, CA: Left Coast Press, 2014); Emily M. Kern, "Archaeology Enters the 'Atomic Age': A Short History of Radiocarbon, 1946–1960," *British Journal for the History of Science* 53, no. 2 (June 2020): 207–27.

7. The half-life of U-238 is approximately 4.5 billion years; the half-life of C-14 is roughly 5,730 years.

8. Martin Rudwick, *Bursting the Limits of Time: The Reconstruction of Geohistory in the Age of Revolution* (Chicago: University of Chicago Press, 2005); Martin Rudwick, *Worlds Before Adam: The Reconstruction of Geohistory in the Age of Reform* (Chicago: University of Chicago Press, 2008).

9. Sarah Dry, *Waters of the World: The Story of Scientists Who Unraveled the Mysteries of Our Seas, Glaciers and Atmosphere and Made the Planet Whole* (Melbourne: Scribe, 2019); John Imbrie and Katherine Palmer Imbrie, *Ice Ages: Solving the Mystery* (Hillside, NJ: Enslow Publishers, 1979); Jean M. Grove, *Little Ice Ages: Ancient and Modern*, vol. 1, 2nd ed. (New York: Routledge, 2004); Alexis Rider, "The Agent of the Most Dire of Calamities," chapter 9 of this volume.

10. See James A. Secord, "Global Geology and the Tectonics of Empire," in *Worlds of Natural History*, ed. H. A. Curry et al. (Cambridge: Cambridge University Press, 2018); Adam Bobbette and Amy Donovan, eds. *Political Geology: Active Stratigraphies and the Making of Life* (Cham, Switzerland: Palgrave Macmillan, 2018).

11. Quoted in Charles Lyell, *Geological Evidences of the Antiquity of Man* (London: John Murray, 1863), 320–22.

12. Lyell, *Geological Evidences*, 322.

13. James Croll, *Climate and Time in Their Geological Relations*, rev. ed. (1875; New York: D. Appleton and Co., 1893).

14. Croll, *Climate and Time*, 341.

15. On Croll's popularity, see Burchfield, *Lord Kelvin*, 121–32. On the rejection of Croll, see Rudwick, *Earth's Deep History*, 174–80; David J. Meltzer, *The Great Paleolithic War: How Science Forged an Understanding of America's Ice Age Past* (Chicago: University of Chicago Press, 2015), 105–6, 161–62.

16. Rudwick, *Earth's Deep History*, 184–85.

17. Croll, *Climate and Time*, 341–43.

18. James Geikie, *The Great Ice Age and Its Relation to the Antiquity of Man*, 3rd ed. (London: E. Stanford, 1894).

19. Geikie, *The Great Ice Age*, 694.

20. Albrecht Penck and Eduard Brückner, *Die Alpen im Eiszeitalter*, 3 vols. (Leipzig, Germany: Chr. Herm. Tauchnitz, 1909).

21. Penck and Brückner, *Die Alpen im Eiszeitalter*, vol. 2, 1169.

22. Milutin Milankovitch, *Théorie mathématique des phénomènes thermiques produits par la radiation solaire* (Paris: Gauthier-Villars et Cie, 1920). See also Dry, *Waters of the World*, 238–41 and James R. Fleming, *Historical Perspectives on Climate Change* (Oxford: Oxford University Press, 1998), 108–10.

23. Lyell, *Geological Evidences*, 207.

24. Lyell, *Geological Evidences*, 207.

25. Trigger, *A History of Archaeological Thought*, 121–29.

26. Gabriel de Mortillet, *Le préhistorique: Antiquité de l'homme* (Paris: G. Reinwald, 1883), 19.

27. De Mortillet, *Le préhistorique*, 20–21. Later, de Mortillet would become involved in the debates over eoliths and the problem of distinguishing naturally flaked from deliberately flaked stone. On the eolith controversy, see Anne O'Connor, "Geology, Archaeology, and 'the Raging Vortex of the "Eolith" Controversy,'" *Proceedings of the Geologists' Association* 114 (2003): 255–62; Marianne Sommer, "Eoliths as Evidence for Human Origins? The British Context," *History and Philosophy of the Life Sciences* 26, no. 2 (2004): 209–41; Matthew R. Goodrum, "The History of Human Origins Research and Its Place in the History of Science: Research Problems and Historiography," *History of Science* 47, no. 3 (2009): 337–57.

28. Trigger, *A History of Archaeological Thought*, 155.

29. De Mortillet, *Le préhistorique*, 183.

30. Penck and Brückner, *Die Alpen im Eiszeitalter*, vol. 3, 1172.

31. W. J. Sollas, "Palaeolithic Races and Their Modern Representatives," *Science Progress in the Twentieth Century* 4, no. 15 (January 1910): 376–92; Marcellin Boule, "Observations sur un silex taillé du Jura et sur la chronologie de M. Penck," *L'Anthropologie* 19 (1908): 1–13.

32. C. E. P. Brooks, "The Meteorological Conditions of an Ice Sheet and Their Bearing on the Desiccation of the Globe," *Quarterly Journal of the Meteorological Society* 40, no. 169 (1914): 53–70; C. E. P. Brooks, *Climate through the Ages* (New York: R. V. Coleman, 1926). On early twentieth-century climatology, see also Dry, *Waters of the World*; Deborah Coen, *Climate in Motion: Science, Empire, and the Problem of Scale* (Chicago: University of Chicago Press, 2018).

33. Henri Breuil, "Les subdivisions du Paléolithique supérieur et leur signification," *Congrès international d'anthropologie et d'archéologie préhistoriques—compte-rendu de la XIVe session, Genève 1912*, vol. 1 (Geneva: Imprimerie Albert Kündig, 1913), 165–238. See also François Bon,

"The Division and Discord of Prehistoric Chronologies," *Res: Anthropology and Aesthetics* 69–70 (Spring–Autumn 2018): 76–84.

34. Breuil, "Les subdivisions du Paléolithique supérieur," 170.

35. Breuil, "Les subdivisions du Paléolithique supérieur," 237–38.

36. F. E. Zeuner, "Pleistocene Chronology of Central Europe," *Geological Magazine* 72 (1935): 350–76.

37. G. C. Simpson, "The Pleistocene Period," review of *The Pleistocene Period*, by Frederick E. Zeuner, *Nature* 156, no. 3973 (December 22, 1945): 730–31.

38. Zeuner, "Pleistocene Chronology," 350.

39. Zeuner, "Pleistocene Chronology," 362.

40. Zeuner, "Pleistocene Chronology," 365.

41. Zeuner, "Pleistocene Chronology," 365.

42. V. G. Childe, "Changing Methods and Aims in Prehistory," *Proceedings of the Prehistoric Society* 1 (1935): 5.

43. Michael Hammond, "The Expulsion of the Neanderthals from Human Ancestry: Marcellin Boule and the Social Context of Scientific Research," *Social Studies of Science* 12, no. 1 (February 1982): 1–36; Stephanie Moser, "The Visual Language of Archaeology: A Case Study of the Neanderthals," *Antiquity* 66 (1992): 831–44; Marianne Sommer, "Mirror, Mirror on the Wall: Neanderthal as Image and 'Distortion' in Early 20th-Century French Science and Press," *Social Studies of Science* 36, no. 2 (April 2006): 207–40.

44. Edward P. F. Rose, Judy Ehlen, and Ursula L. Lawrence, "Military Use of Geologists and Geology: A Historical Overview and Introduction," Geological Society of London, Special Publications 473 (2019): 1–29.

45. E. J. Wayland, *Summary of Progress of the Geological Survey of Uganda for the Years 1919 to 1929* (Entebbe: Government Printer Uganda, 1931), 1. For another example, see R. W. van Bemmelen, "On the Mineral Resources of the Netherlands Indies and Their Industrial Possibilities," trans. J. A. C. Fagginger Auer, in *Science and Scientists in the Netherlands Indies*, ed. Pieter Honig and Frans Verdoorn (New York: Board for the Netherlands Indies, Suriname, and Curaçao, 1945), 5–10.

46. K. A. Davis, "E.J. Wayland C.B.E.—A Tribute," *The Uganda Journal* 31, no. 1 (1967): 1–8.

47. E. J. Wayland, "Rifts, Rivers, Rains, and Early Man in Uganda," *Journal of the Royal Anthropological Institute* 64 (July–December 1934): 333–52.

48. L. S. B. Leakey and J. D. Solomon, "Letter to the Editor: East African Archaeology," *Nature* 124, no. 3114 (July 6, 1929): 9.

49. E. J. Wayland, "African Pluvial Periods and Prehistoric Man," *Man* 29, nos. 87–88 (July 1929): 118–21.

50. Wayland, *Summary*, 38; E. J. Wayland, "Past Climates and Some Future Possibilities in Uganda," *The Uganda Journal* 3, no. 2 (October 1935): 93–118; E. J. Wayland, "Causes of Ice Age," *Geological Magazine* 85, no. 3 (June 1948): 178–81.

51. E. J. Wayland, "Confidential: Chart of Central and East African Prehistory," typescript, Geological Survey Office, Entebbe, Uganda, 1939, Miles Crawford Burkitt Papers, GBR/0012/MS Add.7959, box 4, Cambridge University Library.

52. Before the mid-twentieth century, Asia was widely regarded as the most likely site of the evolutionary origins of humankind. See discussion in brief in Robin Dennell, "From Sangiran to Olduvai, 1937–1960: The Quest for 'Centres' of Hominid Origins in Asia and Africa," in *Studying Human Origins: Disciplinary History and Epistemology*, ed. Raymond Corbey and Wil

Roebroeks (Amsterdam: Amsterdam University Press, 2001), 62–63; Sheela Athreya and Rebecca Rogers Ackermann, "Colonialism and Narratives of Human Origins in Asia and Africa," in *Interrogating Human Origins*, eds. Martin Porr and Jacqueline Matthews (London: Routledge, 2020), 72–95; Emily M. Kern, "Out of Asia: A Global History of the Scientific Search for the Origins of Humankind, 1800–1965" (PhD diss., Princeton, 2018).

53. Edgar B. Howard, "Minutes of the International Symposium on Early Man Held at the Academy of Natural Sciences of Philadelphia, March 17th–20th, 1937, in Celebration of Its One Hundred and Twenty-Fifth Anniversary," *Proceedings of the Academy of Natural Sciences of Philadelphia* 89 (1937): 439–49; Robert McCracken Peck and Patricia Tyson Stroud, *A Glorious Enterprise: The Academy of Natural Sciences of Philadelphia and the Making of American Science* (Philadelphia: Penn Press, 2012): 200–15.

54. "Roundtable on North American Chronology," March 18, 1937, 7, Academy of Natural Sciences Archives, Coll. 422, box 1, folder 4.

55. "Roundtable on North American Chronology," 23.

56. "Roundtable on North American Chronology," 3.

57. Howard, "Minutes of the International Symposium," 442.

58. Sigrid Schmalzer, *The People's Peking Man: Popular Science and Human Identity in Twentieth Century China* (Chicago: University of Chicago Press, 2008); Grace Yen Shen, *Unearthing the Nation: Modern Geology and Nationalism in Republican China* (Chicago: University of Chicago Press, 2014).

59. Wen-Chung Pei (Pei Wenzhong), "An Attempted Correlation of Quaternary Geology, Palaeontology, and Prehistory in Europe and China," Institute of Archaeology Occasional Paper no. 2 (1939): 3–16.

60. Pei, "Attempted Correlation of Quaternary Geology," 3.

61. Pei, "Attempted Correlation of Quaternary Geology," 7.

62. Pei, "Attempted Correlation of Quaternary Geology," 9.

63. Shen, *Unearthing the Nation*, 135–43.

64. George B. Barbour, "The Taiku Deposits and the Problem of Pleistocene Climates," *Bulletin of the Geological Society of China* 10, no. 1 (1931): 71–100; Davidson Black, "Palaeogeography and Polar Shift," *Bulletin of the Geological Society of China* 10, no. 1 (1931): 105–57.

65. See Chang Hsi-Chih, "A Brief Summary of the Tertiary Formations of Inner Mongolia and Their Correlation with Europe and North America," *Bulletin of the Geological Society of China* 10, no. 1 (1931): 301–18.

66. G. C. Simpson, "Some Studies in Terrestrial Radiation," *Memoirs of the Royal Meteorological Society* 2, no. 16 (1928): 69–95.

67. J. S. Lee, "Quaternary Glaciation in the Yangtze Valley," *Bulletin of the Geological Society of China* 13, no. 1 (1934): 15–62; J. S. Lee, "Data Relating to the Study of the Problem of Glaciation in the Lower Yangtze Valley," *Bulletin of the Geological Society of China* 13, no. 1 (1934): 395–422.

68. Lee, "Problem of Glaciation in the Lower Yangtze," 421–22.

69. G. B. Barbour, "Analysis of Lushan Glaciation Problem," *Bulletin of the Geological Society of China* 13, no. 1 (1934): 647–56.

70. J. S. Lee, "Confirmatory Evidence of Pleistocene Glaciation from the Huangshan, Southern Anhui," *Bulletin of the Geological Society of China* 15, no. 3 (1936): 279–94.

71. Pei, "Attempted Correlation of Quaternary Geology," 11.

72. Wayland, "Chart of Central and East African Prehistory," 12.

73. Wayland, "Chart of Central and East African Prehistory," 12.

74. Dry, *Waters of the World*, 238–41.

75. Merrick Posnansky, "Wayland as Archaeologist," *The Uganda Journal* 31, no. 1 (1967): 9–12; P. H. Temple, "E.J. Wayland and the Geomorphology of Uganda," *The Uganda Journal* 31, no. 1 (1967), 13–32.

Chapter Twelve

1. Robert Boyle, *The Christian Virtuoso* (1690) cited in Haileigh Robertson, "Imitable Thunder: The Role of Gunpowder in Seventeenth-Century Experimental Science" (PhD diss., University of York, 2016), 86.

2. Robertson, "Imitable Thunder," 85–86.

3. Robertson, "Imitable Thunder," 104–5.

4. Isabelle Stengers, "Including Nonhumans in Political Theory: Opening Pandora's Box?," in *Political Matter: Technoscience, Democracy and Public Life*, ed. Bruce Braun and Sarah Whatmore (Minneapolis: University of Minnesota Press, 2010), 3–33.

5. Rob Nixon, *Slow Violence and the Environmentalism of the Poor* (Cambridge, MA: Harvard University Press, 2011), 3.

6. Stephen Jay Gould, *Time's Arrow, Time's Cycle: Myth and Metaphor in the Discovery of Geological Time* (Cambridge, MA: Harvard University Press, 1987), 2.

7. See Nigel Clark and Bronislaw Szerszynski, *Planetary Social Thought: The Anthropocene Challenge to the Social Sciences* (Cambridge: Polity, 2021), 19–23.

8. Jan Zalasiewicz et al., "The Working Group on the Anthropocene: Summary of Evidence and Interim Recommendations," *Anthropocene* 19 (2017): 55–60.

9. Jan Zalasiewicz et al., "Petrifying Earth Process: The Stratigraphic Imprint of Key Earth System Parameters in the Anthropocene," *Theory, Culture & Society* 34, nos. 2–3 (2017): 85.

10. Paul Crutzen and John Birks, "The Atmosphere after a Nuclear War: Twilight at Noon," *Ambio* 11, nos. 2–3 (1982): 123.

11. See Joseph Masco, "Bad Weather: On Planetary Crisis," *Social Studies of Science* 40, (2010): 7–40; Paul Edwards, "Entangled Histories: Climate Science and Nuclear Weapons Research," *Bulletin of the Atomic Scientists* 68 (2012): 28–40.

12. See, for example, Eva Lövbrand et al., "Who Speaks for the Future of Earth? How Critical Social Science Can Extend the Conversation on the Anthropocene," *Global Environmental Change* 32 (2015): 211–18.

13. See Clark and Szerszynski, *Planetary Social Thought*, 35–38.

14. Joseph Needham, *Science and Civilisation in China*, vol. 5, *Chemistry and Chemical Technology*, part 7, *Military Technology—The Gunpowder Epoch* (Cambridge: Cambridge University Press, 1986), 1.

15. Needham, *Science and Civilisation in China*, 2.

16. See Clark and Szerszynski, *Planetary Social Thought*, 23–32.

17. Jack Kelly, *Gunpowder: Alchemy, Bombards, and Pyrotechnics: The History of the Explosive that Changed the World* (New York: Basic Books, 2004), vii.

18. Yuk Hui, *The Question Concerning Technology in China: An Essay in Cosmotechnics* (Falmouth, UK: Urbanomic, 2016).

19. Hui, *The Question Concerning Technology in China*, 152.

20. Hui, *The Question Concerning Technology in China*, 153.

21. Peter Lorge, *The Asian Military Revolution: From Gunpowder to the Bomb* (Cambridge: Cambridge University Press, 2008), 1.

22. Lorge, *The Asian Military Revolution*, 8.

23. Needham, *Science and Civilisation in China*, 108.

24. Tonio Andrade, *The Gunpowder Age: China, Military Innovation and the Rise of the West in World History* (Princeton, NJ: Princeton University Press, 2016), 112.

25. Vaclav Smil, *Energy* (Oxford: Oneworld, 2006), 10.

26. Kelly, *Gunpowder*, viii.

27. Kelly, *Gunpowder*, vii, 5–6.

28. Stephen J. Pyne, *World Fire: The Culture of Fire on Earth* (Seattle: University of Washington Press, 1997), 3–7; *Vestal Fire: An Environmental History, Told through Fire, of Europe and Europe's Encounter with the World* (Seattle: University of Washington Press, 1997), 16–17.

29. Lorge, *The Asian Military Revolution*, 42.

30. Andrade, *The Gunpowder Age*, 29.

31. Lorge, *The Asian Military Revolution*, 42.

32. Andrade, *The Gunpowder Age*, 31; see also Nigel Clark, "Vertical Fire: For a Pyropolitics of the Subsurface," *Geoforum* 127 (2021): 364–72.

33. Jack Goody, *Metals, Culture and Capitalism: An Essay on the Origins of the Modern World* (Cambridge: Cambridge University Press, 2012), 207, 258.

34. Needham, *Science and Civilisation in China*, 568.

35. Needham, *Science and Civilisation in China*, 579.

36. Kelly DeVries, "Gunpowder and Early Gunpowder Weapons," in *Gunpowder: The History of an International Technology*, ed. Brenda Buchanan (Bath, UK: Bath University Press, 1996), 131.

37. See Michel Foucault, *Discipline and Punish: The Birth of the Prison* (London: Penguin, 1991), 19.

38. Roy Wolper, "The Rhetoric of Gunpowder and the Idea of Progress," *Journal of the History of Ideas* 31, no. 4 (1970): 597.

39. Simon Werrett, *Fireworks: Pyrotechnic Arts and Sciences in European History* (Chicago: University of Chicago Press, 2010), 59–64.

40. Robertson, "Imitable Thunder," 32–34.

41. Needham, *Science and Civilisation in China*, 16.

42. Lorge, *The Asian Military Revolution*, 5.

43. Lorge, *The Asian Military Revolution*, 17.

44. Lorge, *The Asian Military Revolution*, 81.

45. Lorge, *The Asian Military Revolution*, 4.

46. Stephen Toulmin, *Cosmopolis: The Hidden Agenda of Modernity* (Chicago: University of Chicago Press, 1990), 16–17.

47. Toulmin, *Cosmopolis*, 18–22, 56–62.

48. Toulmin, *Cosmopolis*, 61.

49. Robert Withers, "Descartes' Dreams," *Journal of Analytical Psychology* 53, no. 5 (2008): 691–709.

50. Withers, "Descartes' Dreams," 691–92.

51. Baillet, 1691, cited in Withers, "Descartes' Dreams," 691.

52. Withers, "Descartes' Dreams," 701.

53. Toulmin, *Cosmopolis*, 129–30.

54. Donna Haraway, "Situated Knowledges: The Science Question in Feminism and the Privilege of Partial Perspective," *Feminist Studies* 14, no. 3 (1988): 576, 590.

55. Lewis Mumford, *Technics and Civilization* (Chicago: University of Chicago Press, 1934, 2010), 89.

56. Priya Satia, *Empire of Guns: The Violent Making of the Industrial Revolution* (New York: Penguin Press, 2018), 334.

57. See Foucault, *Discipline and Punish*, 135–36; Mumford, *Technics and Civilization*, 83–84.

58. See DeVries, "Gunpowder and Early Gunpowder Weapons," 122.

59. Andrew Miller, *Now We Shall Be Entirely Free* (London: Sceptre, 2018), 99.

60. Satia, *Empire of Guns*, 395.

61. See Marc-Antoine Croq, "From Shell Shock and War Neurosis to Posttraumatic Stress Disorder: A History of Psychotraumatology," *Dialogues in Clinical Neuroscience* 2, no. 1 (March 2000): 47–55.

62. Cited in Satia, *Empire of Guns*, 332.

63. Lorge, *The Asian Military Revolution*, 180.

64. Lorge, *The Asian Military Revolution*, 7.

65. Lorge, *The Asian Military Revolution*, 21, 177, 162–68.

66. Andrade, *The Gunpowder Age*, 256–58; Lorge, *The Asian Military Revolution*, 168.

67. Needham, *Science and Civilisation in China*, 69.

68. Needham, *Science and Civilisation in China*, 108–17.

69. Needham, *Science and Civilisation in China*, 2–3.

70. Goody, *Metals, Culture and Capitalism*, 165.

71. Goody, *Metals, Culture and Capitalism*, 166.

72. Goody, *Metals, Culture and Capitalism*, 219, 257; Needham, *Science and Civilisation in China*, 39.

73. Kelly, *Gunpowder*, 146.

74. Nigel Clark, "Infernal Machinery: Thermopolitics of the Explosion," *Culture Machine*, 17 (2019), http://culturemachine.net/vol-17-thermal-objects/infernal-machinery/.

75. Kelly, *Gunpowder*, 140–41; Clark, "Infernal Machinery."

76. Nigel Clark and Kathryn Yusoff, "Combustion and Society: A Fire-Centred History of Energy Use," *Theory, Culture & Society* 31, no. 5 (2014): 203–26; Nigel Clark, "Fiery Arts: Pyrotechnology and the Political Aesthetics of the Anthropocene," *GeoHumanities* 1, no. 2 (2015): 266–84.

77. Ilya Prigogine and Isabelle Stengers, *Order Out of Chaos: Man's New Dialogue with Nature* (New York: Bantam Books, 1984), 102.

78. Mumford, *Technics and Civilization*, 88.

79. Needham, *Science and Civilisation in China*, 544–68.

80. Christiaan Huygens, *Oeuvres Completes*, 22 vols. (The Hague: Nijhoff, 1897-1950), 241ff., cited in Needham, *Science and Civilisation in China*, 557.

81. Peter Valenti, "Leibniz, Papin and the Steam Engine: A Case Study of British Sabotage of Science," *American Almanac* (1996), http://members.tripod.com/~american_almanac/papin.htm.

82. Martin van Creveld, "The Rise and Fall of Military Technology," *Science in Context* 7, no. 2 (1994): 332.

83. Kelly, *Gunpowder*, 122–23, 218; Clark, "Vertical Fire," 370.

84. Greta Thunberg, "'Our House Is on Fire': Greta Thunberg, 16, Urges Leaders to Act on Climate," *The Guardian*, January 25, 2019, https://www.theguardian.com/environment/2019/jan/25/our-house-is-on-fire-greta-thunberg16-urges-leaders-to-act-on-climate.

85. Andrade, *The Gunpowder Age*, 45.

86. Walter Benjamin, *Illuminations* (New York: Harcourt, 1968), 84; see also Clark "Vertical Fire," 370.

87. Jan Zalasiewicz, Colin Waters, and Mark Williams, "Human Bioturbation, and the Subterranean Landscape of the Anthropocene," *Anthropocene* 6 (2014): 3–9.

88. Robert Boyle, *Some Considerations Touching the Usefulnesse of Experimental Natural Philosophy*, 2nd ed. (Oxford: Hall & Davis, 1663), pt. 2, essay 5, 14, cited in Needham, *Science and Civilisation in China*, 538.

89. Needham, *Science and Civilisation in China*, 537.

90. Kelly, *Gunpowder*, 229.

91. See Zalasiewicz et al., "Human Bioturbation," 3–4.

92. See Kyle Whyte, "Indigenous Climate Change Studies: Indigenizing Futures, Decolonizing the Anthropocene," *English Language Notes* 55, nos. 1–2 (2017): 153–62.

93. Jacques Derrida, *Writing and Difference* (London: Routledge, 1978), 98.

Chapter Thirteen

1. Hans Joachim Schellnhuber, "Paul Josef Crutzen: Ingeniousness and Innocence," *Proceedings of the National Academy of Sciences* 118, no. 17 (2021), https://doi.org/10.1073/pnas.2104891118.

2. Deborah R. Coen, *Climate in Motion: Science, Empire, and the Problem of Scale* (Chicago: University of Chicago Press, 2018); Perrin Selcer, *The Postwar Origins of the Global Environment: How the United Nations Built Spaceship Earth* (New York: Columbia University Press, 2018); H. A. Curry et al., eds., *Worlds of Natural History* (Cambridge: Cambridge University Press, 2018).

3. Robert A. Davis, "Inventing the Present: Historical Roots of the Anthropocene," *Earth Sciences History* 30: 1 (2011): 63–84.

4. Alison Bashford, "The Anthropocene is Modern History: Reflections of Climate and Australian Deep Time," *Australian Historical Studies* 44, no. 3 (2013): 341–49.

5. I am thankful for a Mellon New Directions Fellowship, which granted me a year to "retrain" in the paleosciences, and to generous colleagues at the University of Michigan for allowing me to participate in their labs, especially Daniel Fisher, Naomi Levin, Benjamin Passey, Raven Garvey, Michael Cherney, Ethan Shirley, and Lauren Pratt.

6. These critiques have merit, particularly of deterministic climate models and Anthropocene discourse. Mike Hulme, "Reducing the Future to Climate: A Story of Climate Determinism and Reductionism," *Osiris* 26, no. 1 (2011): 245–66; Kyle Whyte, "Indigenous Science (Fiction) for the Anthropocene: Ancestral Dystopias and Fantasies of Climate Change Crises," *Environment and Planning E: Nature and Space* 1, nos. 1–2 (2018): 224–42.

7. Geoff Bailey, "Time Perspectives, Palimpsests and the Archaeology of Time," *Journal of Anthropological Archaeology* 26, no. 2 (2007): 198–223.

8. Daniel C. Fisher, "Paleobiology of Pleistocene Proboscideans," *Annual Review of Earth and Planetary Sciences* 46 (2018): 229–60.

9. Richard B. Alley, *The Two-Mile Time Machine: Ice Cores, Abrupt Climate Change, and Our Future* (Princeton, NJ: Princeton University Press, 2000).

10. Jermey Shakun et al., "Global Warming Preceded by Increasing Carbon Dioxide Concentrations during the Last Deglaciation," *Nature* 484, no. 7392 (2012): 53.

11. Paul N. Edwards, *A Vast Machine: Computer Models, Climate Data, and the Politics of Global Warming* (Cambridge, MA: MIT Press, 2010).

12. K. M. Cohen et al., The ICS International Chronostratigraphic Chart, *Episodes* 36, no. 3 (2013; updated 2016–04): 199–204; Felix M. Gradstein et al., *The Geologic Time Scale 2012*,

2 vols. (Amsterdam: Elsevier, 2012); on the history of graphic representations of time, see Daniel Rosenberg and Anthony Grafton, *Cartographies of Time: A History of the Timeline* (Princeton, NJ: Princeton University Press, 2010).

13. Martin Rudwick, *Earth's Deep History: How It Was Discovered and Why It Matters* (Chicago: University of Chicago Press, 2014), 10; emphasis in original.

14. Gradstein et al., "Introduction," *Geologic Time Scale 2012*, 1.

15. Arthur Holmes, *The Age of the Earth* (London: Harper & Brothers, 1913), 18, 175; Joe D. Burchfield, *Lord Kelvin and the Age of the Earth* (London: Macmillan, 1975).

16. G. K. Gilbert, "Rhythms of Geologic Time," *Science* 11, no. 287 (1900): 1001–12; for a more comprehensive discussion of radiometric dating of Quaternary time, see Emily M. Kern, "Earth Time, Ice Time, Species Time," chapter 11 of this volume.

17. W. B. Wright, *The Quaternary Ice Age* (London: Macmillan, 1914).

18. Gerard de Geer, "A Geochronology of the Last 12,000 Years," *Compte Rendu de la XI Session du Congress Geologique International* (Stockholm: P. A. Norstedt & Söner, 1912), 241; E. B. Bailey, "Gerard Jacob de Geer, 1853–1943," *Obituary Notice Fellows of the Royal Society* 4 (1943): 475–81.

19. Steffan Bergwik, "Synchronizing Nature and Culture: Mediating Time in Geochronology and Dendrochronology, 1900–1945," in *Times of History, Times of Nature: Temporalization and the Limits of Modern Knowledge*, ed. Anders Ekström and Staffan Bergwik (New York: Berghahn, 2022), 230–56.

20. Gerard de Geer, "A Thermographical Record of the Late-Quaternary Climate," *Die Veränderungen des Klimas Seit dem Maximum der Letzten Eiszeit*, Dem Execkutivkomitee des 11 Interantionalen Geologenkongresses (Stockholm, 1910), 307.

21. De Geer, "A Thermographical Record," 307.

22. De Geer, "A Geochronology," 252, 253.

23. On the cultural significance of Gondwana's glaciation, see Pratik Chakrabarti, *Inscriptions of Nature: Geology and the Naturalization of Antiquity* (Baltimore: Johns Hopkins University Press, 2020).

24. Gerard de Geer, "Equatorial Palaeolithic Varves in East Africa: Measured in 1929 and 1933 by Erik Nilsson, Teleconnected with the Swedish Time Scale," *Geografiska Annaler* 16, nos. 2–3 (1934): 85; Gerard de Geer, "Geology and Geochronology," *Geografiska Annaler* 16, no. 1 (1934): 1–52; Gerard de Geer, *Geochronologia Suecica Principles* (Stockholm: Almquvist & Wiksells Boktryckeri-A.-B., 1940).

25. E. Antevs, *Late Glacial Correlations and Ice Recession in Manitoba* (Ottawa: Canadian Geological Survey, Mem. 168, 1931).

26. Nils-Axel Mörner, ed., "The Pleistocene/Holocene Boundary: A Proposed Boundary-Stratotype in Gothenburg, Sweden," *Boreas* 5 (1976): 199; V. Nordmann, "Post-glacial Climatic Changes in Denmark," in *Die Veränderungen des Klimas Seit dem Maximum der Letzten Eiszeit* (Stockholm: Generalstabens Litografiska Anstalt, 1910), 313–28.

27. Mörner, "The Pleistocene/Holocene Boundary."

28. Mike Walker et al., "Formal Definition and Dating of the GSSP (Global Stratotype Section and Point) for the Base of the Holocene Using the Greenland NGRIP Ice Core, and Selected Auxiliary Records," *Journal of Quaternary Science* 24, no. 1 (2009): 10.

29. Ronald E. Doel et al., "Strategic Arctic Science: National Interests in Building Natural Knowledge—Interwar Era through the Cold War," *Journal of Historical Geography* 44 (April 2014): 60–80. Calculating deep ice core ages is more complicated than I imply here, involving

skilled microscopic interpretation, isotope chemistry, conductivity tests, temperature measurements, and modeling; see Alley, *The Two-Mile Time Machine.*

30. Sverker Sörlin and Eric Paglia, *The Human Environment: Stockholm and the Rise of Global Environmental Governance* (Cambridge: Cambridge University Press, forthcoming).

31. De Geer, "A Thermographical Record," 307.

32. Anna L. C. Hughes et al., "The Last Eurasian Ice Sheets: A Chronological Database and Time-Slice Reconstruction DATED-1," *Boreas* 45 (January 2016): 1; April S. Dalton, "An Updated Radiocarbon-Based Ice Margin Chronology for the Last Deglaciation of the North American Ice Sheet Complex," *Quaternary Science Reviews* 234 (April 2020): 106–223.

33. J. A. Smith et al., "Early Local Last Glacial Maximum in the Tropical Andes," *Science* 308, no. 5722 (2005): 678–81.

34. Christopher R. Maupin, "Abrupt Southern Great Plains Thunderstorm Shifts Linked to Glacial Climate Variability," *Nature Geoscience* 14 (May 2021): 396–401.

35. Jeremy D. Shakun and Anders E. Carlson, "A Global Perspective on Last Glacial Maximum to Holocene Climate Change," *Quaternary Science Reviews* 29 (2010): 1809.

36. Timothy M. Shanahan et al., "The Time-Transgressive Termination of the African Humid Period," *Nature Geoscience* 8, no. 2 (January 2015): 140–44; Will Steffen et al., *Global Change: Planet under Pressure* (Berlin: Springer, 2004), 86; John E. Kuzbach et al., "African Climate Response to Orbital and Glacial Forcing in 140,000-y Simulation with Implications for Early Modern Human Environments," *Proceedings of the National Academy of Sciences of the United States of America* 117, no. 5 (4 February 2020): 2255–64.

37. Bashford, "The Anthropocene Is Modern History"; Ann C. McGrath and Mary Anne Jebb, eds., *Long History, Deep Time: Deepening Histories of Place* (Canberra: Australian National University Press, 2015).

38. H. E. Wright Jr. et al., eds., *Global Climates since the Last Glacial Maximum* (Minneapolis: University of Minnesota Press, 1993).

39. Richard A. Watson and Herbert E. Wright Jr., "The End of the Pleistocene: A General Critique of Chronostratigraphic Classification," *Boreas* 9, no. 3 (1980): 154, 159.

40. Hollis T. Hedberg, *International Stratigraphic Guide: A Guide to Stratigraphic Classification, Terminology, and Procedure* (New York: John Wiley & Sons, 1976), 76, 73.

41. Rhodes W. Fairbridge, "The Pleistocene-Holocene Boundary," *Quaternary Science Reviews* 1 (1983): 239.

42. Fairbridge, "Pleistocene-Holocene Boundary," 217.

43. Maureen H. Walczack, "Phasing of Millennial Scale Climate Variability in the Pacific and Atlantic Oceans," *Science* 370 (2020): 717.

44. John M. Jaeger and Amelia E. Shevennell, "Steering Iceberg Armadas: The Asian Pacific Tropics Likely Instigated Millennial-Scale Climate Changes," *Science* 370 (2020): 662–63; Alessandro Antonello and Mark Carely, "Ice Cores and the Temporalities of the Global Environment," *Environmental Humanities* 9, no. 2 (2017): 181–203.

45. Robin Sutcliffe Allan, "Geological Correlation and Paleoecology," *GSA Bulletin* 59, no. 1 (January 1948): 5, 6.

46. T. Wayland Vaughan, "Ecology of Modern Marine Organisms with Reference to Paleogeography," *Geological Society of America, Bulletin* 51 (1940): 433–68.

47. Frederic E. Clements, *Plant Succession: An Analysis of the Development of Vegetation* (Washington, DC: Carnegie Institution of Washington, 1916).

48. Von R. Sernander, "Die schwedischen Torfmoore als Zeugen postlazialer Klimaschwan-

kungen," and Gunnar Andersson, "Swedish Climate in the Late-Quaternary Period," in *Die Veränderungen des Klimas Seit dem Maximum der Letzten Eiszeit* (Stockholm: Eleventh International Geological Congress, 1910), 175–294.

49. Melissa Charenko, "Reconstructing Climate: Paleoecology and the Limits of Prediction during the 1930s Dust Bowl," *Historical Studies in the Natural Sciences* 50, nos. 1–2 (2020): 90–128.

50. Holmes, *Age of the Earth*, 176.

51. The inaugural issues of journals devoted to the Quaternary and Holocene celebrate their fundamental interdisciplinarity. See A. L. Washburn, "Editorial: Interdisciplinary Quaternary Research and Environmental History," *Quaternary Research: An Interdisciplinary Journal* 1, no. 1 (1970): 1–2; John A. Matthews, "Editorial," *The Holocene* 1, no. 1 (1991): i–ii.

52. Amos Salvador, ed., *International Stratigraphic Classification, Terminology, and Procedure*, 2nd ed. (Boulder, CO: Geological Society of America and International Union of Geological Sciences, 1994), 7.

53. Gradstein et al., *Geologic Time Scale 2012*, 980.

54. Richard A. Kerr, "A Time War over the Period We Live In," *Science* 319, no. 5862 (January 25, 2008): 402–3.

55. Jun Hu et al., "Assessing Proxy System Models of Cave Dripwater $\delta^{18}O$ Variability," *Quaternary Science Reviews* 254 (2021): 106799.

56. Bailey, "Time Perspectives" 203, 204.

57. Nigel Clark and Bronislaw Szerszynski, *Planetary Social Thought: The Anthropocene Challenge to the Social Sciences* (Cambridge: Polity, 2021); Helge Jordheim, "Introduction: Multiple Times and the Word of Synchronization," *History and Theory* 53, no. 4 (December 2014): 498–518.

58. David Turnbull, "Territorializing/Decolonizing South American Prehistory: Pedra Furada and the Cerutti Mastodon," *Tapuya: Latin American Science, Technology and Society* 2, no. 1 (2019): 137.

59. Julia Kelson, personal communication, June 21, 2021; Kelson's metaphor is an explicit statement of paleoclimatological historicity, as defined in Francis Hartog, *Regimes of Historicity: Presentism and Experiences of Time*, translated by Saskia Brown (New York: Columbia University Press, 2015).

60. Johan Rockström et al., "A Safe Operating Space for Humanity," *Nature* 461 (2009): 472–75.

61. Erle C. Ellis et al., "People Have Shaped Most of Terrestrial Nature for at Least 12,000 Years," *Proceedings of the National Academy of Sciences* 118, no. 17 (2021), e2023483118; Kathleen D. Morrison, "Provincializing the Anthropocene: Eurocentrism in the Earth System," in *At Nature's Edge: The Global Present and Long-Term History*, ed. Gunnel Cederlöf and Mahesh Rangarajan (New Delhi: Oxford University Press, 2018).

62. Kenneth Pomeranz, "Teleology, Discontinuity and World History: Periodization and Some Creation Myths of Modernity," *Asian Review of World Histories* 1, no. 2 (2013): 189–226.

63. Sebastiàn Ureta, Thomas Lekan, and W. Graf von Hardenberg, "Baselining Nature: An Introduction," *Environment and Planning E: Nature and Space* 3, no. 1 (2020): 16.

64. Ureta et al., "Baselining Nature," 15.

65. H. J. Schellnhuber, "'Earth System' Analysis and the Second Copernican Revolution," *Nature* 402, no. S6761 (1999): C19–23; Clive Hamilton and Jacques Grinevald, "Was the Anthropocene Anticipated?," *The Anthropocene Review* 2, no. 1 (2015): 59–72.

Chapter Fourteen

1. Dipesh Chakrabarty, "The Climate of History: Four Theses," *Critical Inquiry* 35, no. 2 (Winter 2009): 197–222.

2. Paul J. Crutzen, "Geology of Mankind," *Nature* 415 (2002): 23.

3. Andreas Malm and Alf Hornborg, "The Geology of Mankind? A Critique of the Anthropocene Narrative," *The Anthropocene Review* 1, no. 1 (2014): 62–69. This critique was evident from the first formulations of the concept, when Paul Crutzen, the popularizer of the Anthropocene, estimated that only a quarter of the world's population had burned enough fossil fuels, fixed enough nitrogen, clear-cut enough forests, and more generally affected earth's processes in ways akin to geological forces.

4. Rob Nixon, "The Anthropocene: The Promise and Pitfalls of an Epochal Idea," in *Future Remains: A Cabinet of Curiosities for the Anthropocene*, ed. Gregg Mitman, Marco Armiero, and Robert S. Emmett (Chicago: University of Chicago Press, 2017), 1–18.

5. Jason Moore, *Capitalism in the Web of Life* (New York: Verso, 2015).

6. Danielle Sands, "Gaia, Gender, and Sovereignty in the Anthropocene," *philoSOPHIA* 5, no. 2 (Summer 2015): 287–307; Laura Pulido, "Racism and the Anthropocene," in *Future Remains: A Cabinet of Curiosities for the Anthropocene*, ed. Gregg Mitman, Marco Armiero, and Robert S. Emmett (Chicago: University of Chicago Press, 2018): 116–28; Janae Davis, Alex A. Moulton, Levi Van Sant, and Brian Williams, "Anthropocene, Capitalocene, . . . Plantationocene? A Manifesto for Ecological Justice in an Age of Global Crises," *Geography Compass* 13, no. 5 (May 2019): e12438.

7. Pratik Chakrabarti, "Gondwana and the Politics of the Deep Past," *Past & Present* 242, no. 1 (February 2019): 119–53.

8. Pulido, "Racism and the Anthropocene," 116.

9. Kevin J. Francis, "'Death Enveloped All Nature in a Shroud': The Extinction of Pleistocene Mammals and the Persistence of Scientific Generalists" (PhD diss, University of Minnesota, 2002), 16–112.

10. Paul S. Martin, "Pleistocene Ecology and Biogeography of North America," in *Zoogeography*, ed. Carl L. Hubbs, The American Association for the Advancement of Science 51 (Baltimore: Horn-Shafer Company, 1958): 413.

11. Discoveries of projectile points at Clovis, NM, in 1927 and Folsom, NM, in 1929 extended human antiquity in the Americas to the Pleistocene, see David J. Meltzer, *The Great Paleolithic War: How Science Forged an Understanding of America's Ice Age Past* (Chicago: University of Chicago Press, 2015). On early versions of the Bering Strait hypothesis, see Ernst Antevs, "The Spread of Aboriginal Man to North America," *Geographical Review* 25, no. 2 (April 1935): 302–9.

12. Martin, "Pleistocene Ecology," 413.

13. Paul S. Martin, "Africa and Pleistocene Overkill," *Nature* 212 (October 1966): 339–42.

14. Walter Sullivan, "'Overkill' of Animals Laid to Huntsmen in 9000 B.C.," *The New York Times*, February 13, 1972, 62; Paul S. Martin to Steven P. Christman, June 25, 1985, Paul S. Martin Papers, MS442, box 2, folder 7, Special Collections, University of Arizona Library (hereafter cited as Martin MSS).

15. James E. Mossiman and Paul S. Martin, "Simulating Overkill by Paleoindians," *American Scientist* 63, no. 3 (May–June 1975): 304–16.

16. Donald K. Grayson and David J. Meltzer, "A Requiem for North American Overkill," *Journal of Archaeological Science* 30, no. 5 (2003): 590.

17. Steven P. Christman to Paul S. Martin, June 18, 1985, Martin MSS, box 2, folder 7.

18. Erika Lorraine Milam, *Creatures of Cain: The Hunt for Human Nature in Cold War America* (Princeton, NJ: Princeton University Press, 2019), esp. 79–124. See also Matt Cartmill, *A View of Death in the Morning: Hunting and Nature through History* (Cambridge, MA: Harvard University Press, 1993), esp. 1–14, 189–210.

19. Richard Barry Lee and Irven DeVore, eds., *Man the Hunter* (New Brunswick, NJ: Aldine Transaction, 1968).

20. Sherwood L. Washburn and C. S. Lancaster, "The Evolution of Hunting," in *Man the Hunter*, ed. Richard Barry Lee and Irven Devore (New Brunswick, NJ: Aldine Transaction, 1968), 303.

21. Washburn and Lancaster, "The Evolution of Hunting," 299.

22. L. Binford et al., "Primate Behavior and the Evolution of Aggression," in *Man the Hunter*, ed. Richard Barry Lee and Irven Devore (New Brunswick, NJ: Aldine Transaction, 1968), 339–44.

23. Martin had used the term "Paleo-Indian" to describe the hunters as early as 1963. See Paul S. Martin, *The Last 10,000 Years: Fossil Pollen Study of the American Southwest* (Tucson: University of Arizona Press, 1963).

24. Paul S. Martin, "The Discovery of America," *Science* 179, no. 4077 (March 1973): 972.

25. In 1984, Stephen Whittington and Bennett Dyke created new models "with different values more likely to be acceptable to human ecologists and archeologists." Their bibliography includes critiques from the 1970s. Martin similarly summarized these critiques in a 1982 letter. See Stephen L. Whittington and Bennett Dyke, "Simulating Overkill: Experiments with the Mosimann and Martin Model," in *Quaternary Extinctions: A Prehistoric Revolution*, ed. Paul S. Martin and Richard G. Klein (Tucson, AZ: University of Arizona Press, 1984): 451–65; Paul S. Martin to Marjolaine Boutin-Sweet, September 7, 1982, Martin MSS, box 2, folder 7.

26. Mosimann and Martin, "Simulating Overkill by Paleoindians," 104–13.

27. Mosimann and Martin, "Simulating Overkill by Paleoindians," 311.

28. "Classes Shortened, Symposium Opens," *The Idaho Argonaut* 78, no. 44, (March 14, 1969): 1; "Symposium Available on Television," *Idahonian*, March 13, 1969.

29. "Martin and Wallrich Advocate Non-violence," *The Idaho Argonaut* 78, no. 44 (March 14, 1969), 6.

30. Ladd Hamilton, "Non-violence Plea Coupled with Theory Man 'Killer Ape,'" *Lewiston Morning Tribune*, March 14, 1969.

31. See, for example, "Ice-Age Man Guilty of 'Overkill' of Large Mammals, Scientist Says," *The Toledo Times*, February 15, 1972; "Man Always a Natural Killer," *Syracuse Herald Tribune*, March 4, 1972; "Idea that Man Killed Off Mammoths Gaining Favor," *Chicago Tribune*, March 3, 1972.

32. "Man Wiped Our Giant Beasts, Scientist Says," *Los Angeles Times*, March 19, 1972.

33. Lisa Nagaoka, Torben Rick, and Steve Wolverton, "The Overkill Model and Its Impact on Environmental Research," *Ecology and Evolution* 8, no. 19 (October 2018): 9683–96.

34. Elizabeth Kolbert, *The Sixth Extinction: An Unnatural History* (New York: Henry Holt and Company, 2014), 229–30; Yuval Noah Harari, *Sapiens: A Brief History of Humankind* (New York: HarperCollins, 2015); J. R. McNeil and Peter Engelke, *The Great Acceleration: An Environmental History of the Anthropocene since 1945* (Cambridge, MA: The Belknap Press of Harvard University Press, 2014), 208.

35. Nagaoka et al., "Overkill Model and Its Impact"; Donald K. Grayson and David J. Meltzer, "Clovis Hunting and Large Mammal Extinction: A Critical Review of the Evidence," *Journal of*

World Prehistory 16, no. 4 (2002): 313–59; Donald K. Grayson and David J. Meltzer, "Revisiting Paleoindian Exploitation of Extinct North American Mammals," *Journal of Archaeological Science* 56 (April 2015): 177–96; David J. Meltzer, "Pleistocene Overkill and North American Mammalian Extinctions," *Annual Review of Anthropology* 44 (October 2015): 33–53; Jennifer Raff, *Origin: A Genetic History of the Americas* (New York: Twelve, 2022).

36. Grover S. Krantz, "Human Activities and Megafaunal Extinctions," *American Scientist* 58, no. 2 (1970): 166.

37. Robert L. Kelly and Mary M. Prasciunas, "Did the Ancestors of Native Americans Cause Animal Extinctions in Late-Pleistocene North America?," in *Native Americans and the Environment: Perspectives on the Ecological Indian*, ed. Michael E. Harkin and David Rich Lewis (Lincoln, NB: University of Nebraska Press, 2007): 104–5. On archaeologists' reluctance to accept human hunting as the sole cause of the extinctions, see Nagaoka et al., "Overkill Model and Its Impact."

38. Paul S. Martin to David Burney, December 23, 1986, Martin MSS, box 2, folder 2.

39. Martin, "The Discovery of America," 969; italics in original.

40. Shepard Krech III, *The Ecological Indian: Myth and History* (New York: W.W. Norton, 1999), 36.

41. Grayson and Meltzer, "Requiem for North American Overkill," 585. Others have critiqued what they see as Grayson and Meltzer's "dogmatic dismissal of competing scientific hypotheses." See Stuart Fiedel and Gary Haynes, "A Premature Burial: Comments on Grayson and Meltzer's 'Requiem for overkill,'" *Journal of Archaeological Science* 31 (2004): 121–31.

42. Brian W. Dippie, *The Vanishing American: White Attitudes and US Indian Policy* (Middletown, CT: Wesleyan University Press, 1982); Berry Brewton, "The Myth of the Vanishing Indian," *Phylon* 21 (1960): 51–57; Martin Barker and Roger Sabin, *The Lasting of the Mohicans: History of an American Myth* (Jackson, MS: University of Mississippi Press, 1995).

43. Kent Redford, "The Ecologically Noble Savage," *Cultural Survival Quarterly* 15 (1991): 46–48.

44. Finis Dunaway, "Gas Masks, Pogo, and the Ecological Indian: Earth Day and the Visual Politics of American Environmentalism," *American Quarterly* 60, no. 1 (2008): 67–99; Finis Dunaway, *Seeing Green: The Use and Abuse of American Environmental Images* (Chicago: University of Chicago Press, 2015), 79–95.

45. Paul Nadasdy, "Transcending the Debate over the Ecologically Noble Indian: Indigenous Peoples and Environmentalism," *Ethnohistory* 52, no. 2 (2005): 291–331.

46. Diamond frames his interests in history and geography around World War II. He remembers tracking army movements and seeing images of concentration camps as a boy. He then studied in London in the 1950s, where he saw signs of the blitz. See Jared Diamond, "About Me," accessed June 5, 2021, http://www.jareddiamond.org/Jared_Diamond/About_Me.html.

47. Jared M. Diamond, "Man the Exterminator," *Nature* 298 (1982): 789. See also Jared M. Diamond, "Extinctions, Catastrophic and Gradual," *Nature* 304 (1983): 396–97.

48. Jared M. Diamond, "The American Blitzkrieg: A Mammoth Undertaking," *Discover* (June 1987): 82.

49. Vine Deloria Jr., *Custer Died for Your Sins: An Indian Manifesto* (Norman: University of Oklahoma Press, 1969).

50. Vine Deloria Jr., *Red Earth, White Lies: Native Americans and the Myth of Scientific Fact* (New York: Scribner, 1995), 112.

51. Diamond, "The American Blitzkrieg," 84.

52. Deloria, *Red Earth, White Lies*, 128–29.

53. Jack D. Forbes, *Apache, Navaho, and Spaniard* (Norman: University of Oklahoma Press, 1960); Deloria, *Custer Died for Your Sins*, 1–2.

54. Jennifer Nez Denetdale, "Planting Seeds of Ideas and Raising Doubts about What We Believe: An Interview with Vine Deloria, Jr.," *Journal of Social Archaeology* 4, no. 2 (2004): 137.

55. Deloria quoted in Nez Denetdale, "Planting Seeds of Ideas," 133.

56. Deloria quoted in Nez Denetdale, "Planting Seeds of Ideas," 140.

57. Deloria, *Red Earth, White Lies*, 19.

58. Deloria, *Red Earth, White Lies*, 140.

59. Deloria, *Red Earth, White Lies*, 128.

60. George Nicholas, "When Scientists 'Discover' What Indigenous People Have Known For Centuries," *Smithsonian Magazine*, February 21, 2018, https://www.smithsonianmag.com/science-nature/why-science-takes-so-long-catch-up-traditional-knowledge-180968216/; Rick Budhwa, "Correlations between Catastrophic Paleoenvironmental Events and Native Oral Traditions of the Pacific Northwest" (PhD diss., Simon Fraser University, 2002); Patrick Nunn, *The Edge of Memory: Ancient Stories, Oral Tradition and the Post-glacial World* (London: Bloomsbury, 2018).

61. Krech similarly dismissed Deloria and other native thinkers who were "invested in the Ecological Indian" but who wrote historically framed inventories of Europeans. Krech, *The Ecological Indian*, 303. See also Michael G. Doxtater, "Indigenous Knowledge in the Decolonial Era," *American Indian Quarterly* 28, no. 3/4 (Summer–Autumn 2004): 618–33.

62. Amos Esty, "An Interview with Paul S. Martin," *American Scientist*, 2009, accessed May 19, 2019, https://www.americanscientist.org/bookshelf/pub/an-interview-with-paul-s-martin (web page no longer available).

63. In Martin's 2005 book, he presented different views: he said that blaming particular groups was not helpful. Instead, he claimed that "the proposal that near-time extinctions in some critical way involve people, our species, *Homo sapiens*, requires at least a modicum of cultural sensitivity. Certainly no one can pass judgment, from long after the fact, on the peoples who first discovered and inhabited new lands. Their achievements are truly remarkable. It is one thing to note synchronicity of arrival of first pioneering prehistoric people in various corners of the planet and the concurrent extinction of many native animals; it is another to make a judgment. It would be absurd to assign blame to the progeny of Paleolithic Europeans or of the First Americans for the extinction of the Old World or New World mammoths, to Australian Aborigines for the end of the diprotodonts, or to the New Zealand Maoris for eliminating the moa. It is important to remember that the extinctions of near time occurred worldwide. To the extent that responsibility is assigned, it belongs to our species as a whole. This may be an even more disturbing thought for many." Paul S. Martin, *Twilight of the Mammoths: Ice Age Extinctions and the Rewilding of America* (Berkeley: University of California Press, 2007), 54.

64. Gesa Mackenthun, "Sacred Pact or Overkill? Human-Bison Relations in North American Mythologies," in *An Eclectic Bestiary: Encounters in a More-than-Human World*, ed. Birgit Spengler and Babette B. Tischleder (Bielefeld: transcript Verlag, 2019): 199.

65. Philip J. Deloria, *Playing Indian* (New Haven, CT: Yale University Press, 1998).

66. Douglas Fisher, "The Myth of the Ecological Indian," *Toronto Sun*, February 23, 2000.

Chapter Fifteen

1. Ann Stoler, "Imperial Debris: Reflections on Ruins and Ruination," *Cultural Anthropology* 23, no. 2 (2008): 194.

2. Carl Martius, *Von dem Rechtzustande unter den Ureinwohnern Brasiliens. Eine Abhandlung* (Munique/Leipzig: Friedrich Fleischer, 1832).

3. Stoler, "Imperial Debris," 193.

4. Raphael Uchôa, "Contextualizing the 'American Race' in the Atlantic: The Case of Carl von Martius and His German and Iberian Sources," *Lychnos: An Annual for History of Ideas and Science* (2019): 91–109.

5. Uchôa, "Contextualizing the 'American Race.'"

6. Stoler, "Imperial Debris," 195.

7. See, for example, Manuela Carneiro, ed. *História dos índios no Brasil* (São Paulo: Companhia das Letras, 1992).

8. Davi Kopenawa and Bruce Albert, *A queda do céu: Palavras de um xamã yanomami* (São Paulo: Cia das Letras, 2019), 24.

9. Ailton Krenak, *Ideas to Postpone the End of the World*, trans. Anthony Doyle (Toronto: House of Anansi Press, 2020); Ailton Krenak, *Life Is Not Useful*, trans. Alex Brostoff and Jamille Pinheiro Dias (Cambridge: Polity, 2023); Davi Kopenawa and Bruce Albert, *The Falling Sky: Words of a Yanomami Shaman*, trans. Nicholas Elliot and Alison Dundy (Cambridge, MA: Harvard University Press, 2013).

10. In this chapter, "Amerindians" refers to "Indigenous peoples of the Americas." Eduardo Viveiros de Castro, *Metafísicas Canibais: Elementos para uma Antropologia Pós-estrutural* (São Paulo: Cosac Naify, 2015), 72.

11. Ailton Krenak, *Ideias Para Adiar O Fim Do Mundo* (São Paulo: Cia Das Letras, 2020).

12. See, for example, Staffan Muller-Wille and Christina Brandt, eds. *Heredity Explored: Between Public Domain and Experimental Science, 1850–1930* (Cambridge, MA: MIT Press, 2016); Pratik Chakrabarti, *Inscriptions of Nature: Geology and the Naturalization of Antiquity* (Baltimore: Johns Hopkins University Press, 2020); Sadiah Qureshi, "Dying Americans: Race, Extinction and Conservation in the New World," in *From Plunder to Preservation: Britain and the Heritage of Empire, 1800–1950*, ed. Astrid Swenson and Peter Mandler (Oxford: Oxford University Press, 2013), 269–88.

13. Carl Martius, "O Estado do Direito entre os Autóctones do Brasil," *Instituto Histórico e Geográfico de São Paulo* 11 (1907): 63.

14. Martius, "Estado do Direito," 63.

15. Uchôa, "Contextualizing the 'American Race,'" 91–94.

16. Uchôa, "Contextualizing the 'American Race,'" 91–94.

17. See, for example, James Tracy, *The Political Economy of Merchant Empires: State Power and World Trade, 1350–1750* (Cambridge: Cambridge University Press, 1997).

18. See Marcelo Basile, "Império Brasileiro: Panorama Político," in *História Geral do Brasil*, ed. Maria Linhares (Rio de Janeiro: Elsevier, 2000), 175–291.

19. Carl Martius, *Viagem pelo Brasil: 1817–1820*, vol. 3, trans. L. Lahmeyer (São Paulo: Edusp, [1831] 1981), 320.

20. Carl Martius, *Frei Apolônio: Um Romance do Brasil*, trans. Erwin Theodor (São Paulo: Brasiliense, [1831] 1992), 92.

21. Martius, "Estado do Direito," 61.

22. See Martin Rudwick, *Worlds Before Adam: The Reconstruction of Geohistory in the Age*

of Reform (Chicago: University of Chicago Press, 2008); Paolo Rossi, *The Dark Abyss of Time: The History of the Earth & the History of Nations from Hooke to Vico* (Chicago: University of Chicago, 1984).

23. This quotation and preceding ones, Martius, "Estado do Direito," 63.

24. Martius, "Estado do Direito," 63.

25. Martius, "Estado do Direito," 64.

26. Martius, "Estado do Direito," 65.

27. Robert Richards, *The Romantic Conception of Life: Science and Philosophy in the Age of Goethe* (Chicago: University of Chicago Press, 2002), 221.

28. Phillip Sloan, "The Gaze of Natural History," in *Inventing Human Science: Eighteenth-Century Domains*, ed. Christopher Fox, Roy Porter, and Robert Wokler (Berkeley: University of California Press, 1995), 136.

29. Sloan, "The Gaze of Natural History," 135–36.

30. See, for example, Dain Borges, "'Puffy, Ugly, Slothful and Inert': Degeneration in Brazilian Social Thought, 1880–1940," *Journal of Latin American Studies* 25, no. 2 (1993): 235–56; Philip L. Kohl, Irina Podgorny, and Stefanie Gänger, eds. *Nature and Antiquities: The Making of Archaeology in the Americas* (Tucson: The University of Arizona Press, 2014).

31. Edward Tylor, "The Degeneracy of Man," *Nature* 10, no. 243 (1874): 146–47.

32. The first publication appeared in Carl Friedrich Philipp von Martius, *Von dem Rechtzustande unter den Ureinwohnern Brasiliens. Eine Abhandlung* (Leipzig: Friedrich Fleischer, 1832). A comprehensive summary of the text was published in English by Rev. George Cecil Renouard as "On the State of Civil and Natural Rights among the Aboriginal Inhabitants of Brazil," *The Journal of the Royal Geographical Society of London* 2 (January 1832): 191–227.

33. Tylor, "Degeneracy of Man," 147.

34. Martius, "Estado do Direito," 24.

35. See, for example, Robert Ginsberg, *The Aesthetics of Ruins* (Amsterdam: Rodopi, 2004).

36. Carl Martius, "A Ethnographia da America, Especialmente do Brasil," *Instituto Histórico e Geográfico de São Paulo* 9 (1905): 534–62.

37. Carl Martius, *Natureza, doenças, medicina e remédios dos índios Brasileiros* (São Paulo: Companhia Editora Nacional, 1939), 286.

38. Carl Martius, "The Natural History, the Diseases, the Medical Practice, and the Materia Medica of the Aborigines of Brazil," *Calcutta Journal of Natural History and Miscellany of the Arts and Sciences in India* 6 (1845): 1–33.

39. Samuel Morton, "Inquiry into the Distinctive Characteristics of the Aboriginal Race of America," *Calcutta Journal of Natural History and Miscellany of the Arts and Sciences in India* 6 (1845): 117–49.

40. Morton, "Inquiry into the Distinctive Characteristics," 124.

41. Morton, "Inquiry into the Distinctive Characteristics," 124.

42. See, for example, John Douglas Bishop, "Locke's Theory of Original Appropriation and the Right of Settlement in Iroquois Territory," *Canadian Journal of Philosophy* 27, no. 3 (1997): 311–37.

43. See, for example, Kavita Philip, "Imperial Science Rescues a Tree: Global Botanic Networks, Local Knowledge and the Transcontinental Transplantation of Cinchona," *Environment and History* 1, no. 2 (1995): 173–200.

44. Clements Markham, ed. *Expeditions into the Valley of the Amazons, 1539, 1540, 1639*, trans. Clements R. Markham (Cambridge: Cambridge University Press, 2010).

45. Markham, "Expeditions," lxiii.

46. Martius, "Ethnographia da America," 561.

47. Martius, "Ethnographia da America," 561.

48. See, for example, Anthony Pagden, *The Burdens of Empire: 1539 to the Present* (Cambridge: Cambridge University Press, 2015).

49. Martius, *Viagem*, 281.

50. Kopenawa and Albert, *The Falling Sky*, 347–48.

51. See, for example, Alan Robson Alexandrino Ramos et al. "Mercúrio nos Garimpos da Terra Indígena Yanomami e Responsabilidades," *Ambiente & Sociedade* 23 (2020): 1–22.

52. Bruce Albert, "O ouro canibal e a queda do céu: Uma crítica xamânica da economia política da natureza (Yanomami)," in *Pacificando o branco: Cosmologias do contato no norte-Amazônico*, ed. Alcida Rita Ramos and Bruce Albert (Marseille: IRD Éditions, 2018), 239–74.

53. Kopenawa and Albert, *The Falling Sky*, 349.

54. Ailton Krenak, *A vida não é útil* (São Paulo: Cia das Letras, 2020), 10.

55. Ailton Krenak, *Ideias Para Adiar O Fim Do Mundo* (São Paulo: Cia Das Letras, 2020), Kindle.

56. Viveiros de Castro, *Idées pour retarder la fin du monde* (Bellevaux: Dehors, 2020), 60.

57. Déborah Danowski and Eduardo Viveiros de Castro, *Há mundo por vir? Ensaio sobre os medos e o fins* (Santa Catarina: Cultura e Barbárie, 2017), 12.

58. Claude Lévi-Strauss, *Tristes trópicos* (São Paulo: Companhia das Letras, 1996), 51.

59. See, for example, da Cunha, ed., *História dos índios no Brasil.*

60. See Martius, "Ethnographia da America," 550.

61. See Martius, "Ethnographia da America," 559.

62. Krenak, *Ideias Para Adiar.*

Chapter Sixteen

1. Alfred Wegener, *Die Entstehung der Kontinente und Ozeane* (Braunschweig, Germany: Druck und Verlag von Friedr. Vieweg & Sohn, 1915); Mott T. Greene, *Alfred Wegener* (Baltimore: Johns Hopkins University Press, 2015).

2. Alexander du Toit, *Our Wandering Continents* (Edinburgh: Oliver & Boyd, 1937); Suryakanthie Chetty, *Africa Forms the Key: Alex Du Toit and the History of Continental Drift* (Basingstoke, UK: Macmillan, 2021).

3. Naomi Oreskes, *The Rejection of Continental Drift: Theory and Method in American Earth Science* (New York: Oxford University Press, 1999); Naomi Oreskes and Homer Le Grand, eds., *Plate Tectonics: An Insider's History of the Modern Theory of the Earth* (Boulder, CO: Westview Press, 2001); H. R. Frankel, *The Continental Drift Controversy*, 4 vols. (Cambridge: Cambridge University Press, 2012); see also, James A. Secord, "Global Geology and the Tectonics of Empire," in *Worlds of Natural History*, ed. Helen Anne Curry et al. (Cambridge: Cambridge University Press, 2018).

4. Sabin Zahirovic, "Gondwana break-up," https://www.earthbyte.org/gondwana-breakup-and-the-plate-tectonic-evolution-of-the-tethyan-oceans (with thanks to Sabin Zahirovic). See also GPlates, http://www.gplates.org/docs.html.

5. Comments on "Gondwanaland Breakup," Mawson's Huts Foundation, September 3, 2019, YouTube video, https://www.youtube.com/watch?v=s9gHLs7QeTw [video no longer available].

6. Sumathi Ramaswamy, *The Lost Land of Lemuria: Fabulous Geographies, Catastrophic Histories* (Berkeley: University of California Press, 2004).

7. Ramaswamy, *The Lost Land of Lemuria*, 1–2.

8. Marita T. Bradshaw et al., "Out of Gondwana," in *Shaping a Nation: The Geology of Australia*, ed. Richard Blewett (Canberra: ANU Press, 2012).

9. Alison Bashford, Pratik Chakrabarti, and Jarrod Hore, "Towards a Modern History of Gondwanaland," *Journal of the British Academy* 9, 6 (2021): 5–26.

10. An international team is researching and writing this modern history of Gondwanaland: Alessandro Antonello, Alison Bashford, Pratik Chakrabarti, Saul Dubow, and Jarrod Hore. See http://gondwanaland.net.

11. "I will at once adopt instead the name GONDWANA series or system, to be understood in the same wide sense as when we speak of the Jurassic or Silurian series or system. The name was proposed some years ago by Mr. Medlicott, and has since been more or less current on the survey; it has been once used in print by Mr. H.F. Blanford in his little work on the Physical Geology of India." Ottakar Feistmantel, "Notes on the Age of Some Fossil Floras in India," *Records of the Geological Survey of India* IX, part 2 (1876): 28. See also W. Blanford, "Dr Oskar Mantel's Paper on the Gondwana Series," *Geological Magazine* 4, no. 4 (1877): 189–90; Pratik Chakrabarti, "Gondwana and the Politics of Deep Past," *Past & Present* 242, no. 1 (2019): 119–53; A. Leviton and M. L. Aldrich, "Contributions of the Geological Survey of India, 1851–1890, to the Concept of Gondwanaland," *Earth Sciences History* 31, no. 2 (2012): 247–69; Pratik Chakrabarti, *Inscriptions of Nature: Geology and the Naturalization of Antiquity* (Baltimore: Johns Hopkins University Press, 2020); Robert A. Stafford, "Annexing the Landscapes of the Past: British Imperial Geology in the Nineteenth Century," in *Imperialism and the Natural World*, ed. John MacKenzie (Manchester: Manchester University Press, 1990).

12. Bhangya Bhukya, *The Roots of the Periphery: A History of the Gonds of Deccan India* (New Delhi: Oxford University Press, 2017); Asoka Kumar Sen, *Indigeneity, Landscape and History: Adivasi Self-Fashioning in India* (New Delhi: Routledge 2017); W. van Schendel, "The Dangers of Belonging: Tribes, Indigenous Peoples and Homelands in South Asia," in *The Politics of Belonging in India: Becoming Adivasi*, ed. Daniel J. Rycroft and Sangeeta Dasgupta (Abingdon: Routledge, 2011), 19–43.

13. Mayuri Patankar, "'Gondwana'/'Gondwanaland' as a Homeland of the Gonds: Storytelling in the Paintings of Gond Pilgrims," *Summerhill* 22, no. 2 (2016): 39–48.

14. Eyre Chatterton, *The Story of Gondwana* (London: Pitman, 1916). See also R. V. Russell and Rai Bahadur Hira Lāl, *The Tribes and Castes of the Central Provinces of India* (London: Macmillan and Co., 1916).

15. Patankar, "'Gondwana'/'Gondwanaland,'" 39–48.

16. Patankar, "'Gondwana'/'Gondwanaland,'" 43.

17. Akash Poyam [Akash K. Prasad], "Gondwana Movement in Post-colonial India: Exploring Paradigms of Assertion, Self-Determination and Statehood," in *Social Work in India: Tribal and Adivasi Studies—Perspectives from Within*, ed. bodhi s. r. (Kolkata: adivaani, 2016), 131–66; also cited as *Journal of Tribal Intellectual Collective India*, ser. 4, 3, no. 1 (September 21, 2017): 37–45.

18. Poyam, "Gondwana Movement in Post-colonial India"; V. R. Mandala, "The Making and Unmaking of the Gonds," *Global Environment* 10, no. 2 (2017): 421–81.

19. Indrajit Singh, *The Gondwana and the Gonds* (Lucknow, India: Universal Publishers, 1944), vi.

20. Sangeeta Dasgupta, "Adivasi Studies: From a Historian's Perspective," *History Compass* 16, no. 10 (2018): e12486.

21. Akash Poyam, "10 Things You Need to Know about 'Gondwana State' Demand," Adivasi

Resurgence, December 18, 2015, http://adivasiresurgence.com/2015/12/18/10-things-you-need-to-know-about-gondwana-state-demand/.

22. For Mukerjee's environmentally, even physiologically, determinist anticolonialism, see Alison Bashford, "Anti-colonial Climates: Physiology, Ecology, and Global Population, 1920s–50s," *Bulletin of the History of Medicine* 86 (2012): 595–626.

23. Radhakamal Mukerjee, "Foreword," in Singh, *The Gondwana and the Gonds*, i-iv.

24. Singh, *The Gondwana and the Gonds*, 2.

25. Singh, *The Gondwana and the Gonds*, 197–98.

26. Lahar Singh, *Gondwana: A Journey to the Centre of India* (Kolkata: Writers Workshop India, 2009).

27. "About the Author," Writers Workshop India, accessed September 7, 2021, https://www.writersworkshopindia.com/books/fiction/novel/gondwana-a-journey-to-the-centre-of-india/.

28. Tahir Shah, *Beyond the Devil's Teeth: Journeys in Gondwanaland* (London: Octagon Press, 1995), 4–7.

29. Tellurian [Charles Albert Fenner], "Gondwana Land," *The Australasian* (Melbourne), November 28, 1931, 40. Fenner was sailing to London to the centenary meeting of the British Association for the Advancement of Science. Thanks to Jarrod Hore for this reference.

30. Tellurian, "Gondwana Land," *The Australasian*, 40. For Lemuria and Australia, see Frank Bongiorno, "Aboriginality and Historical Consciousness: Bernard O'Dowd and the Creation of an Australian National Imaginary," *Aboriginal History* 24 (2000): 39–58.

31. Tellurian, "Gondwana Land," *The Australasian*, 40.

32. Nicholas Ruddick, *The Fire in the Stone: Prehistoric Fiction from Charles Darwin to Jean M. Auel* (Middletown, CT: Wesleyan University Press, 2009).

33. Martin J. S. Rudwick, *Earth's Deep History: How It Was Discovered and Why It Matters* (Chicago: University of Chicago Press, 2014).

34. Juliette Huxley, *Wild Lives of Africa* (New York: Harper & Row, 1963), 137–39.

35. J. B. Rowley, *Trapped in Gondwana* (Oklahoma City: Potoroo Press, 2013).

36. Arthur Conan Doyle, *The Lost World* (London: Hodder & Stoughton, 1912), chap. XI.

37. Adrian Desmond, *The Ape's Reflexion* (London: Quartet, 1979).

38. Craig Robertson, *Song of Gondwana* (Melbourne: Penguin, 1989).

39. "Nominations to the World Heritage List: Convention Concerning the Protection of the World Cultural and Natural Heritage," *Gondwana Rainforests of Australia: Advisory Body Evaluation (IUCN)*, UNESCO, April 1986, 20, https://whc.unesco.org/en/list/368/documents/.

40. "Gondwana Rainforests of Australia," *Department of the Environment, Water, Heritage and the Arts* (Coffs Harbour, Australia: NSW National Parks and Wildlife Service 2021), https://www.environment.gov.au/system/files/pages/0e4f8d74-98ce-45b4-be3f-e681f4c4e6fe/files/gondwana-factsheet.pdf.

41. Craig Robertson, "Gondwana: How It Got Its Name," *The Study*, March 2008, https://www.thestudy.net.au/projects/gondwana-name.html.

42. Margaret Andrew, *Flight to Antarctica* (Cambridge: Burlington Press, 1985), 15–18.

43. Radhakamal Mukerjee, *The Oneness of Mankind* (London: Macmillan, 1965).

44. Rolf Schmitt, *Gondwanaland: Eine Geschichte für Kinder und solche, die wieder Kinder werden wollen* [Gondwanaland: A Story for Children and Those Who Want to Become Children Again], illus. Michael Schmitt (Bonn: Simon Verlag, 1985), 10.

45. Schmitt, *Gondwanaland*, 25, 31–2.

46. Stuart Cooke, "Echoes of Gondwana," *Meanjin* 75, no. 1 (2016): 142–49.

47. Rachael Mead, "Rachel Mead Reviews Stuart Cooke," *Cordite Poetry Review*, April 10, 2017, http://cordite.org.au/reviews/mead-cooke/.

48. Stuart Cooke, "Toward an Ethological Poetics: The Transgression of Genre and the Poetry of the Albert's Lyrebird," *Environmental Humanities* 11, no. 2 (2019): 302; see also Stuart Cooke and Peter Denney, eds., *Transcultural Ecocriticism: Global, Romantic and Decolonial Perspectives* (London: Bloomsbury, 2021).

49. M. Pearce and R. Louis, "Mapping Indigenous Depth of Place," *American Indian Culture and Research Journal* 32, no. 3 (2008) 107–26.

50. Stuart Cooke, "Fire Was in the Reptile's Mouth: Towards a Transcultural Ecological Poetics," *Landscapes: The Journal of the International Centre for Landscapes and Language* 7, no. 1 (2016): https://ro.ecu.edu.au/landscapes/vol7/iss1/17/.

51. Mead, "Rachael Mead Reviews Stuart Cooke."

52. Stuart Cooke, interview by Zalehah Turner, *Rochford Street Review*, January 15, 2017, https://rochfordstreetreview.com/2017/01/15/fallen-myrtle-trunk-by-stuart-cooke-zalehah-turner-interviews-the-winner-of-the-new-shoots-poetry-prize-2016/; see also Stuart Cooke, "Fallen Myrtle Trunk," *Lyre* (Crawley, Australia: UWA Publishing, 2019).

53. For planetary localities, see Cooke and Denney, *Transcultural Ecocriticism*, 18.

54. Excerpts from Nathaniel Tarn, "Gondwana," *Gondwana and Other Poems* (New York: New Directions, 2017), 3–12. Copyright 2017 by Nathaniel Tarn. Reprinted by permission of New Directions Publishing Corp.

55. Michael Saler, "Modernity and Enchantment," *American Historical Review* 111, no. 3 (2006): 692–716.

56. Ramaswamy, *The Lost Land of Lemuria*, 5.

57. "Coal Reserves," Government of India, Ministry of Coal, August 31, 2021, https://coal.gov.in/major-statistics/coal-reserves.

58. Matthew D. Merrill, "GIS Representation of Coal-Bearing Areas in Antarctica: U.S. Geological Survey Open-File Report 2016–1031," (Reston, VA: US Geological Survey, 2016).

59. Nigel Clark, *Inhuman Nature: Sociable Life on a Dynamic Planet* (London: Sage, 2011); Jeffrey Jerome Cohen, *Stone: An Ecology of the Human* (Minneapolis: University of Minnesota Press, 2015).

Index